The Navigation Control Manual

The Navigation Control Manual

Second edition

A. G. Bole, Extra Master Mariner, FRIN, FNI

W. O. Dineley, Extra Master Mariner, M Phil.

C. E. Nicholls, Master Mariner, BSc., MNI

Newnes
An imprint of Butterworth-Heinemann Ltd
Linacre House, Jordan Hill, Oxford OX2 8DP

PART OF REED INTERNATIONAL BOOKS

OXFORD LONDON BOSTON
MUNICH NEW DELHI SINGAPORE SYDNEY
TOKYO TORONTO WELLINGTON

First published 1987
Second edition 1992

British Library Cataloguing in Publication Data

Bole, A. G.
Navigation Control Manual. – 2Rev. ed
I. Title
623.89

ISBN 0 7506 0542 1

Typeset by Vision Typesetting, Manchester
Printed and bound in Great Britain by Thomson Litho, Scotland

Contents

Preface to the second edition

The years that have passed since the publication of the first edition of *The Navigation Control Manual* have seen the *Future* Global Maritime Distress and Safety System become *The* Global Maritime Distress and Safety System with the entering into force, in February 1992, of the GMDSS amendments to Solas 74. The effects of these amendments on the management of communications and SAR procedures form an important part of the updating of this book.

Attention has also been given to recent amendments to the Collision Avoidance Rules, changes in the requirements for ship manoeuvring data, rearrangement of the ALRS and the increased availability of GPS.

As with any text of this nature, regulations quoted in the first edition have changed and M Notices have been superseded. Extracts from such official publications have all been brought up to date.

W. O. Dineley
14 February 1992

Introduction

Navigation control is the exercise of command decision-making based on all the data available. This includes the ability to evaluate navigational information, derived from all sources, in order to make and implement command decisions for collision avoidance and for directing the safe navigation of the ship.

A full appreciation of system errors and a thorough understanding of the operational aspects of modern navigation systems and effective bridge team work procedures are essential.

It is not intended to repeat here the theory upon which the various navigation instruments are based, other than where it is directly relevant to the topic under consideration. Where a more complete understanding is required or some detailed revision of basic theory is felt necessary, reference should be made to the many existing texts which already cover this topic.

In general, the 'break-in' point to a particular instrument will be where the raw readout of data from the instrument is obtained.

It might seem desirable that a manual should be all embracing and cover the subject in such as way that it is the only publication necessary. However, it is intended that this book will guide you to and through the welter of publications (official and otherwise) which, under normal circumstances, will be carried on board – that is, it will draw together all the elements of navigation control.

Acknowledgements

The authors wish to express their gratitude to:

The International Maritime Organisation for permission to reproduce the various extracts from Resolutions adopted by the Assembly;

The Controller of Her Majesty's Stationery Office for permission to reproduce extracts of Crown copyright material from 'M' Notices, Statutory Instruments, The Survey of Merchant Shipping Navigational Equipment, and the Guide to the Planning and Conduct of Sea Passages;

The British Ports Association for permission to reproduce the extract from Vessel Traffic Services in British Ports;

Lloyds of London Press for permission to reproduce the extract from the Handbook for Radio Operators;

Racal-Decca Marine Navigation Ltd, for permission to reproduce the Decca Navigator Marine Data Sheets;

The Commanding Officer, Omega Project Office, Washington, USA for permission to reproduce the three figures from the Omega Navigation System-User's Guide;

Ms Dorothy Hatchett of IMC Ltd, for her helpful comments based on a reading of Chapter 19;

Captain P. T. Owen of Liverpool Polytechnic for his helpful comments based on a reading of Chapter 22;

Families and friends without whose assistance, support and understanding this undertaking would never have been completed.

Part I
Data Sources

1 The compass

1.1 The magnetic compass

In recent years the magnetic compass has tended to play a lesser role in the immediate navigation of the vessel, but because of its independence of power supplies etc. it has remained an essential element in the ship's overall navigational equipment. It is still a legal requirement that it be carried (see Section 10.3) and, in spite of the trend towards total reliance on the gyro compass, it is important that the magnetic compass error is checked and logged

1 At least once per watch
2 At each change of course.

The methods more commonly available for checking compass error are

1 Celestial observations, i.e. azimuth and amplitude
2 Transit bearings of shore objects.

It is of course good practice to check the magnetic against the gyro compass, not to determine error but rather as a first warning that any change in the already noted difference in reading would indicate that an error was developing in one of the instruments.

1.1.1 Magnetic variation

Since the earth's magnetic and geographic poles do not coincide, the compass needle will indicate the direction to the magnetic pole, which will differ from the direction of the geographic or true pole. This difference in direction is termed magnetic variation and in some navigable areas can exceed some 25°. Information regarding magnetic variation is printed at the centre of the compass roses on navigational charts. Since its value may vary over the area of the chart, it is important to use the value on the compass rose nearest to the vessel's position or even to interpolate between values where there is a marked change. A special chart (no. 5374) giving worldwide lines of magnetic variation is also available.

Magnetic variation undergoes continual change and the published values are all related to some particular date, e.g. variation of 6° 05′ W (1979), decreasing about 4′ annually. It is essential that before use the published value is updated for the current year.

1.1.2 Deviation

After correction of the compass for the effect of the ship's magnetism, any residual effects will cause the compass needle to deviate from the direction of magnetic north. These residual errors are termed deviation and should be tabulated on a 'deviation card' by the compass adjuster at the time the vessel is swung. As a consequence of determining compass error, i.e. by the application of variation, the deviation component can be evaluated. Deviations should be recorded in the compass log, with the deviation card being updated as necessary and any significant changes investigated.

The deviation card should be readily available on the bridge at all times.

It should be noted that a table of low values for deviation does not necessarily indicate correct compensation on a worldwide basis. Thus it is essential that the values of deviation are recorded because much can be learned from an analysis of the logged values.

1.1.3 *Maintenance, service and correction*

Careful monitoring of compass error will indicate when the vessel will need to be swung and the correctors readjusted. In the absence of changes in the vessel's structure or of special cargoes (e.g. scrap iron), or of significant changes in the logged deviations, the vessel should only need to be swung at intervals of approximately ten years and even then should only require some minor adjustment of the correctors which have already been placed.

The positions, sizes, etc. of the correcting magnets should be logged at the time the vessel is swung so that in the event of some accident to the compass, which results in their displacement, they can be restored to their original positions.

When a spare compass bowl is carried, it is important that it is stowed clear of magnetic fields which might otherwise affect it. Since the corrector magnets placed in the binnacle are to compensate for the ship's magnetism, at a properly corrected compass position the magnetic field should be the same as would be experienced at that place in the absence of ship, binnacle, etc. Thus it would seem that one compass bowl is interchangeable (assuming this is mechanically possible) with another, but some care needs to be exercised.

The two most common sources of problems which occur with magnetic compasses are:

Bubbles in the liquid These should be removed according to the maker's instructions, and here the importance of topping up with the correct fluid cannot be overstressed.
Worn pivot This can result in the card 'sticking', and should be tested at intervals. No attempt should be made to repair it: the complete bowl should be replaced and the faulty bowl returned to the makers.

1.2 The gyro compass

In recent years the gyro compass has tended to come into its own, mainly because of its high reliability, compact size, minimal siting constraints, small errors, and ability to run repeaters and to provide a heading reference for other navigational equipment.

Unfortunately, the good points of the gyro compass have tended to result in it being taken for granted, but it is essential that it is monitored in exactly the same way as any other electronic navigational aid.

1.2.1 *Course speed and latitude error*

The gyro compass is controlled so as to align its axis at 90° to what it senses to be the earth's direction of rotation. The vessel's velocity, when combined with the earth's rotation, will result in an error which is dependent on the vessel's course, speed and latitude. In normal navigable latitudes (<70° N or S) this error is quite small (less than 2°) and should be applied by means of the facilities provided on the particular compass. Unfortunately, different compass manufacturers provide different means of compensation. These vary from nothing, in which case the error is determined from tables and the 'course to steer' adjusted accordingly; through mechanical compensators, which adjust the lubber line; to digital techniques, which have to be set with speed and latitude. Although the errors may be small it is important that the necessary adjustments are made:

1 Regularly when changing speed; and
2 Daily when significantly changing latitude.

1.2.2 *Starting, settling and stopping controls*

With modern gyro compasses, there should be no need to keep starting and stopping the compass. With the many compasses available today it is important to follow the manufacturer's instructions, but this should involve little more than switching on. Some require a warming-up period and also a period to allow the compass to settle, i.e. align with the meridian, although in the latter case a 'slew' or fast align facility is usually provided. In any case, it is good practice to have the gyro running well before it is needed. After switching on and alignment have been completed, and immediately prior to use, the speed and latitude controls should be set and all the repeaters aligned with the master.

1.2.3 Routine maintenance

Modern gyro compasses are essentially sealed units and should require no maintenance, but again the manufacturer's handbook should be consulted. Older gyro compasses do require routine maintenance, but here it is essential to attend some form of training course before anything more than basic cleaning and oiling is undertaken.

1.2.4 Error log

As with magnetic compasses, it is important that the compass error is checked regularly, i.e. at least once per watch and at each change of course, and then logged along with course, speed and latitude at the time.

1.3 Autopilots

Steering the vessel has long been seen as wasteful of labour time; some form of automatic steering device has been available since approximately 1907. The autopilot is primarily for use on long ocean passages, although the precision and sophistication of today's adaptive autopilots permit their use under more demanding steering conditions. One consequence of the provision of autopilots is that the traditional steering skills of the helmsman or quartermaster are being lost. Whereas it was the practice to call for a helmsman in bad weather or where precision conning was required, it has been found (e.g. during refuelling and transferring stores at sea during the Falklands War) that very few of the crew are capable of the task. It is essential therefore to ensure that there are sufficient crew members capable of competent steering when the occasion demands and that the skill is maintained by regular practice, even if this means a spell on the wheel each day.

1.3.1 Checking, setting and change-over procedures

Again, there is variation in the controls and facilities provided on the autopilot and consoles in the sophistication of the functions which the system is intended to perform, so that reference to the manufacturer's manual is essential.

1.3.2 Basic controls

Setting This is usually by means of a compass card which is set to the 'demanded' heading. The difference between the demanded and actual heading is then sensed and the difference removed by altering the course of the ship.
Rudder angle This usually sets or limits the maximum rudder angle which the autopilot may apply.
Sensitivity This sets the amount by which the vessel is allowed to be off course before correction is applied. Too small a setting will result in excessive and unnecessary working of the steering gear and added drag due to rudder reaction, while too large a setting can result in long periods off course and therefore poor track keeping.
Weather helm When steering in heavy weather with the wind and sea at an angle to the vessel's heading, there is a tendency for the vessel's head to be turned in a particular direction. The effect of this can be offset by maintaining some permanent value of rudder angle; this angle is set using 'weather helm' after a period of trial and error.

1.3.3 The adaptive autopilot

This is intended to remove the need for the operator to assess the correct control settings. In effect the autopilot 'learns' the vessel's handling characteristics in calm weather and then, when the weather deteriorates, can distinguish between those errors in heading due to the weather and those due to the vessel's normal handling characteristics. It has been found that in heavy weather the vessel's head deviates very little *relative to the water*; therefore the adaptive autopilot concerns itself primarily with those errors in heading resulting from normal handling characteristics, so reducing the work and hence drag of the rudder.

1.4 Course recorders

These, when provided, can perform a useful course/autopilot monitoring function. They provide a paper trace of course against time and have an off-course alarm facility which can be set to warn if the vessel strays more than the set number of degrees off course. When linked to a magnetic/fluxgate compass it can prove extremely useful in detecting a slow off-course drift in the gyro controlling the autopilot. This would not be the case where the alarm is linked to the same gyro as the autopilot.

1.5 Effect of errors on dead reckoning

1.5.1 Constant errors

It must be recognized that the course as ordered may differ from the course made good, due to:

1 Wind, i.e. leeway
2 Tidal stream or ocean current
3 Periods off course for collision avoidance
4 Careless or inaccurate steering.

When this difference is known to exist, it should be taken into account when working up the ship's estimated position (see Section 20.1).

1.5.2 Random errors

It may be that during bad weather it is only possible to steer within say 2° of the desired course, with the error at any instant being random within the 2°. If the position is then worked up by dead reckoning (DR), the cross-track error should be borne in mind.

Example
If a vessel has departed from a known position, what would be the uncertainty in position when the log reads 420 nautical miles (nm) if the random course error is estimated to be ±2° and the random log error is estimated to be ±2 per cent (Figure 1.1)?

estimated cross-track error = 420 tan 2°
= ±14.66 nm
estimated along-track error = 2% of 420
= ±8.4 nm

When making landfall after a period without a reliable fix, it is important that considerations of this sort be taken into account. The area of uncertainty is dependent upon:

1 Random course error
2 Random speed error
3 Distance run since the last fix.

1.6 Extracts from official publications

Within Sections 1.6.1 to 1.6.7, the paragraph numbering of the original publications has been retained.

The International Maritime Organization is abbreviated IMO. It was formerly the Intergovernmental Maritime Consultative Organization (IMCO).

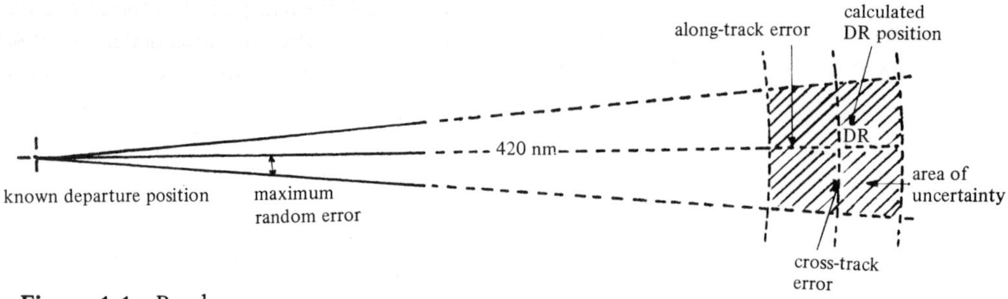

Figure 1.1 Random errors

1.6.1 IMO Resolution A382(X): Performance Standards for Magnetic Compasses

1 Definitions

1.1 A magnetic compass is an instrument designed to seek a certain direction in azimuth and to hold that direction permanently, and which depends, for its directional properties, upon the magnetism of the earth.

1.2 The standard compass is a magnetic compass used for navigation, mounted in a suitable binnacle containing the required correcting devices and equipped with a suitable azimuth reading device.

1.3 The steering compass is a magnetic compass used for steering purposes mounted in a suitable binnacle containing the required correcting devices.

Note If the transmitting image of a sector of the standard compass card of at least 15° to each side of the lubber mark is clearly readable for steering purposes at the main steering position, both in daylight and artificial light according to subparagraph 7.1, the standard compass can also be regarded as the steering compass.

2 Compass card

2.1 The compass card should be graduated in 360 single degrees. A numerical indication should be provided every ten degrees, starting from north (000°) clockwise to 360°. The cardinal points should be indicated by the capital letters N, E, S and W. The north point may instead be indicated by a suitable emblem.

2.2 The directional error of the card, composed of inaccuracies in graduation, eccentricity of the card on its pivot and inaccuracy of orientation of the card on the magnetic system, should not exceed 0.5° on any heading.

2.3 The card of the steering compass should clearly be readable in both daylight and artificial light at a distance of 1.4 m. The use of a magnifying glass is permitted.

3 Materials

3.1 The magnets used in the directional system and the corrector magnets for correcting the permanent magnetic fields of the ship should have a high coercivity of at least 11.2 kA/m.

3.2 Material used for correcting induced fields should have a low remanence and coercivity.

3.3 All other materials used in the magnetic compass and in the binnacle should be non-magnetic, so far as reasonable and practicable and such that the deviation of the card caused by these materials should not exceed $(9/H)°$, where H is the horizontal component of the magnetic flux density in microtesla (μT) at the place of the compass.

4 Performance

The magnetic compass equipment should operate satisfactorily and remain usable under the operational and environmental conditions likely to be experienced on board ships in which it is installed.

5 Constructional error

5.1 With the compass rotating at a uniform speed of 1.5° per second and a temperature of the compass of 20 °C ± 3 °C, the deflection of the card should not exceed $(54/H)°$ (H being defined as in subparagraph 3.3).

5.2 The error duc to friction should not exceed $(3/H)°$ at a temperature of 20 °C ± 3 °C (H being defined as in subparagraph 3.3).

5.3 With a horizontal component of the magnetic field of 18 μT the half period of the card should be at least 12 seconds after an initial deflection of 40°. The time taken to return finally to within ±1° of the magnetic meridian should not exceed 60 seconds after an initial deflection of 90°. Aperiodic compasses shall comply with the latter requirements only.

6 Correcting devices

6.1 The binnacle should be provided with devices

for correcting semicircular and quadrantal deviation due to:

.1 The horizontal components of the ship's permanent magnetism
.2 Heeling error
.3 The horizontal component of the induced horizontal magnetism
.4 The horizontal component of the induced vertical magnetism.

6.2 The correcting devices provided in subparagraph 6.1 should ensure that no serious changes of deviation occur under the influence of the environmental conditions described in paragraph 4 and particularly considerable alteration of magnetic latitude. Sextantal and deviations of higher order should be negligible.

7 Construction

7.1 Primary and emergency illumination should be installed so that the card may be read at all times. Facilities for dimming should be provided.

7.2 With the exception of the illumination, no electrical power supply should be necessary for operating the magnetic compass.

7.3 In the case where an electrical reproduction of the indication of the standard compass is regarded as a steering compass, the transmitting system should be provided with both primary and emergency electrical power supply.

7.4 Equipment should be constructed and installed in such a way that it is easily accessible for correcting and maintenance purposes.

7.5 The compass, binnacle and azimuth reading device should be marked to the satisfaction of the administration (the government of each country applying the rules).

7.6 The standard compass should be suspended in gimbals so that its verge ring remains horizontal when the binnacle is tilted up to 40° in any direction, and so that the compass cannot be dislodged under any condition of sea or weather. Steering compasses suspended in gimbals should meet the same requirements. If they are not suspended in gimbals they should have a freedom of the card of at least 30° in all directions.

7.7 Material used for the manufacture of magnetic compasses should be of sufficient strength and be to the satisfaction of the administration.

8 Positioning

8.1 The magnetic compass equipment should be installed if practicable and reasonable on the ship's centreline. The main lubber mark should indicate the ship's heading with an accuracy of ±0.5°.

8.2 The standard compass should be installed so that from its position the view is as uninterrupted as possible, for the purpose of taking horizontal and celestial bearings. The steering compass should be clearly readable by the helmsman at the main steering position.

8.3 The magnetic compasses should be installed as far as possible from magnetic material. The minimum distances of the standard compass from any magnetic material which is part of the ship's structure should be to the satisfaction of the administration. The following diagram gives general guidelines to indicate the minimum desirable distances from the standard compass. The minimum desirable distances for the steering compass may be reduced to 65 per cent of the values given by the diagram provided that

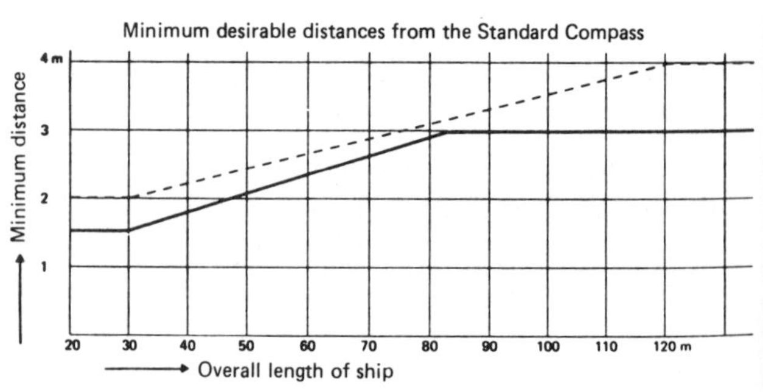

—— Uninterrupted fixed magnetic material

- - - - End parts of fixed magnetic material such as top edges of walls, partitions and bulkheads, extremities of frames, girders, stanchions, beams pillars and similar steel parts. Magnetic material subject to movement at sea such as davits, ventilators, steel doors, etc. Large masses of magnetic material with variable fields such as funnels.

no distance is less than 1 m. If there is only a steering compass the minimum distances for the standard compass should be applied as far as practicable.

8.4 The distance of the magnetic compass from electrical or magnetic equipment should be at least equal to the safe distance specified for the equipment and be to the satisfaction of the administration.

1.6.2 UK Statutory Instrument 1984, no. 1203

Part I Magnetic compass installation

Magnetic compass performance standards
10 Every magnetic compass installation required to be provided shall comply with the performance standards adopted by the organization.

The magnetic compass installation
11(1) Except in the case of ships having Passenger Certificates of Class IV, VI or VI(A), the magnetic compass installation shall comprise:

(a) A standard magnetic compass, fitted on the centreline of the ship and mounted on a binnacle;
(b) A steering magnetic compass, fitted on the centreline of the ship and mounted on a binnacle, unless heading information provided by the standard compass required under (a) is made available and is clearly readable by the helmsman at the main steering position;
(c) Adequate means of communication between the standard compass position and the normal navigation control position; and
(d) Means for taking bearings as nearly as practicable over an arc of the horizon of 360°.

(2) In the case of ships having Passenger Certificates of Class IV, VI or VI(A) the magnetic compass installation shall comprise one efficient magnetic compass at the steering position.

Adjustment of magnetic compasses
12 Each of the magnetic compasses referred to in regulation 11(1) shall be properly adjusted and its table or curve of residual deviations shall be available at all times.

Spare magnetic compass
13 A spare magnetic compass, interchangeable with the standard compass, shall be carried in every ship of 150 tons and over to which these regulations apply, unless a steering compass mentioned in regulation 11(1)(b) is fitted.

Emergency steering position
14 On ships of 150 tons and over which are provided with emergency steering positions, arrangements shall be made to supply heading information to such positions.

Part II Gyro compass installation

Gyro compass performance standards
15 Every gyro compass installation required to be provided shall comply with the performance standards adopted by the organization and shall, in addition, comply with the relevant performance specifications issued by the Department of Transport.

Siting of gyro compass installation
16(1) The master compass shall be installed with its fore-and-aft datum line parallel to the ship's fore-and-aft datum line to within ±0.5°.
(2) The compass card of the master compass, or a repeater of the heading information, shall be sited so that it is clearly readable by the helmsman when steering the ship.
(3) Where provided, repeaters used for taking visual bearings shall be installed with their fore-and-aft datum lines parallel to the ship's fore-and-aft datum line to within ±0.5°.
(4) The master compass shall be sited so as to avoid, where practicable, excessive errors being caused to the gyro compass installation due to the ship rolling, pitching or yawing.
(5) Where, in a gyro compass installation fitted on or after 1 September 1984, failure of one repeater could cause an error in any other repeater, a readily accessible means shall be provided for isolating each repeater output from the master compass.

Provision of gyro repeaters
17 On ships of 1600 tons or over a gyro repeater or

gyro repeaters shall be provided and shall be suitably placed for taking bearings as nearly as practicable over an arc of the horizon of 360°.

1.6.3 IMO Resolution A424(XI): Performance Standards for Gyro Compasses

1 Introduction

1.1 The gyro compass required by Regulation 12 of Chapter V of the International Convention for the Safety of Life at Sea, 1974, should determine the direction of the ship's head in relation to geographic (true) north.

1.2 The equipment should comply with the following minimum performance requirements.

2 Definitions

For the purpose of these performance standards the following definitions apply:

.1 The term 'gyro compass' comprises the complete equipment and includes all essential elements of the complete design.

.2 The 'true heading' is the horizontal angle between the vertical plane passing through the true meridian and the vertical plane passing through the ship's fore-and-aft datum line. It is measured from true north (000°) clockwise through 360°.

.3 The compass is said to be 'settled' if any three readings taken at intervals of 30 minutes, when the compass is on a level and stationary base are within a band of 0.7°.

.4 The 'settle point heading' is the mean value of ten readings taken at 20 minute intervals after the compass has settled as defined in paragraph 2.3.

.5 The 'settle point error' is the difference between settle point heading and true heading.

.6 The other errors to which the gyro compass is subject are taken to be the difference between the observed value and the settle point heading.

3 Method of presentation

The compass card should be graduated in equal intervals of one degree or a fraction thereof. A numerical indication should be provided at least at every ten degrees, starting from 000° clockwise through 360°.

5 Accuracy

5.1 *Settling of equipment*
5.1.1 When switched on in accordance with the manufacturer's instructions, the compass should settle within 6 hours in latitudes of up to 60°.

5.1.2 The settle point error as defined in paragraph 2.5 at any heading and at any latitude up to 60° should not exceed ±0.75° × secant latitude, where heading indications of the compass should be taken as the mean of ten readings at 20 minute intervals, and the root mean square value of the differences between individual heading indications and the mean should be less than 0.25° × secant latitude. The repeatability of settle point error from one run-up to another shall be within 0.25° × secant latitude.

5.2 *Performance under operational conditions*
5.2.1 When switched on in accordance with the manufacturer's instructions, the compass should settle within 6 hours in latitudes of up to 60° when rolling and pitching with simple harmonic motion of any period between 6 and 15 seconds, a maximum angle of 5°, and a maximum horizontal acceleration of 0.22 m/s^2.

5.2.2 The repeatability of the settle point error of the master compass shall be within ±1° × secant latitude under the general conditions mentioned in chapter 1.1 of this publication, paragraphs 3.1 and 4, and including variations in magnetic field likely to be experienced in the ship in which it is installed.

5.2.3 In latitudes of up to 60°:

.1 The residual steady state error, after correction for speed and course influences at a speed of 20 knots, shall not exceed ±0.25° × secant latitude.

.2 The error due to a rapid alteration of speed of 20 knots should not exceed ±2°.

.3 The error due to a rapid alteration of course of 180° at a speed of 20 knots should not exceed ±3°.

.4 The transient and steady state errors due to the ship rolling, pitching and yawing, with simple harmonic motion of any period between 6 and 15 seconds, maximum angle of 20°, 10° and 5° respectively, and maximum horizontal acceleration not exceeding 1 m/s^2, should not exceed 1° × secant latitude.

5.2.4 The maximum divergence in reading between the master compass and repeaters under all operational conditions should not exceed ±0.5°.

Note When the compass is used for purposes other than steering and bearing, a higher accuracy might be necessary. To ensure that the maximum error referred to in subparagraph 5.2.3.4 is not exceeded in practice, it will be necessary to pay particular attention to the siting of the master compass.

9 Construction and installation

9.1 The master compass and any repeaters used for taking visual bearings should be installed in a ship with their fore-and-aft datum lines parallel to the ship's fore-and-aft datum line to within ±0.5°. The lubber line should be in the same vertical plane as the centre of the card of the compass and should be aligned accurately in the fore-and-aft direction.

9.2 Means should be provided for correcting the errors induced by speed and latitude.

9.3 An automatic alarm should be provided to indicate a major fault in the compass system.

9.4 The system should be designed to enable heading information to be provided to other navigational aids such as radar, radio direction finder and automatic pilot.

1.6.4 The Survey of Merchant Shipping Navigational Equipment (HMSO)

5.1 Siting of gyro compass units

5.1.1 The master unit should be installed on a firm horizontal base using a wooden or similar mounting pad. It should be as free from vibration as is practicable and have adequate space around the unit for access and ventilation. Adequate ventilation is particularly important for units which incorporate cooling fans.

5.1.2 The fore-and-aft line of master units and binnacles must align with, or be parallel with, the fore-and-aft axis of the ship and compass cards should be clearly visible so that accurate reading of ship's heading is readily available.

5.1.3 Where the gyro compass installation provides steering information, the master unit or repeater, as applicable, should be installed adjacent to the steering wheel, in such a position that the ship's heading can easily be read by the helmsman.

5.1.4 Bearing repeaters, particularly those on the bridge wings for taking bearings, should be rigidly mounted, and protective covers should be available and in place whenever any exposed repeaters are not in use.

5.1.5 Bridge wing repeaters should be installed in positions which provide the maximum possible unobstructed view of the horizon.

5.2 Operational checks

5.2.1 The compass should be allowed to settle and the readings then checked to ensure that they are within the tolerances given in IMO resolution A424(XI), and all repeaters should be aligned with the master unit to within ±0.5°.

5.2.2 All repeaters should be checked to ensure that they follow the master unit accurately, smoothly and in the correct sense, by deflecting the master unit a small amount, about ±10°.

5.2.3 The mechanical parts of bridge wing repeaters should be checked to ensure that they are in a good condition and capable of operating satisfactorily.

1.6.5 IMO Resolution A342(IX): Performance Standards for Automatic Pilots

Automatic pilot equipment aboard a seagoing vessel should comply with the following minimum operational requirements in addition to the general requirements contained in Assembly Resolution A.281(VIII).

1 General

1.1 Within limits related to a ship's manoeuvrability the automatic pilot, in conjunction with its source of heading information, should enable a ship to keep a preset course with minimum operation of the ship's steering gear.

1.2 The automatic pilot equipment should be capable of adapting to different steering characteristics of the ship under various weather and loading conditions, and provide reliable operation under prevailing environmental and normal operational conditions.

2 Changing over from automatic to manual steering and vice versa

2.1 Changing over from automatic to manual steering and vice versa should be possible at any rudder position and be effected by one, or at the most two, manual controls, within a time lag of 3 seconds.

2.2 Changing over from automatic to manual steering should be possible under any conditions, including any failure in the automatic control system.

2.3 When changing over from manual to automatic steering, the automatic pilot should be capable of bringing the ship to the preset course.

2.4 Change-over controls should be located close to each other in the immediate vicinity of the main steering position.

2.5 Adequate indication should be provided to show which method of steering is in operation at a particular moment.

3 Alarm signalling facilities

3.1 A course monitor should be provided which actuates an adequate 'off-course' audible alarm signal after a course deviation of a preset amount.

3.2 The information required to actuate the course monitor should be provided from an independent source.

3.3 Alarm signals, both audible and visual, should be provided in order to indicate failure or a reduction in the power supply to the automatic pilot or course monitor, which would affect the safe operation of the equipment.

3.4 The alarm signalling facilities should be fitted near the steering position.

4 Controls

4.1 The number of operational controls should be minimized as far as possible and they should be designed to preclude inadvertent operation.

4.2 Unless features for automatic adjustments are incorporated in the installation, the automatic pilot should be provided with adequate controls for operational use to adjust effects due to weather and the ship's steering performance.

4.3 The automatic pilot should be designed in such a way as to ensure altering course to starboard by turning the course setting control clockwise. Normal alterations of course should be possible by one adjustment only of the course setting control.

4.4 Except for the course setting control, the actuation of any other control should not significantly affect the course of the ship.

4.5 Additional controls at remote positions should comply with the provisions of these performance standards.

5 Rudder angle limitation

Means should be incorporated in the equipment to enable rudder angle limitation in the automatic mode of operation. Means should also be available to indicate when the angle of limitation has been reached.

6 Permitted yaw

Means should be incorporated to prevent unnecessary activation of the rudder due to normal yaw motion.

1.6.6 UK DTp Merchant Shipping Notice M.1102: Operational Guidance for Officers in Charge of a Navigational Watch

Note The guidelines were first produced by the Ministry of Transport, then by the DoT. They are currently published by the DTp.

Periodic checks of navigational equipment

9 Operational tests of shipboard navigational equipment should be carried out at sea as frequently as practicable and as circumstances permit, in particular when hazardous conditions affecting navigation are expected: where appropriate these tests should be recorded.

10 The officer of the watch should make regular checks to ensure that:

(a) The helmsman or the automatic pilot is steering the correct course
(b) The standard compass error is determined at least once a watch and, when possible, after any major alteration of course; the standard and gyro compasses are frequently compared and repeaters are synchronized with their master compass
(c) The automatic pilot is tested manually at least once a watch
(d) The navigation and signal lights and other navigational equipment are functioning properly.

Automatic pilot

11 The officer of the watch should bear in mind the necessity to comply at all times with the requirements of regulation 19, chapter V of the International Convention for the Safety of Life at Sea 1974. He should take into account the need to station the helmsman and to put the steering into manual control in good time to allow any potentially hazardous situation to be dealt with in a safe manner. With a ship under automatic steering it is highly dangerous to allow a situation to develop to the point where the officer of the watch is without assistance and has to break the continuity of the look-out in order to take emergency action. The change-over from automatic to manual steering and vice versa should be made by, or under the supervision of, a responsible officer.

Electronic navigational aids

12 The officer of the watch should be thoroughly familiar with the use of electronic navigational aids carried, including their capabilities and limitations.

13 The echo sounder is a valuable navigational aid and should be used whenever appropriate.

1.6.7 UK DTp Merchant Shipping Notice M.1471: Use of Automatic Pilot

1 There have been many casualties in which a contributory cause has been the improper use of, or over-reliance upon, the automatic pilot. Collisions have occurred where one and sometimes both vessels have been on automatic steering with no proper lookout being kept; strandings and other casualties have occurred where automatic steering systems have been in use in restricted waters and a person has not been immediately available to take the wheel; casualties have also happened because watchkeepers were not familiar with the procedure or precautions necessary when changing over from the automatic pilot to manual steering.

2 Attention is drawn to the possible inability of an automatic pilot to closely maintain set headings when a ship is making low speed and/or in heavy seas. The performance of some automatic steering systems is very dependent upon correct control settings suited to the prevailing conditions of ship speed, displacement, and sea state particularly. Use of the automatic pilot must be restricted to conditions within the designed parameters of the automatic control system.

3 If shipowners do not use all the control options which may be incorporated by the various manufacturers into a control console, positive measures should be taken to prevent redundant control settings being used inadvertently, and the labelling arrangements should be amended accordingly.

4 Certain requirements on the use of the automatic pilot are included in Regulation 4 of The Merchant Shipping (Automatic Pilot and Testing of Steering Gear) Regulations 1981 (SI 1981 No. 571) which is reproduced as an appendix to this notice. Masters, skippers and watchkeeping officers should be aware of these requirements as well as the general need to ensure that arrangements are adequate for maintaining a safe navigational watch, as described in Merchant Shipping Notice M.1102.

5 Masters, skippers and all watchkeeping personnel must be familiar with the procedure for changing over from steering with the automatic pilot to hand steering (e.g. through a telemotor) and must ensure that sufficient time is allowed for the operation. Clear instructions must be provided at the control console, and special attention should be given to the procedure when joining a ship because it will vary depending on the particular equipment installed. The operations manual should be kept on the bridge and be readily available to masters, skippers and navigation watchkeeping personnel.

6 Some steering gear control systems enable alignment to be maintained between the helm and the steering gear at all times, irrespective of whether the automatic pilot is or has been used. Where the design does not include this provision, suitable measures should be taken immediately before and after the changeover to ensure that the helm and steering gear are aligned.

7 Attention is drawn to the need to test the manual steering. Paragraph 10(c) on page 3 of M.1102 recommends that the automatic pilot should be 'tested manually at least once a watch', while Regulation 4(4) in the appendix to this notice requires that, whilst the vessel is on passage and continuously using the automatic pilot, the manual steering gear be tested at least once a day. To comply with the former recommendation, the manual steering over-ride alter course control incorporated in the automatic-pilot console should be operated once every watch. To comply with the latter requirement, the wheel (or equivalent) steering should be engaged at least once every day and the ship steered by hand. It is strongly recommended that a roster system should be employed so that all persons recognized and qualified for the purpose of steering take a turn at this task. They should steer for a sufficient period for them to maintain their proficiency, including manoeuvring the vessel thus gaining experience in the vessel's response to helm orders.

Appendix: Extract from The Merchant Shipping (Automatic Pilot and Testing of Steering Gear) Regulations 1981 (SI 1981, no. 571)

Use of the automatic pilot: regulation 4

1 The master shall ensure that an automatic pilot, where fitted, shall not be used in areas of high traffic density, in conditions of restricted visibility, or in any other hazardous navigational situation, unless it is possible to establish manual control of the ship's steering within 30 seconds.

2 Before entering any area of high traffic density, and whenever visibility is likely to become restricted or some other hazardous navigational situation is likely to arise, the master shall arrange, where practicable, for the officer of the watch to have available without delay the services of a qualified helmsman who shall be ready at all times to take over the manual steering.

3 The change-over from automatic to manual steering and vice versa shall be made by, or under the supervision of, the officer of the watch, or, if there is no such officer, the master.

4 The master shall ensure that the manual steering gear is tested (a) after continuous use of the automatic pilot for 24 hours and (b) before entering any areas where navigation demands special caution.

2 The log

Logs of one form or another have been with us from the earliest days of navigation and are probably the most imprecise instruments we have today.

2.1 The water reference log

As its name implies, the speed and/or distance displayed is related to the vessel's movement 'through the water', i.e. it is completely independent of tidal stream or current. It should be borne in mind that water reference logs are generally single axis devices, i.e. they only give distance 'along track' so that the effect of leeway could result in errors (Figure 2.1).

2.1.1 *Impeller/rotor-based logs*

These suffer from 'slip', which can vary with weather conditions, speed and, when it is a towed log, with the freeboard of the vessel. As a result the slip, usually expressed as a percentage, is imprecisely known and results in an inaccurate knowledge of speed/distance.

Logs of this type which are towed need to be handed (taken in) when the vessel is likely to manoeuvre, e.g. in fog, while those extending on a short shaft below the hull need to be retracted in shallow water. In both cases, speed indication is lost at a time when it could be most useful.

2.2 The ground reference log

With this type of log, the Doppler effect on a signal reflected from the sea-bed is used to determine speed 'over the ground' and hence distance. The transducers may be in single or dual axis arrays.

With a single axis array, it is only the along-track component of velocity which is measured. It is important to remember here that, when the vessel is working in tide, this reading is neither the distance through the water nor that over the ground, and some careful thought is required in calculating the course to steer to counteract a tide.

The outputs from a dual axis Doppler log will be both *a* and *b* in Figure 2.2, from which course and distance (over the ground) may be determined.

There is a maximum limit (under normal circumstances) to the depth of water in which ground lock can be achieved. Therefore in deeper water the log may only be locked to the watermass, in which case the ground reference advantages normally attributed to this type of log will be lost. It should also be noted that in shallower water there is a minimum depth at which water lock is possible. This is of considerable importance when the log output is fed to a true motion radar display, or ARPA (see Sections 5.4.2, 5.4.3 and 5.8.1).

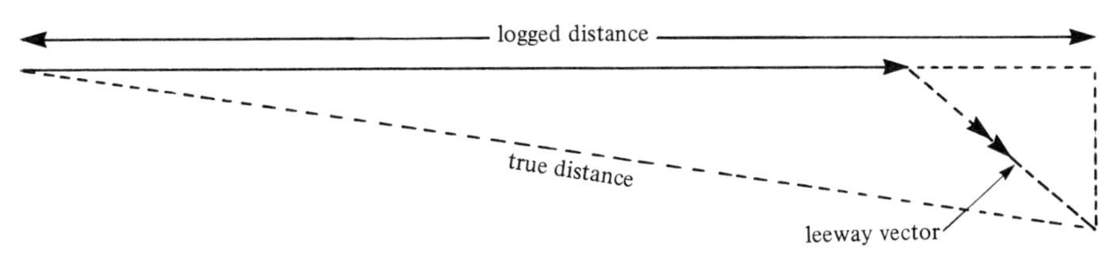

Figure 2.1 The effect of leeway

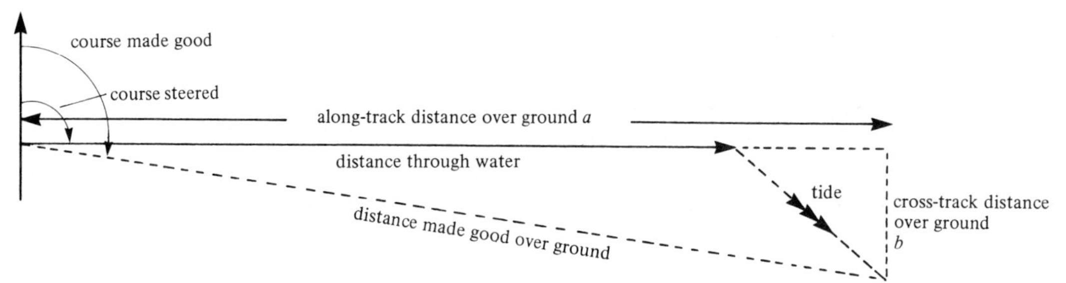

Figure 2.2 The effect of tide

Some modern Doppler logs claim to be able to indicate simultaneously the speed/distance over the ground and through the water, thus giving two independent readouts. This is achieved by 'locking' to the watermass in a 'window' which can be set so that one is certain that the response is coming from a depth less than the depth of water at the time. However, it is important to remember that ground lock will be lost when the water becomes too deep. It is then interesting to compare the two readings, which are now being referenced to the watermass at two different levels.

Ground reference logs usually have some means of calibration, but here it is necessary to refer to the operating/installation manual since it would appear that no two logs are alike. Once completed, calibration should remain valid for a long period.

2.3 Errors

The effects of errors in course and speed on dead reckoning position have been dealt with in Section 1.5.

2.4 Extracts from official publications

Within Sections 2.4.1 and 2.4.2, the paragraph numbering of the original publications has been retained.

2.4.1 IMO Resolution A478(XII): Performance Standards for Devices to Indicate Speed and Distance

1 Introduction

1.1 Devices to indicate speed and distance required by Regulation 12, Chapter V, of the 1974 SOLAS Convention, as amended, are intended for general navigational use to provide information on the distance run and the forward speed of the ship, through the water or over the ground. The equipment should function at forward speeds up to the maximum speed of the ship and in water of depth greater than 3 metres beneath the keel.

1.2 In addition to the recommended general requirements for electronic navigational aids the equipment should conform to the following minimum performance standards.

2 Methods of presentation

2.1 Speed information may be presented in either analogue or digital form. Where a digital display is used, its incremental steps should not exceed 0.1 knots. Analogue displays should be graduated at least every 0.5 knots and be marked with figures at least every 5 knots. If the display can present the speed of the ship in both forward and reverse directions, the direction of movement should be indicated unambiguously.

2.2 Distance run information should be presented in digital form. The display should cover the range from 0 to not less than 9999.9 nautical miles and the incremental steps should not exceed 0.1 nautical miles. Where practicable, means should be provided for resetting a readout to zero.

2.3 The display should be easily readable by day and by night.

2.4 Means should be provided for feeding distance run information to other equipment fitted on board. The information should be in the form of one contact closure or the equivalent for each 0.005 nautical miles run.

2.5 If equipment is capable of being operated in either the 'speed through the water' or 'speed over the ground' modes, mode selection and mode indication should be provided.

3 Accuracy of measurement

3.1 Errors in the indicated speed, when the ship is operating free from shallow water effect and from the effects of wind, current and tide, should not exceed 5 per cent of the speed of the ship, or 0.5 knots, whichever is the greater.

3.2 Errors in the indicated distance run, when the ship is operating free from shallow water effect and from the effects of wind, current and tide, should not exceed 5 per cent of the distance run by the ship in one hour or 0.5 nautical miles in each hour, whichever is greater.

3.3 If the accuracy of devices to indicate speed and distance run can be affected by certain conditions (e.g. sea state and its effects, water temperature, salinity, sound velocity in water, the depth of water under the keel, heel and trim of ship), details of possible effects should be included in the equipment handbook.

4 Roll and pitch

The performance of the equipment should be such that it will meet the requirements of these standards when the ship is rolling up to plus or minus 10 degrees and pitching up to plus or minus 5 degrees.

5 Construction and installation

5.1 The system should be so designed that neither the method of attachment of parts of the equipment to the ship nor damage occurring to any part of the equipment which penetrates the hull could result in the ingress of water to the ship.

5.2 Where any part of the system is designed to extend from and retract into the hull of the ship, the design should ensure that it can be extended, operated normally and retracted at all speeds up to the maximum speed of the ship. Its extended and retracted positions should be clearly indicated at the display position.

2.4.2 UK Statutory Instrument 1984, no. 1203

Part V Speed and distance measuring installation

Speed and distance measuring equipment performance standards

29 Every speed and distance measuring installation required to be provided shall comply with the performance standards adopted by the IMO and shall, in addition, comply with the relevant performance specifications issued by the Department of Transport.

Siting of speed and distance measuring installation

30(1) Where applicable, the transducer unit of the speed and distance measuring installation shall be sited so as to avoid, where practicable, the vicinity of all underwater openings in, or projections from, the hull, such as plugs, anodes or other transducers, so that satisfactory overall performance is achieved.

(2) Where a towed log is fitted, the position of the log register shall be selected so that the log line and its rotator when streamed are as clear as is practicable from disturbed water in the close vicinity of the ship and so that the rotation of the log line is not impeded by any part of the ship or its equipment.

(3) The display shall, where practicable, be sited on the bridge in a position to facilitate easy access and viewing and where the effect of any lighting necessary for the equipment does not interfere with the keeping of an effective look-out.

3 The echo sounder

Some form of sounding system has been in operation since man first ventured out to sea. In these days of sophisticated sensors he ignores the data from this source at his peril.

3.1 System introduction

The basic principle upon which the system is based is that if the velocity of sound in water is known (or assumed) and the time interval between the transmission of a pulse and the reception of an echo from the sea-bed is measured, then the depth of water can be determined.

3.1.1. The basic system

To obviate the need for a clock as such and to facilitate the provision of a recorded trace, a rotating stylus triggers the transmitter as it passes 'zero depth' and the amplified echo signal causes a mark to be left on the recording paper (Figure 3.1).

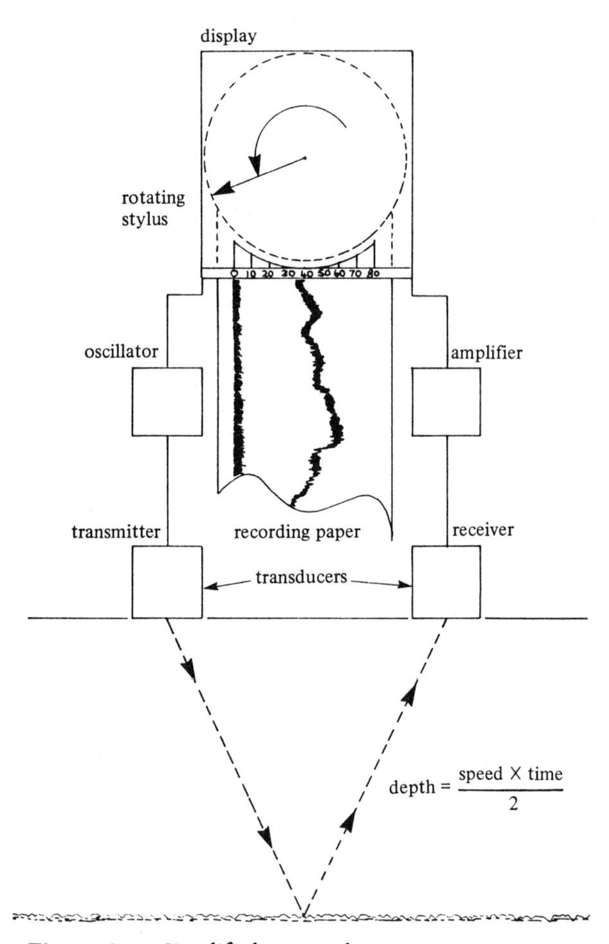

Figure 3.1 Simplified system diagram

3.1.2 Data output format

In its most simple and up-to-date format this may take the form of a digital display of depth in either metres, fathoms or feet, but equipment which meets the various specifications (see Section 3.4.1) is required to present the information in graphical form, indicating immediate depth and a visible record of soundings. It is also possible to give an output in colour on a video display unit as well as synthesized speech.

In order to overcome the dangers inherent in the phasing of range scales, all range scales should commence at zero. However, care must be taken using older equipment to ensure that the correct depth is being displayed by always working from zero depth into deeper scales when searching for the sea-bed, and not just looking for a return at the depth at which it is expected.

3.2 Setting-up procedure

Prior to use, the back plate or platen should be cleaned and the roll of recording paper fitted. The equipment should be switched on; a range scale commencing at zero and covering the expected sounding should be selected, e.g. 0–100 metres. The gain (or sensitivity) control should be adjusted until there is a clear trace from the sea-bed. In the absence of a response, the maximum range scale having a zero should be selected and the gain set so that noise speckles are just marking the paper. Controls for paper speed (if provided) should be set at slow to permit trace integration of weak responses while making initial detection. Alarms (if provided) may also be set. Where a trace is already available, care should be taken in setting the gain control so as to avoid multiple traces which can result from signals bouncing between sea-bed and the hull (see Figure 3.4).

Great care should be exercised when setting the gain control and using phased range scales. In any event, the neon indicator on the stylus should never be switched off.

3.3 Corrections before chart comparison

The depth indicated on the echo sounder is usually the depth of water below the transducer (on some ships there may be more than one transducer). Thus it is essential to know just where that transducer is situated, since both heel and trim can be important when navigating in channels where the depth of water is limited.

Some echo sounders provide for a correction for draft to be inserted, which means that the echo sounder will now indicate the depth of water. While this can have advantages in special circumstances (e.g. in hydrographic surveying) its use should be avoided for normal navigation, since if the sounder was left in the 'water depth' mode and the output was assumed to be depth under the keel, stranding could result.

3.3.1 *The effect of draft and rise of tide*

Before the echo sounder reading is compared with the charted depth, corrections for draft and height of tide will need to be applied.

Consider Figure 3.2. If the vessel is in the position indicated by other fixing systems, then

$$\text{charted depth} = \text{echo sounder reading} + \text{draft} - \text{height of tide}$$

Alternatively, in an assumed position,

$$\text{echo sounder reading} = \text{charted depth} + \text{height of tide} - \text{draft}$$

3.3.2 *Errors in indicated depth*

In echo sounding, the basic assumption is that the velocity of sound in water is known and is constant. Changes in salinity, temperature and pressure from that assumed when the instrument was originally calibrated will result in errors in depth indication.

Incorrect stylus speed will also result in an incorrect indication of depth. For example, the stylus should move from the zero mark to the 100 metres mark on

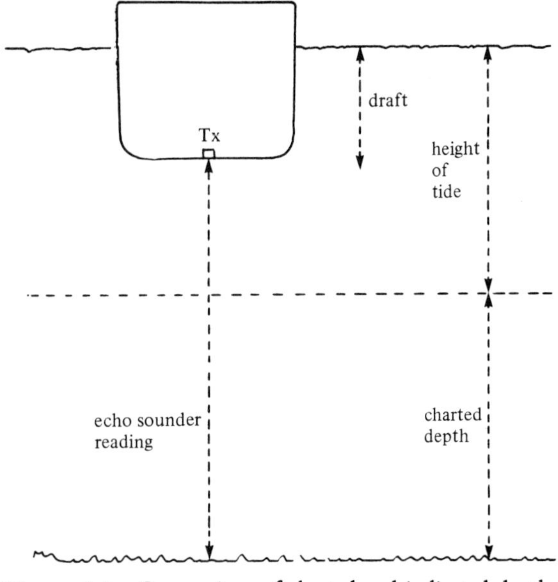

Figure 3.2 Comparison of charted and indicated depth

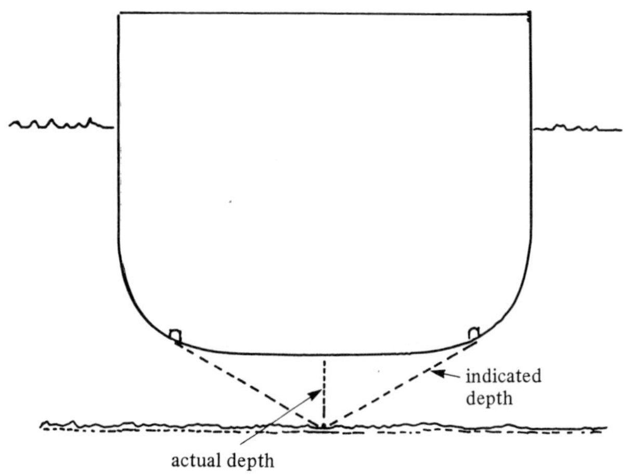

Figure 3.3 Pythagorean error

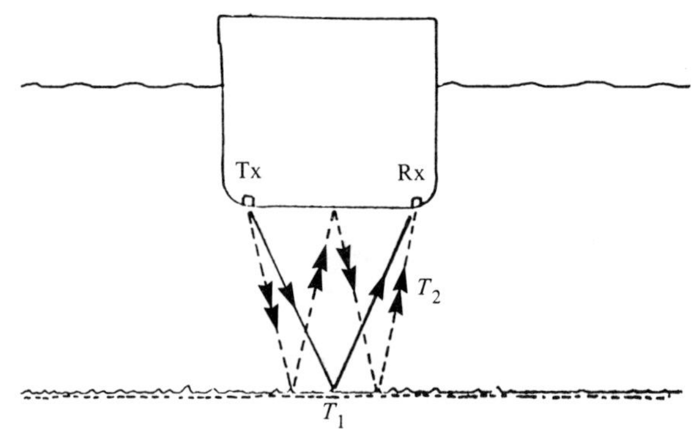

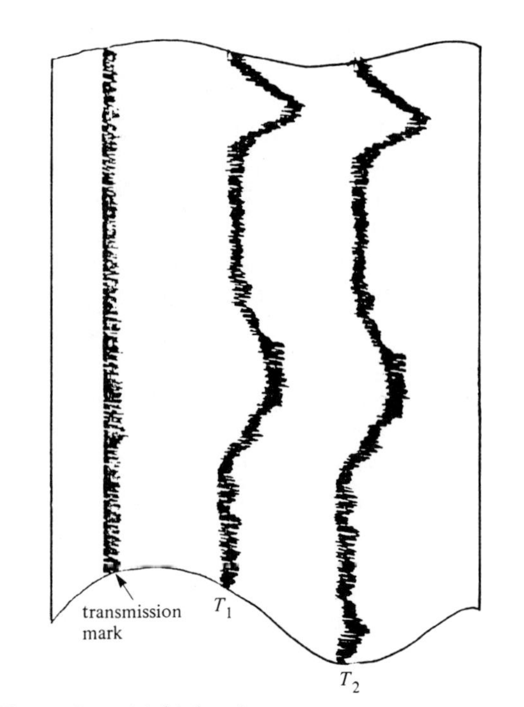

Figure 3.4 Multiple echoes

the scale in the same time as it takes for the acoustic pulse to travel to the sea-bed at 100 metres and back. If the rotation rate of the stylus is too fast then at all depths it will indicate that the water is deeper than it really is, the error being proportional to the depth and the ratio of rotation rates.

Pythagorean error is likely in shallow water where the transducers are widely spaced (Figure 3.3). In addition, in shallow water a high gain setting can result in multiple echoes which, if care is not exercised, can give an incorrect reading of under-keel depth (Figure 3.4).

Air bubbles trapped under the hull as a result of rolling, pounding or working the propeller astern can, because of the high acoustic impedence of the bubbles, result in the dispersion of all the transmitted signal and hence the inability to obtain a reading.

In very deep water the stylus may have completed more than 360° of rotation and, if again over the calibrated scale when the echo signal returns, will indicate a depth far less than the true depth.

3.4 Extracts from official publications

Within Sections 3.4.1–3.4.3, the paragraph numbering of the original publications has been retained.

3.4.1 IMO Resolution A224(VII): Performance Standards for Echo Sounding Equipment

1 Introduction

1.1 The echo sounding equipment required by Regulation 12 of Chapter V, as amended, should provide reliable information on the depth of water under a ship to aid navigation.

1.2 The equipment should comply with the following minimum performance requirements.

2 Range of depths

Under normal propagation conditions the equipment should be capable of measuring any clearance under the transducer between 2 metres and 400 metres.

3 Range scales

3.1 The equipment should provide a minimum of two range scales, one of which, the deep range, should cover the whole range of depth, and the other, the shallow range, one-tenth thereof.

3.2 The scale of display should not be smaller than 2.5 mm per metre depth on the shallow range scale and 0.25 mm per metre depth on the deep range scale.

4 Method of presentation

4.1 The primary presentation should be a graphical display which provides the immediate depth and a visible record of soundings. Other forms of display may be added but these should not affect the normal operation of the main display.

4.2 The record should, on the deep range scale, show at least 15 minutes of soundings.

4.3 Either by marks on the recording paper, or by other means, there should be a clear indication when the paper remaining is approximately 10 per cent of the length of the roll.

6 Pulse repetition rate

The pulse repetition rate should not be slower than 12 pulses per minute.

7 Accuracy of measurement

Based on a sound speed in water of 1500 metres per second, the allowable tolerance on the indicated depth should be:

either: ±1metre on the shallow range scale
±5metres on the deep range scale

or: ±5 per cent of the indicated depth,

whichever is the greater.

8 Roll and pitch

The performance of the equipment should be such that it will meet the requirements of these performance standards when the ship is rolling ±10° and/or pitching ±5°.

3.4.2 UK Statutory Instrument 1984, no. 1203

Part IV Echo sounder installation

Echo sounder performance standards
27 Every echo sounder installation required to be provided shall comply with the performance standards adopted by the IMO and shall, in addition, comply with the relevant performance specification issued by the Department of Transport.

Siting of echo sounder installation
28(1) The transducer unit or units of such echo sounder installation shall be sited so as to avoid, where practicable, the vicinity of all underwater openings in, or projections from, the hull, such as plugs, anodes or other transducers, so that satisfactory overall performance is achieved.
(2) The echo sounder graphical display shall, where practicable, be sited on the bridge in a position to facilitate easy access and viewing, and where the effect of any lighting necessary for the equipment does not interfere with the keeping of an effective look-out.

3.4.3 The Survey of Merchant Shipping Navigational Equipment

3.1 Siting of the transducer

3.1.1 One of the most important considerations to be taken into account when installing echo sounder equipment is the selection of the transducer position.

The ideal position is one in which there is 'solid' water free from aeration beneath the transducer, and where the effect of surface, engine and propeller noise are at a minimum. There are, however, few positions in a ship which are suitable in every respect, and moreover a position found to be satisfactory in one design of ship will not necessarily produce equally good results in another.

3.1.2 The principal source of aeration is the bow wave created by a moving ship. This wave rises some way up the stem, curls over, and then is forced down beneath the ship, taking a quantity of air with it. The resultant bubble stream normally starts about a quarter length of the ship from the stem, and divides about three-quarters of the length from the bow. The bubble stream varies in form and intensity according to the speed, draught, shape of bow and hull, and the trim of the ship as well as the sea state. These factors must be taken into account when siting the transducer. In particular, in the case of a ship with a bulbous bow, the only satisfactory forward site may be within the bulb, although the consequence of physical damage has to be recognized.

3.1.3 To avoid aeration, a position at the forepeak is desirable but it may be unsatisfactory in a ship with a light draught forward, especially in bad weather conditions. In addition, the hull shape may make fitting difficult. In a laden ship of normal design a position within a quarter of the ship's length from the stem will often be found to give satisfactory results. On small vessels damage may occur due to pounding and care should be taken when siting the transducer. An aft position may be more suitable than one forward. Care should be taken, however, to site the receiving transducer a sufficient distance from the propellers to avoid the effects of noise or aeration. When separate transmitting and receiving transducers are fitted, they should be sufficiently separated to prevent interaction between them but the separation should be as small as possible to ensure accurate soundings in shallow water. Positions either side of the keel are sometimes found to be satisfactory.

3.1.4 Other factors which should be borne in mind when fitting transducers are as follows:

3.1.4.1 It is desirable to install the transducer in a horizontal position. In some cases fitting with a slight projection from the hull will be desirable to avoid the effects of aeration at the hull surface. If the transducer projects from the hull it will be necessary to 'fair off' this projection.

3.1.4.2 If, in exceptional circumstances, a windowed transducer has to be used, the window should be acoustically thin, so that the range of the equipment will not be adversely affected.

3.1.4.3 Siting near and particularly aft of obstructions such as the forward propeller, bow thruster, water intake pipes, drain plugs and external speed measuring devices should be avoided.

3.1.4.4 To minimize the effects of roll and pitch a position near to the centreline should be chosen when practicable.

3.1.4.5 When appropriate, care should be taken to minimize interference between echo sounders and Doppler speed devices.

3.1.5 Information regarding the location of the transducers should be kept on the vessel with the equipment handbook.

3.2 Siting of displays

3.2.1 It is recommended that a display be sited in the wheelhouse as close as practicable to the place from which the ship is normally navigated and in a position to facilitate easy access, viewing and servicing, and where the effect of any lighting necessary for the equipment does not interfere with the keeping of an effective look-out.

3.2.2 As echo sounders are normally operated for long periods, adequate ventilation is essential. When positioning the graphical display sufficient space should be allowed around the sides of the display unit to ensure that it is provided with proper ventilation to avoid overheating and the effect of fumes from some types of dry recorder paper.

4 The radio direction finder

The radio direction finder (DF) was one of the earliest electronic navigational aids which was required to be carried. Although somewhat unpopular because of poor positional accuracy, the equipment was comparatively cheap, simple in concept, and well provided with shore-based transmitting stations worldwide. The system has remained virtually unchanged since its inception.

4.1 System introduction

A receiver using a double loop aerial is able to obtain the bearing of a shore-based transmitting station whose position is indicated on the navigational chart. By repeating the process with a number of other stations, a series of position lines (bearings) can be obtained and used to fix the ship's position.

Transmitting stations are indicated on the chart by the symbols

RC ◎ or Aero RC ◎

and details (call sign, service transmission frequency and format, range, grouping and precise position) are promulgated in the Admiralty List of Radio Signals (ALRS), vol. 2, as are Radiobeacon Diagrams.

Shortly after installation, the equipment should have been calibrated and a calibration table completed (see Figure 4.1). The sections relating to the state of the vessel at the time of calibration should have been completed, e.g. derricks topped or lowered, and prior to using the direction finder the vessel should be returned to this condition.

Although there are many different makes of direction finder, the basic operations in obtaining a bearing are much the same for all:

1 Identify suitable stations using the charts (to ensure that the bearings will cut at acceptable angles) and ALRS, vol. 2 (to ensure that the stations chosen have sufficient range).
2 Tune the equipment to the appropriate frequency and identify the station by its call sign.
3 Determine the relative bearing and then use the 'sense' control to resolve any ambiguity.
4 Correct the bearing for:
   (a) Calibration error;
   (b) Ship's head at the time the bearing was taken; and
   (c) Half convergency, if the bearing is to be laid off on a Mercator chart.
5 Repeat to obtain at least two more bearings.
6 Lay off the bearings on the chart.
7 Consider the reliance which can be placed on the position, taking account of the possibility of errors (see Section 4.5), the angle of cut, the distance from the stations and the quality of the aural nulls (if used).

4.2 Setting-up procedure

Having ensured that the vessel is in the same condition as it was when the DF was calibrated (including hoisting the sense aerial), and having decided on suitable stations, switch on and tune in to the appropriate frequency. When marine beacons are being used (those in the frequency band 285 kHz to 315 kHz), switch to the bandspread range covering these frequencies. Select manual (and wide bandwidth) and tune for maximum signal. Rotate the bearing pointer to ensure that it is not already on a null. Change to narrow bandwidth and fine tune. When tuned in, identify the station by its call sign. (It may be necessary to switch in the beat-frequency oscillator (BFO) control if the transmission classification is A0, A1 or A02A and sometimes A2*.)

4.3 Obtaining the bearing

Rotate the bearing pointer until a null (i.e. no signal) is detected, at which time the signal strength meter should have dropped to its lowest reading: note the bearing. Rotate the bearing pointer to approximately the reciprocal of the first bearing and repeat the procedures: again note the bearing. In most cases there will be no uncertainty as to the correct bearing, but if there is then use the sense control (and the stub bearing pointer) to resolve the ambiguity. Normally with the sense control switched on and the stub pointer set to each of the two bearings already obtained, the bearing associated with the lowest reading on the meter is correct. The sense procedure should only be used to determine which of the two bearings *already obtained* is the correct one and should *not* be used in its own right to obtain a bearing.

If an auto facility is available for obtaining the bearing, it is essential to confirm station identity after tuning to the correct frequency. On selection of auto the bearing pointer should then move automatically to the correct bearing. Auto has a particular advantage when working with grouped beacons in that no retuning is required and so the bearing pointer should indicate the bearing of each station in the group (as it transmits) sequentially. Note that it is also possible to identify a beacon in a group by the time slot in which it is transmitting.

4.4 Correction of the relative bearing

This is probably most easily explained by means of an example.

Example

While a vessel was heading 043° (T) in DR position lat. 59° 00′ N, long. 8° 00′ W, the relative bearing on Sule Skerry radiobeacon (lat. 59° 05.1′ N, long. 4° 24.3′ W) was 034° (relative). Using the calibration table (Figure 4.1), determine the bearing to lay off on a Mercator chart.

relative bearing = 034 degrees
correction from calibration curve = + 13.5 degrees

corrected relative bearing = 047.5
ship's head = 043 (T)

true great circle bearing = 090.5 degrees
half convergency = 1.5

True (Mercator) bearing = 092 degrees

The half convergency is obtained from nautical tables or from the graph in NP 282.

4.5 Error sources and accuracy

4.5.1 *Quadrantal error*

Perhaps the main source of error in DF bearings is re-radition from the ship's structure (see Sections 4.8.2 and 4.8.3). Residual errors are determined by means of calibration. The procedure is akin to the swinging of a ship to correct the compass and draw up a deviation table, but in this case visual and DF bearings are taken of a shore-based radio transmitter as the vessel steams in a circle. The difference between the visual and radio bearing is graphed against relative (radio) bearing.

4.5.2 *Night effect*

When bearings are obtained in the hour before sunrise and the hour after sunset, the effect of skywave

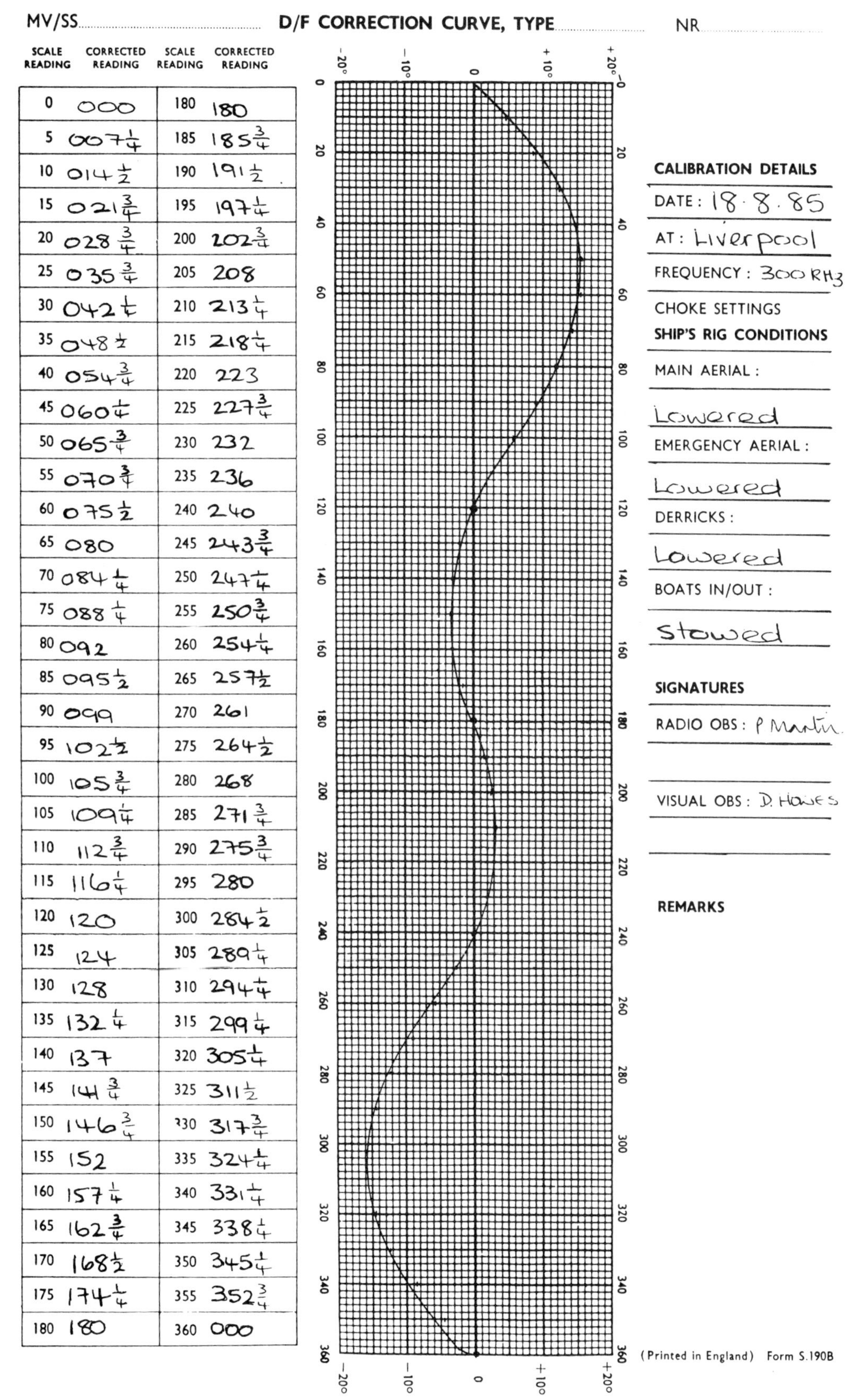

MV/SS.................... **D/F CORRECTION CURVE, TYPE**.................... NR....................

SCALE READING	CORRECTED READING	SCALE READING	CORRECTED READING
0	000	180	180
5	007¼	185	185¾
10	014½	190	191½
15	021¾	195	197¼
20	028¾	200	202¾
25	035¾	205	208
30	042¼	210	213¼
35	048½	215	218¼
40	054¾	220	223
45	060¼	225	227¾
50	065¾	230	232
55	070¾	235	236
60	075½	240	240
65	080	245	243¾
70	084¼	250	247¼
75	088¼	255	250¾
80	092	260	254¼
85	095½	265	257½
90	099	270	261
95	102½	275	264½
100	105¾	280	268
105	109¼	285	271¾
110	112¾	290	275¾
115	116¼	295	280
120	120	300	284½
125	124	305	289¼
130	128	310	294¼
135	132¼	315	299¼
140	137	320	305¼
145	141¾	325	311½
150	146¾	330	317¾
155	152	335	324¼
160	157¼	340	331¼
165	162¾	345	338¼
170	168½	350	345¼
175	174¼	355	352¾
180	180	360	000

CALIBRATION DETAILS

DATE: 18.8.85

AT: Liverpool

FREQUENCY: 300 kHz

CHOKE SETTINGS

SHIP'S RIG CONDITIONS

MAIN AERIAL: Lowered

EMERGENCY AERIAL: Lowered

DERRICKS: Lowered

BOATS IN/OUT: Stowed

SIGNATURES

RADIO OBS: P Martin

VISUAL OBS: D. Howes

REMARKS

(Printed in England) Form S.190B

Figure 4.1 Calibration curve

contamination of the groundwave is particularly indeterminate; the received composite signal is subject to fading, and the null is likely to blur. Obtaining bearings at these times should be avoided or, if that is not possible, the bearings showing signs of fading, blur, wide or erratic movement of the null should not be relied upon. Skywaves can affect the groundwave signals throughout the night but tend to result in less instability of the bearings.

4.5.3 *Land effect*

When the signal path changes from over land to over water there is a change in velocity of propagation and hence a refraction effect. It can be seen from Figure 4.2 that the vessel's plotted position is displaced. The amount of displacement is increased the further inland the transmitting station is situated.

Effects similar to night effect can occur as a result of signals, reflected from coastal features, if they arrive out of phase with the direct wave. Included in the entry in ALRS, vol. 2 for a particular beacon will be details of any unreliable sectors due to these causes.

4.5.4 *Half convergency*

The path of radiowaves over the earth's surface is a great circle, and so the bearing as measured by DF is a great circle bearing. Before this can be laid off on a Mercator chart it must be corrected for half convergency. The values are tabulated in nautical tables and graphed in NP282. The application of the correction is always to shift the bearing at the ship towards the equator (Figure 4.3).

4.6 Shore-based direction finding stations

It is possible in some areas of the world to request a number of shore stations to obtain the bearing of the ship. The bearings are then used to fix the position of the ship. The ship will be requested to transmit at a specific frequency. This service normally attracts a fee. Some years ago in the UK, MF/DF shore stations were withdrawn. Some ten coastguard stations have now been fitted with VHF/DF equipment whose prime purpose is to allow the coastguard to obtain the position of vessels transmitting distress messages by VHF. Details of the service are contained in ALRS, vol. 2. This service should not be confused with the VHF radio lighthouses which can be used for obtaining a bearing by listening to special transmissions on VHF channel 88.

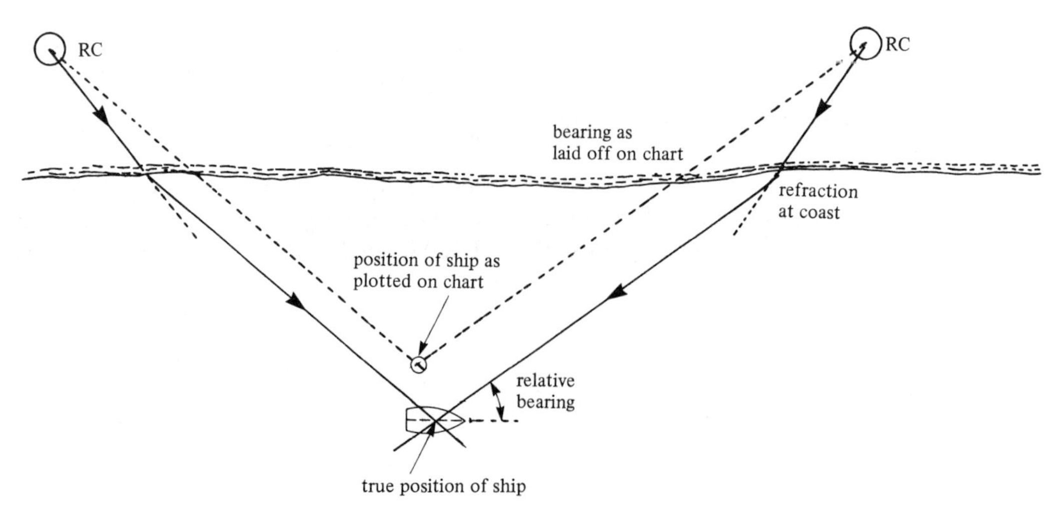

Figure 4.2 The displacement of position due to land effect

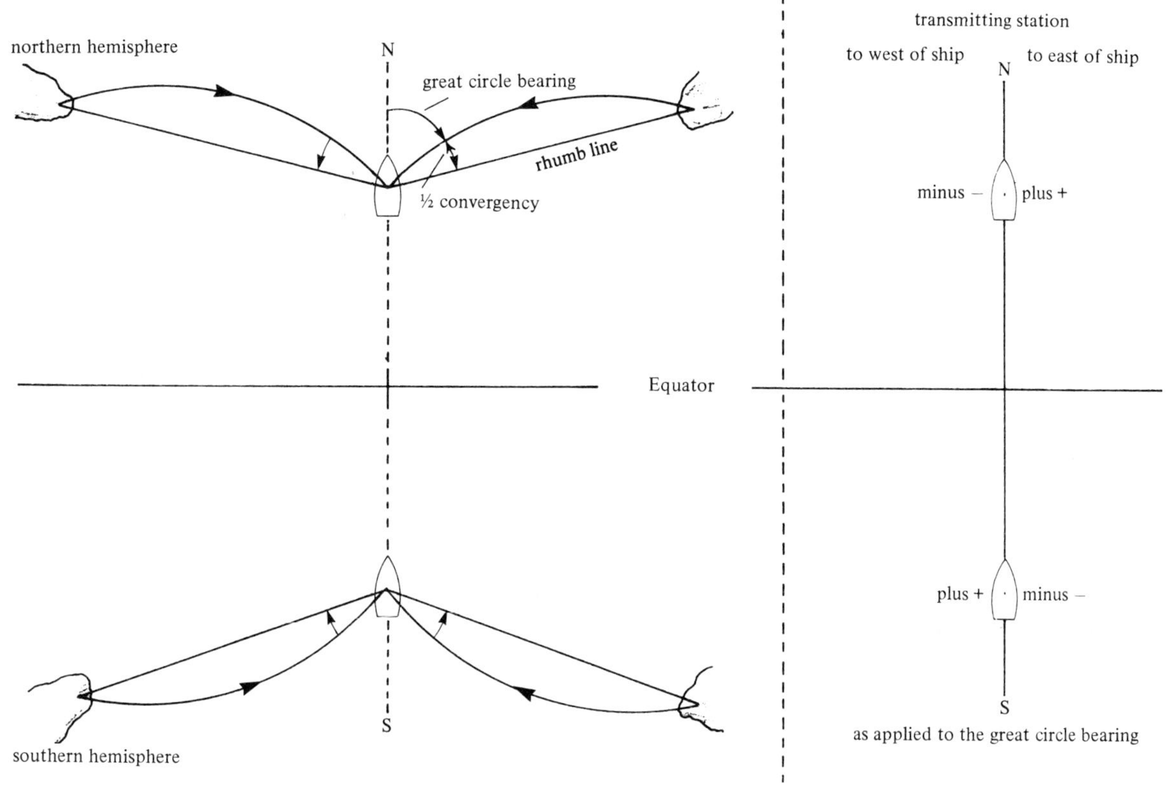

Figure 4.3 The application of half convergency

4.7 Fixing position

4.7.1 The effect of distance from the transmitting station

Some DF stations have a range of some 200 nm, and it is important to consider the effect of bearing error of just ±1° on the position obtained at a range of 200 nm. As Figure 4.4 shows, with a good angle of cut an error circle of some 3.5 nm radius can result. For typical accuracy of position see Section 10.2.

4.7.2 Grouped beacons

Since 1 April 1992, the grouping of beacons around the United Kingdom has been discontinued; also all UK Marine Radio Beacons require the use of BFO.

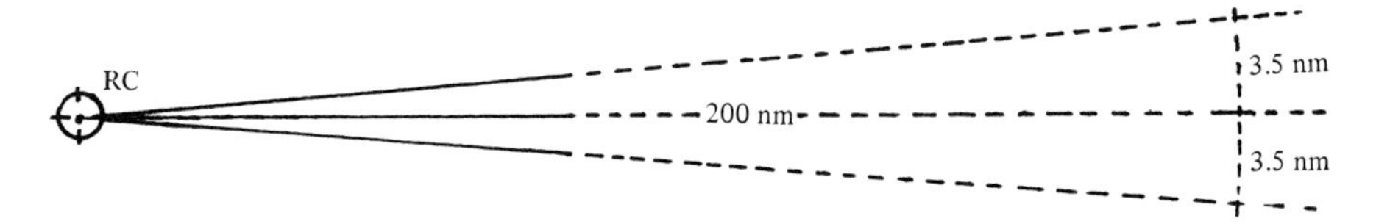

Figure 4.4 The effect of a 1° bearing error at 200 nm

4.8 Extracts from official publications

Within Sections 4.8.1–4.8.3, the paragraph numbering of the original publications has been retained.

4.8.1 IMO Resolution A223(VII): Performance Standards for Radio Direction Finding Systems

1 Introduction

1.1 The direction finding equipment required by Regulation 12 of Chapter V is to indicate both bearing and sense of radio transmissions in the frequency bands specified in paragraph 2 of this Recommendation.

1.2 In addition to the provisions of Regulation 11 of Chapter IV, as amended, the equipment should comply with the following minimum performance requirements.

2 Frequency ranges and classes of emission

The equipment should be capable of receiving signals of classes of emission A1, A2 and A2H in the frequency range 255 kHz to 525 kHz and A1, A2, A2H, A3 and A3H in the frequency range 2167 kHz to 2197 kHz.

3 Selectivity

The selectivity should be such as to allow a bearing to be taken readily without interference from other radio transmissions on frequencies more than 2 kHz from the desired signal.

4 Signal identification

4.1 Means of audio monitoring should be provided regardless of the method used for direction finding.

4.2 The equipment should be suitable for use with headphones. A loudspeaker, if provided, should be capable of being rendered inoperative by simple means.

5 Bearing indication

Means should be provided to indicate the bearing of the desired transmission. Such indication should be capable of being easily, rapidly and precisely resolved within 0.25 degrees.

6 Bearing accuracy

6.1 The instrumental accuracy in taking relative bearings should be within ±1°. This requirement should be met at all frequencies in the frequency bands specified in paragraph 2 of this recommendation and throughout the whole 360 degrees of azimuth at field strength values between 50 μV/m and 50 mV/m.

Note The instrumental accuracy referred to above does not include the operational accuracy attainable in service, which should be determined for each installation taking into account paragraphs (a)(iv), (a)(v) and (a)(vii) of regulation 12 of chapter IV. In particular the operational accuracies in the 2 MHz band should be sufficient for homing purposes.

6.2 Preset facilities to correct the quadrantal error should normally be provided for the frequency band 255–525 kHz.

7 Manual controls and their operation

7.1 A tuning scale or indicator should be provided, calibrated to indicate directly the carrier frequency of the signal to which the equipment is intended to be tuned.

7.2

(a) If a tuning scale is provided, at all points in its range, 1 mm should correspond to not more than 2.5 kHz in the frequency range 255–525 kHz.
(b) The maritime distress frequencies should be prominently marked.
(c) Where other means of frequency indication are provided, the resolution should be at least 1 kHz.

7.3 All controls should be of such size and location as to permit normal adjustments to be easily performed, and should be easy to identify and use.

7.4 The sense switch, if fitted, should be of a non-locking type.

8 Operational availability

The equipment should be ready for operation within 60 seconds of switching on.

11 Special requirements for different methods of direction finding

11.1 Aural minimum method:

(a) With a field strength sufficient to ensure a signal/noise ratio of at least 50 dB, a change in the setting of the bearing indicator of 5° in either direction from the position of minimum output should cause the audio-frequency output to increase by not less than 18 dB. Similarly, a change of 90° in either direction should cause an increase of not less than 35 dB.
(b) The equipment should be provided with a minimum-clearing control giving a noticeable minimum of the output at all settings.
(c) The sense should be determined with reference to the lower output.
(d) The sense ratio in the frequency ranges 255–525 kHz and 2167–2197 kHz should be 15 dB and 10 dB, respectively.
(e) The automatic gain control, if provided, should be rendered inoperative automatically when the equipment is used for bearing determination.

11.2 Other methods:

(a) There should be means of indicating that the receiver gain and signal strength are sufficient to enable a correct bearing to be taken.
(b) With a field strength of 1 mV/m the indicated bearing should not change by more than 1° when the receiver is detuned to a point where the indication referred to in subparagraph 11.2(a) shows that the signal strength is just sufficient to take a bearing.
(c) For any signal of strength sufficient to give a bearing indication, there should be no observable change of indicated bearing when the beat-frequency oscillator is switched on.
(d) Fluctuations of the indicated bearing caused by any servo mechanism should not exceed ±0.5° from the mean value.
(e) If, after identifying a station the bearing of which is required, it is necessary to check or alter the adjustment of any control as part of the process of direction finding, this check and adjustment should be capable of being made within 10 seconds.

12 Miscellaneous

12.1 The equipment should be protected from excessive voltages induced in the aerials.

4.8.2 UK Statutory Instrument 1984, no. 1203

Part VI Direction finder installation

Direction finder performance standards

31 Every direction finder installation required to be provided shall comply with the performance standards adopted by the IMO and shall, in addition, comply with the relevant performance specifications issued by the Department of Trade and Industry (now Department of Transport).

Interference with reception

32(1) Radio antennas installed in any ship required to be fitted with a direction finder installation which rise above the base of, and are within 17 metres horizontal distance of, the loop antennas of the direction finder installation shall be isolated whenever bearings are being obtained by the direction finder installation: provided that such radio antennas which do not cause significant errors in the accuracy of the bearings obtained by the direction finder installation need not be isolated.

(2) Any ship fitted on or after 25 May 1980 with a direction finder installation shall be provided with a communal antenna system for all broadcast receivers in respect of which it is impracticable to erect efficient and properly installed antennas which:

(a) Are outside a radius of 17 metres from the direction finder antennas; or
(b) Do not rise above the base of the direction finder antennas; or
(c) Can be lowered quickly and stowed easily when the direction finder is in use.

Siting of direction finder installation

33(1) The direction finder shall be so sited that efficient determination of radio bearings by means of the direction finder will not be affected by extraneous noises.

(2)(a) The direction finder antenna system shall be mounted in such a manner that the efficient determination of radio bearings by means of the direction finder will be affected as little as possible by the proximity of antennas, derricks, wire halyards and other large metal objects.

(b) Adequate precautions shall be taken to protect the cables connecting the direction finder antenna system with the receiver forming part of the direction finder installation from the ingress of water and from damage, including any which might be caused by excess heat.

(3) The direction finder installation shall, where practicable, be mounted so as to prevent the performance and reliability of the installation being adversely affected by vibration and so that the installation will not, whilst in service, normally be subject to greater vibration than that specified in the relevant performance specification for the climatic and durability testing of maritime radio equipment, issued by the Department of Trade and Industry (now Department of Transport).

Means of communication

34(1) In every ship required to be fitted with a direction finder installation an efficient two-way means of calling and voice communication shall be provided between the receiver forming part of the direction finder and the position from which the ship is normally navigated.

(2) In every such ship an efficient means of signalling shall be provided for use when calibrating or taking check bearings of the direction finder installation between the receiver forming part of the direction finder installation and the place on the ship from which visual bearings are taken.

(3) If the direction finder installation is not installed in the ship's radiotelegraph operating room, and radio antennas on the ship are required by regulation 32(1) of these regulations to be isolated, means shall be provided at the direction finder operating position to indicate when such antennas are isolated.

Calibration

35(1) The master of every ship required to be fitted with a direction finder installation shall cause the direction finder installation to be calibrated in accordance with this regulation as soon as practicable after it has been installed in the ship and whenever any change is made in the position of the direction-finder antenna system.

(2)(a) The direction-finder installation shall be calibrated by two persons, one being experienced in the taking of radio bearings and the other experienced in the taking of visual bearings. The calibration shall be carried out by taking simultaneous radio and visual bearings of a transmitter, and such bearings shall be taken at intervals of not greater than 5 degrees throughout 360 degrees on a frequency between 283.5 kHz and 315 kHz.

(b) Calibration tables and curves, which enable radio bearings obtained by the direction finder installation to be adjusted to within two degrees of the correct bearing, shall be prepared on the basis of the bearings taken in accordance with paragraph (2)(a) of this regulation.

(c) Following satisfactory calibration and the preparation of calibration tables and curves, a Certificate of Calibration of Direction Finder shall be completed in the form specified in schedule 2 to these regulations.

(d) On each occasion that an arrangement of cargo carried above deck level varies significantly from an arrangement in respect of which the direction finder installation has been calibrated, check bearings shall be taken, if practicable, to determine whether any substantial inaccuracy in the direction finder installation is being caused by the arrangement of cargo. Where substantial errors are found, further check bearings shall be taken to establish a correction curve.

(3) The master of every such ship shall cause the calibration tables and curves prepared in accordance

SCHEDULE 2 Regulation 35(2)*(c)*

Certificate of Calibration of Direction-Finder

We, the undersigned, hereby certify that we have today—

(a) calibrated, in accordance with Part VI of the Merchant Shipping (Navigational Equipment) Regulations 1984, the direction-finder installed in the s.s. / m.v. ....

(b) handed to the master of that ship tables of calibration corrections;

(c) adjusted the said direction-finder so that the readings taken thereby, when corrected with such tables, differ from the correct bearings by no more than plus or minus two degrees.

We hereby further certify that the master of the said ship has been furnished with a list or diagram indicating the position, at the time of such calibration, of the antennas and of all moveable structures on board the ship which might affect the accuracy of the direction-finder.

....Radio Observer

....Visual Observer

....Date

with the foregoing provisions of this regulation to be verified by means of not less than four check bearings in each quadrant:

(a) At intervals not exceeding 12 months; and
(b) Whenever any change is made in any structure or fitting on deck or in any rigging or antenna above deck which is likely to affect the accuracy of the direction finder.

If such verification shall show that the calibration tables or curves are substantially inaccurate, the master of the ship shall cause the direction finder to be recalibrated as soon as practicable in the manner specified in paragraphs in (2) and (3) of this regulation.
(4) In addition, bearings shall be taken in each quadrant, where practicable at intervals not exceeding 12 months, on a frequency at about 500 kHz. These bearings should not be substantially inaccurate after being corrected by use of the calibration curves.

Records of calibration and verification
36 The master of every ship required to be fitted with a direction finder installation shall cause the following records to be kept in a place accessible to any person operating the direction finder, and to be available for inspection at any reasonable time by a surveyor of ships:

(a) A list or diagram indicating the position, on the most recent occasion on which the direction finder was calibrated, of the antennas and all movable structures on board the ship which might affect the accuracy of the direction finder;
(b) The calibration tables and curves which were prepared on the most recent occasion on which the direction finder was calibrated;
(c) A certificate of calibration signed by the persons making the calibration relating to the most recent occasion on which the direction finder was calibrated; and
(d) A record, in the form specified in schedule 3 to these regulations, of check bearings taken for the verification of calibration, the bearings being numbered in the order in which they were taken.

4.8.3 The Survey of Merchant Shipping Navigational Equipment

4.1 Siting

4.1.1 The loop antenna system should be mounted as near as practicable to the centreline of the ship, and as far as possible from the antennas of other radio equipment, and from large movable metal objects such as derricks and wire halyards.

SCHEDULE 3

Regulation 36(*d*)

Record of Check-Bearings taken by means of the Direction-Finder

(1)	Serial Number of Bearings	
(2)	Date	
(3)	Times (GMT(UTC) and ship's)	
(4)	Latitude (Ship's Approximate Position)	
(5)	Longitude (Ship's Approximate Position)	
(6)	Distance from Transmitter	
(7)	Direction-Finder Bearing of (Name and frequency)	
(8)	Direction-Finder Relative Bearing Correct for QE	
(9)	Ship's Head by Compass 0/360°	
(10)	Total Compass Error	
(11)	½ Convergency Applied	
(12)	Ship's Head Corrected (True)	
(13)	True Bearing by Direction-Finder [Col. (8) and Col. (12)]	
(14)	True Bearing by Visual Check or Calculation (whether Visual or Calculation to be indicated; if by Calculation, the method to be stated)	
(15)	Correction required to make Col. (13) equal Col. (14) (indicating whether − or +)	
(16)	Signature of Observer or Observers	

4.1.2 Fixed metallic structures may give rise to considerable errors depending on their distance from the loop antenna system. The latter should be placed not less than two metres from such structures which rise above the base of the loop. The best results are obtained when the distribution of the metallic structure is symmetrical in relation to the loop.

4.1.3 Wire stays should be broken by insulators if they are so close to the loop antenna that the accuracy of the direction finder is likely to be prejudiced. Long wire stays should be broken into lengths not exceeding six metres.

4.1.4 It should be ensured that antenna feeder cables are adequately protected against physical damage and, where necessary, from the effect of excess heat such as funnel efflux. It should also be ensured that cable joints are adequately protected against the ingress of water, particular attention being given to joints where the cables enter the deck-head of the space in which the direction finder is installed. Arrangements may also be necessary at the foot of a pedestal to permit egress of condensed water and to permit suitable ventilation of the pedestal and loop system.

4.1.5 Metal antenna pedestals and any protecting metalwork for feeders should be bonded to earth.

4.1.6 Surveyors should ensure that insulation on antennas is clean and free from paint; particular attention should be paid to metallic loop antenna because changes of impedance in the loop system can introduce significant bearing errors.

4.1.7 An efficient two-way means of calling and voice communication must be provided between the direction finder receiver position and the bridge from which the ship is navigated. There must also be an efficient means of signalling by either a gong, buzzer or other similar means which provides a single sharp aural indication, between the DF receiver position and the ship's standard compass or gyro compass repeater for use when calibrating for taking check bearings.

4.2 Calibration

4.2.1 The direction finder must be calibrated after installation in the ship, whenever the position of the loop antenna system is changed, or whenever any alteration is made to any structure or fitting on deck which is likely to affect the accuracy of the equipment. In each instance calibration should be completed before commencement of the first voyage after the installation or alteration. Where this is impracticable, calibration should be carried out during the early stage of the first voyage using the first calibration facilities which become available.

4.2.2 Before calibration starts the apparatus should be carefully checked for mechanical and electrical performance.

4.2.2.1 Other antennas in the ship, including those provided for Omega, Decca Navigator, satellite navigation and broadcast receivers, can cause serious errors to radio bearings, especially when rising above the base of, and near to, the loop antenna and when connected to transmitters or receivers which are tuned to or near to the frequency on which radio bearings are being taken.

4.2.2.2 All wire and self-supported antennas which have any part rising above the base of the loop antenna and are within 17 metres of it, should be kept isolated during calibration and whenever the direction finder is actually in use unless and until they have been individually tested and found not to introduce any substantial error. Such tests should be made on first installation by noting the maximum effect on radio bearings when the antenna is alternatively connected to and isolated from equipment tuned to a frequency which is as near as possible to the frequency on which the test radio bearings are being taken.

4.2.2.3 Other antennas, such as satellite terminals and radar scanners, should be treated as fixed metallic structures as described in paragraph 4.1.2.

4.2.2.4 The antennas used for maintaining the safety watch, i.e. those to which the radiotelegraph auto alarm and radiotelephone distress frequency watch receiver are normally connected, should, where practicable, be erected so that they are below the base of the loop antenna or outside a radius of 17 metres from the loop antenna. If this is impracticable the antennas should be tested as in paragraph 4.2.2.2, and if it is found that serious error to radio bearings is being caused the shipowner or his representative should be requested to display warning notices at the direction finder installation regarding isolation of these antennas.

4.2.2.5 The haphazard erection of broadcast antennas can be a serious source of error in radio bearings. All broadcast receivers should be attached to either

(a) A communal antenna system; or
(b) Antennas which do not rise above the base of the loop antenna; or
(c) Antennas which are outside a 17 metre radius from the loop antenna.

Any broadcast antennas other than those mentioned in (b) and (c) above should always be dealt with in the manner described in paragraph 4.2.2.2.

4.2.2.6 The calibration data should include a list of all antennas showing their condition (i.e. whether isolated or connected) and also the position and condition of any movable deck structures and deck cargo etc. so that, as far as is practicable, the same conditions may obtain when the direction finder is used navigationally.

4.2.2.7 When new antennas are fitted or existing ones are resited and come within the above conditions as to height and distance from the loop antenna, they should be tested immediately as described and particulars and condition for use entered in the record.

4.2.3 Calibration should be carried out clear of land so as to minimize the effects of coastal refraction and, whenever practicable, during the period between two hours after sunrise and two hours before sunset. It may be carried out by either:

(a) Swinging the ship in relation to a fixed radio station; or
(b) Receiving signals transmitted by an auxiliary ship while circling the stationary ship.

It should be noted that unless the auxiliary ship uses either a vertical transmitting antenna or a T antenna in which the horizontal limbs are substantially symmetrical with respect to the vertical limb, large and variable errors in bearing may be experienced.

4.2.4 When possible the fixed radio stations referred to in paragraph 4.2.3(a) should be a special calibration station or a maritime radiobeacon transmitting a suitable and adequately strong signal. The siting conditions of other radio stations may affect the accuracy of bearings.

4.2.5 The ship should be so placed that throughout the period of calibration the position of the antenna of the calibrating transmitter can be seen. The distance between the calibrating transmitter and the direction finder should not be less than 1 mile.

4.2.6 Visual and radio bearings should be taken simultaneously at intervals of not more than 5° in all arcs in which visual bearings are possible. The visual bearings should, where practicable, be taken from positions which do not introduce parallax error into the calculations. The radio bearings must be taken on a frequency within the band 285–315 kHz (radiobeacon band). In each instance the visual and radio bearings must be recorded.

4.2.7 Curves and correction tables must be prepared from the details of visual and radio bearings taken during the calibration period. These curves and tables and data as required by paragraph 4.2.2.6 must be kept on board for the use of any person operating the direction finder. The recorded radio bearings when corrected from the curves should not differ from the correct bearings by more than 2°.

4.2.8 Details of established calibration facilities are contained in Admiralty List of Radio Signals, vol. 2.

4.3 Check bearings

4.3.1 The curves and tables must be verified at least once a year and whenever any changes are made in any structure or fitting on deck which is likely to affect the accuracy of the equipment by taking at least four check bearings in each quadrant. During these checks, the effect on radio bearings of antennas falling within the limits detailed in paragraph 4.2.2.2 should be confirmed.

4.3.2 Ships carrying cargo above deck level in an arrangement which differs significantly from one for which tables and curves are carried should, if practicable, take check bearings early in the voyage. Check bearings should be taken by all ships as often as practicable on all frequencies liable to be used for the taking of navigational bearings. The usual method of checking the accuracy of calibration data is to take a series of radio bearings and visual bearings while the ship is on passage in the vicinity of a radiobeacon or coast radio station. If the check bearings disclose errors which differ from those indicated by the calibration curve, it should be remembered that these curves are not always purely quadrantal in shape; the new readings, therefore, should not be used to amend the calibration curve in unchecked sectors although they may be useful as a general guide. Useful check bearings may also be taken on other ships at sea.

4.3.3 Check bearings must be taken in each quadrant at least once a year on a frequency of about 500 kHz to ensure that bearings taken on the radiotelegraph distress frequency are not substantially inaccurate when corrected by the tables and curves. However, when taking check bearings on frequencies above 315 kHz, it should be borne in mind that the tables and curves may not be very accurate for such frequencies.

4.3.4 If at any time check bearings show significant discrepancies from the data obtained at the last calibration, the direction finder should be used with extreme caution and should be fully recalibrated at the earliest opportunity.

4.4 Radio navigational bearings

When the direction finder is used for taking radio bearings it is important that the movable deck structures and antennas of the other radio equipment in the ship should be in the same position and condition as when the direction finder was calibrated.

4.5 Calibration records

4.5.1 The following records must be kept on board the ship in a place accessible to any person operating the direction finder, and available for inspection by the surveyor:

(a) A list or diagram indicating the position and condition at the last occasion on which the direction finder was calibrated of the antennas on board the ship and of all movable structures which might affect the accuracy of the direction finder (see paragraph 4.2.2.1);
(b) The calibration tables and curves which were prepared as a result of the last calibration (see paragraphs 4.2.7 and 4.3.1);
(c) A certificate of calibration, in the form specified in the regulations, signed by the persons who made the last calibration;
(d) A record, in the form specified in the regulations, of all check bearings taken for verification purposes, and of any other information which might be useful in showing whether or not the equipment gives satisfactory performance (see paragraphs 4.2.7 and 4.3.1). The bearings must be entered and numbered in the order in which they are taken.

5 Radar

5.1 System introduction

The intrinsic property of a marine radar system is the ability to detect certain objects (referred to as targets) which are present in the vicinity of a radar equipped vessel and hence to determine the range and bearing of such targets. Any other data obtained from the radar system is deduced from these two basic quantities. It is assumed that the reader has studied the detailed principles which underlie the operation of the system. This introduction is intended merely to summarize the application of such principles as an initial step in considering the use of radar data in navigation control.

Figure 5.1 shows a simplified block diagram of a marine radar system.

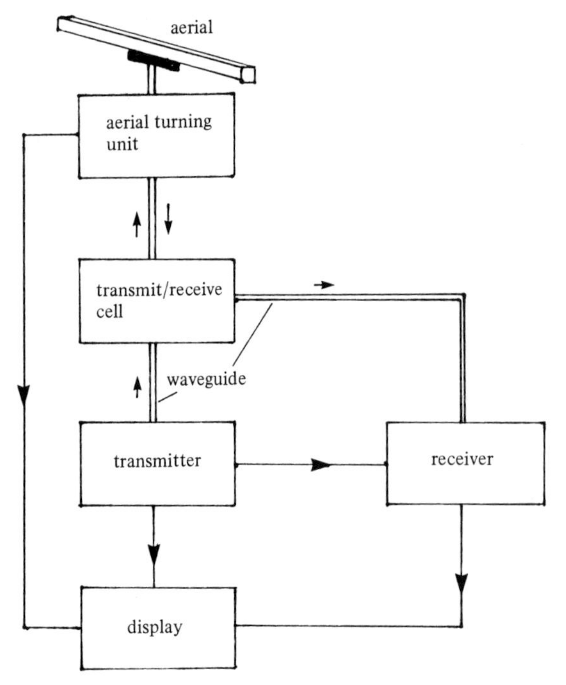

Figure 5.1 A simplified block diagram of a marine radar system

5.1.1 *The transmitter*

This unit generates short, powerful pulses of radio energy. All civil marine radar systems operate in either the X band (3 cm wavelength) or in the S band (10 cm wavelength). From 1 February 1995 there is a requirement for one radar to operate in the X band (see Section 10.3.1(g) and (h)). The number of pulses transmitted in one second may vary from about 500 to 3500, and this quantity is known as the pulse repetition frequency (PRF). Normally a higher rate is employed when the shorter range scales are selected. The pulse length may vary from about 0.05 to 1 μs, and this value too is likely to vary with range scale selection. Longer pulses are employed when the longer range scale are selected.

At the same time as each pulse is transmitted, a trigger (or synchronizing pulse) is sent to the receiver and display. This initiates the timing circuitry which makes it possible to measure the range of the targets from the elapsed time between transmission of pulse and reception of echo.

5.1.2 *The antenna*

Commonly referred to as the scanner, this device ensures that the pulses are transmitted in a beam which is narrow (0.5°–2.0°) in a horizontal sense but wide (approximately 20°) in a vertical sense. The returning echoes are detected within the same limits, thus making it possible to determine the bearing of

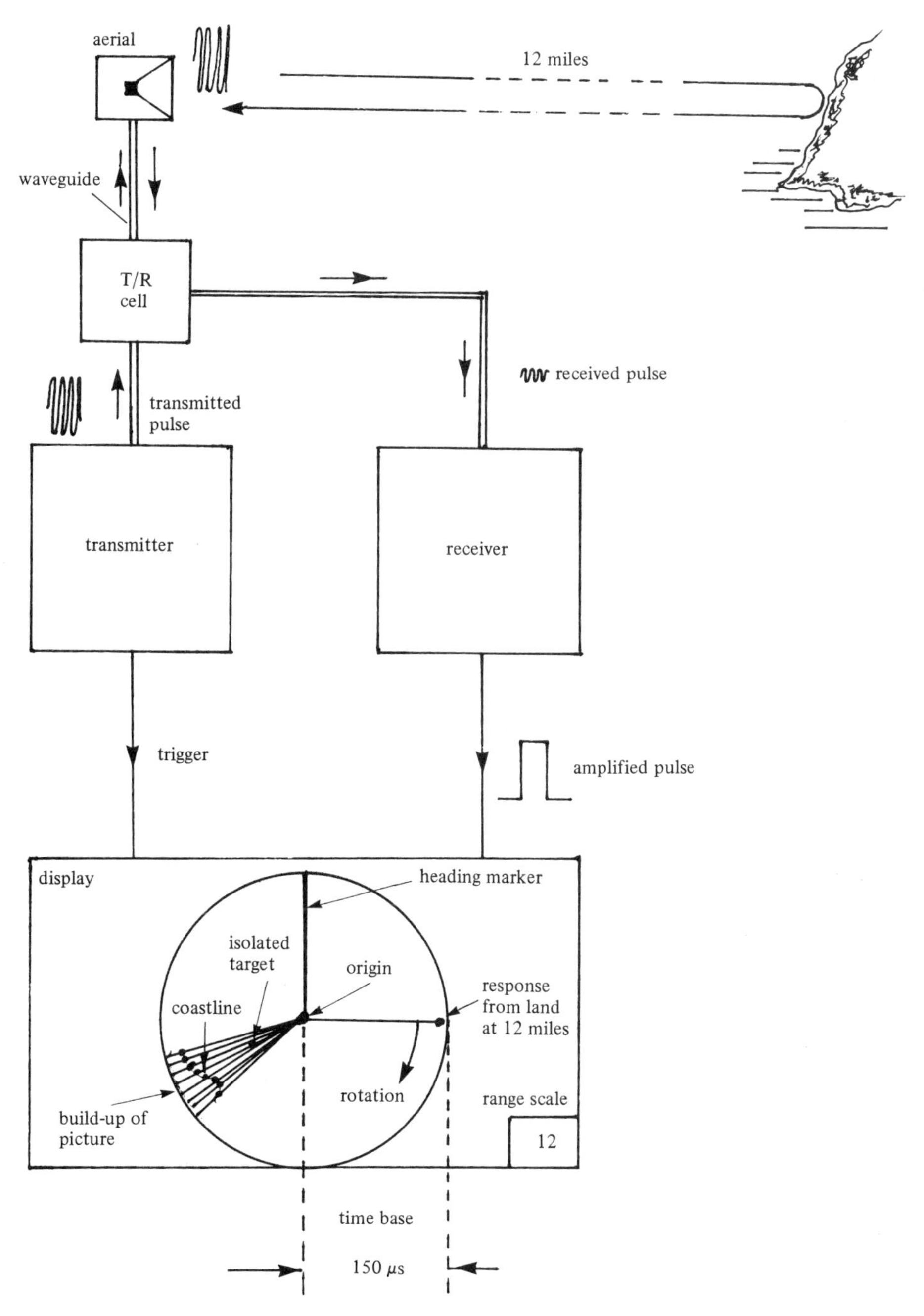

Figure 5.2 The build-up of the radar picture

targets. By rotating the antenna, the axis of the horizontal beam covers the full 360° of azimuth in about 3 seconds.

5.1.3 The receiver

This unit detects the extremely weak pulses captured by the antenna. It processes the pulse such that it is of suitable power and form to produce a visible response on the display.

5.1.4 The display

The essential feature of the display is a cathode ray tube having a circular screen, the inside of which is coated with a long persistence phosphor. An echo is displayed as a bright spot whose distance from the origin is proportional to the range of the target. The echo is painted once every rotation of the antenna and the persistence of the screen ensures that it remains visible until the next paint is produced. Traditionally, the radar picture has been displayed in real time, but since the early 1970s synthetic pictures have become progressively more common. With the advent of rasterscan radars, it seems that in many applications the real time picture may well disappear forever.

As mentioned before, an essential feature of the radar display is that the echo should appear at a distance from the origin which is proportional to the range of the target. Traditionally, this has been achieved as illustrated in Figure 5.2.

Consider the antenna to be stationary with the axis of the beam in the horizontal plane, at 90° to the ship's head. Let there be a detectable target at a range of 12 miles and the display be switched to the 12 mile range scale. At the instant that the pulse leaves the antenna, a spot of light leaves the origin and sets off towards the edge of the screen, in a direction which represents 090° (relative). The speed of the spot is set (by range scale selection) to be such that the time taken for it to transit the radius of the screen (the timebase) is the same as that for a radar pulse to travel the actual distance represented by the radius, and back, in this case 24 miles (12 miles there, 12 miles back).

The speed of the spot is literally faster than the eye can see, taking only 150 μs to travel from centre of screen to edge. As a result, it appears as a radial line. In a correctly set up display, the brilliance is adjusted such that the line (or trace) is barely visible because the returning echo will be indicated by an increase in the brightness of the spot.

In the example chosen, the echo will return just as the spot is reaching the edge of the screen, and will thus produce a bright spot, representing, in the chosen example, a range of 12 miles. Where there is a detectable target at 6 miles, its echo would return when the spot is half-way to the edge of the screen and would thus indicate an echo at 6 miles. In summary, the essential features are:

1 Each radial line commences at the instant of transmission.
2 The spot speed is related to the speed of the radar pulse by the range scale selected.
3 The returned echo brightens the spot.

From this it can be seen that the range is deduced by time measurements and that an echo is displayed at the instant the timing is completed. The radar pulse makes its journey to and from the target in the same time that the spot is writing the line on the screen. For this reason, this type of display is referred to as a real time display. Until a few years ago all civil marine radar pictures were produced in this way, and hence the expression 'real time' was seldom used. More recently synthesized displays, in which the information is stored and retrieved, have been developed, and these will be considered in Section 5.1.5.

In the foregoing example, one line only was considered. As the antenna rotates, successive radial lines are drawn. Taking average values of PRF (say 1200) and antenna rotational speed (say 20 rpm), it can be shown that the picture is built up of about 3600 radial lines, i.e. roughly ten lines per degree. As a result, a plan view is built up, which shows the targets in the correct angular relationship one to another. A rotating trace display is sometimes referred to as a plan position indicator (PPI) since the area around the vessel is displayed in plan.

When the radar beam crosses the fore-and-aft line in the forward direction, several traces are fully

Table 5.1 Terminology associated with traditional and stored radar pictures

	Terminology	*Origin*
Group A	Real time picture	On any range scale each radial line is written in the time taken by radiowaves to travel twice the maximum range of that scale
	Analogue picture	The amplitudes of displayed echoes may lie anywhere in a continuous range from noise level to saturation level, and within such limits are proportional to the strength of the received signal
Group B	Synthetic picture	The displayed echoes are generated artificially by a computer
	Digital picture (or video)	The radar data is stored in, and retrieved from, a digital computer
	Retimed video	The radar data is read into memory in real time but read out subsequently at a different rate
	Quantized video	The displayed echoes have one or more discrete amplitudes (i.e. their range is not continuous)
	Bright display	The displayed echoes are regenerated in such a way as to facilitate viewing in a wide range of ambient light conditions

Both terms in group A are commonly used to describe the traditional radar picture.

All terms in group B are commonly used to describe the modern stored picture.

brightened to give a distinctive heading marker. Hence it is possible to determine the bearing of the target by measuring the angle between the heading marker and the line (or lines) on which the target lies.

5.1.5 The synthetic picture

It is becoming common practice to store the picture information in a computer memory, process the data and then regenerate a synthetic picture. In promotional and technical literature such pictures are described by a wide variety of terms. It is not always clear to the user that, although the terms have differing origins, they essentially describe the same process (Table 5.1).

5.1.5.1 Range words

The simple principle of storage is that the information on each radial line is stored separately in binary code, and in this form it is known as a range word.

Figure 5.3 portrays one line of bearing on which there lies a buoy, a small vessel and a large vessel. In the interests of simplicity, let any clutter be successfully suppressed such that the real time picture (waveform A) would consist of three echoes of differing amplitude set against a background of receiver noise. This is the signal which would produce the traditional analogue picture on the cathode ray tube. Where a synthetic picture is required, the signal is used as input to a threshold detector. Such a device gives no output if the input is less than a given value (known as the threshold), and a fixed level pulse output if the input is above threshold. This is illustrated by waveform B. If the threshold can be set just at the top of the noise, then the output of the detector will indicate the presence (high) or absence (low) of targets on that particular line.

As waveform B is generated, it is used to set binary digits in a computer memory device known as a switch register. In this simple example, 12 elements of memory have been allocated to represent 12 miles of range and the information has been stored as binary digits (bits). It should be noted that the range of each target is implicit in the range word. If a 'one' appears in element number 6, it signifies the presence of a target

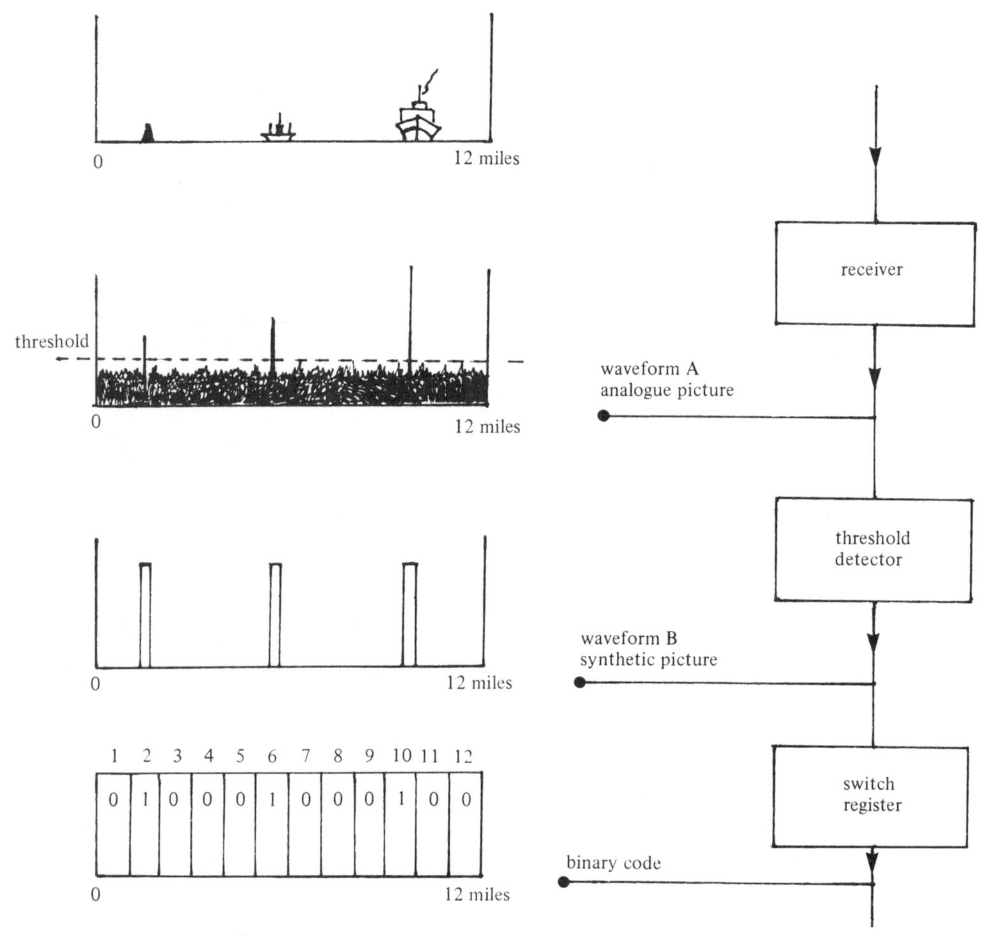

Figure 5.3 Storage of range

whose range lies between 5 and 6 miles. In a practical system, some 1200 elements would be allocated to represent 12 miles. This will give storage with a range resolution of some 20 metres and targets may well register in several consecutive range cells. Further, some systems employ two (or more) thresholds and store two (or more) range words in parallel per line, so that, on regeneration, two (or more) brightness levels of artificial echo can be produced.

5.1.5.2 Bearing words

Having established the principle of storing one line, it can be seen that, in theory, range data for the entire picture can be stored by repeating the process for each radial line of the picture and hence producing a matrix of say 1200 × 3600 elements of memory. Because each horizontal line in the matrix represents a radial line on the PPI which differs in azimuth from its predecessor by approximately 0.1°, bearing information will be implicit in the storage format if the lines are stored sequentially and labelled on an incremental basis. The labels, which are called bearing words, may be produced by a device known as a shaft encoder which generates an incremental binary number as it rotates through one revolution in synchronism with the antenna. In practice, it is common to use a 4096 line format because this represents the number of unique

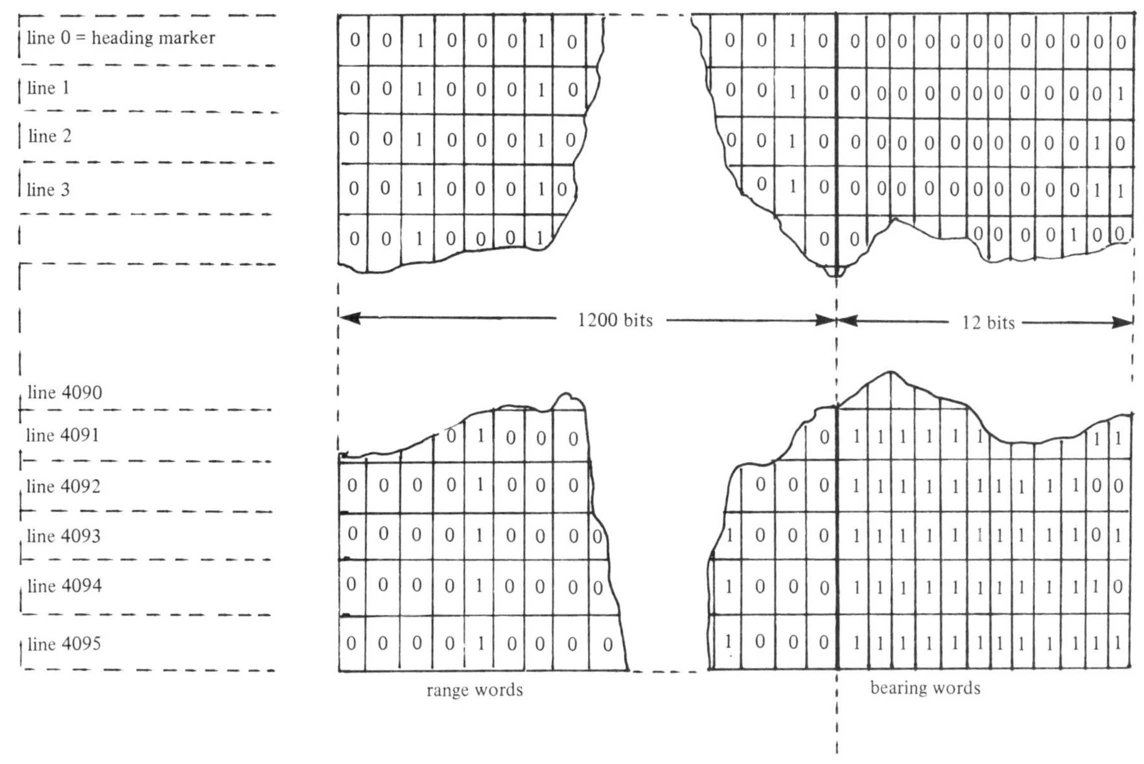

Figure 5.4 Memory matrix

codes that can be produced from a 12 element binary word (Figure 5.4).

5.1.5.3 Regeneration

From the foregoing, it is apparent that in theory the entire range and bearing data of a conventional marine radar picture can be stored in a matrix of some 1212 columns and 4096 lines. The range of each target is implicit in the column that it occupies, while the bearing is determined by the line on which it occurs.

Once the data has been stored it can be processed and retrieved at a time and at a rate which is convenient, i.e. a synthetic picture can be regenerated (Figure 5.5). The time delay due to detection, storage and retrieval is, as far as the navigator is concerned, negligible.

The amount of memory required to store one complete rotation of the antenna is considerable, there being nearly 5 million bits of information. In applications where simple processing is undertaken, it is only necessary to maintain a small sector of the data in memory at any given time, previously stored data being overwritten as the antenna rotates. Where more complex processing is required, such as scan conversion (see Section 5.1.7), the full matrix will have to be stored.

The significant principle is that artificial echoes, of a chosen shape and size, are computer generated to indicate the presence or absence of targets on each line of bearing. There is considerable variation in the way in which individual radar manufacturers have utilized the availability of regenerated pictures. For a detailed treatment of this topic see *Radar and ARPA Manual.*

5.1.6 Real and synthetic pictures compared

Synthetic pictures have many features which compare

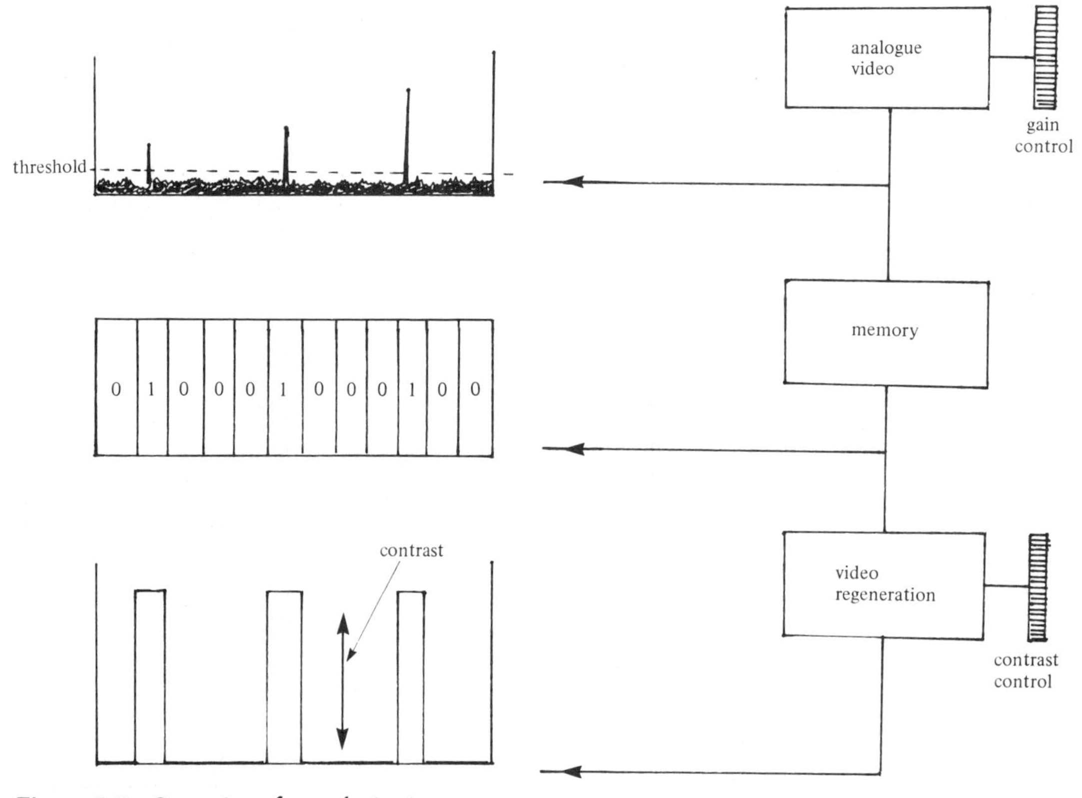

Figure 5.5 Generation of a synthetic picture

favourably with the traditional real time analogue picture. However, they also have limitations, and it must be borne in mind that the basic rules of adjustment and interpretation learned for real time pictures cannot invariably be applied to synthetic pictures (see particularly the discussions on gain, tuning and contrast in Sections 5.2.4.4, 5.2.4.5 and 5.2.4.3).

5.1.6.1 Probability of detection

Because the displayed responses are computer generated, it is possible to make them very obvious. Even echoes just at the lower limit of detection can be enhanced so that they appear as bright rectangular pulses rather than dim, barely visible spikes. This makes a positive contribution to the probability of detection of targets. It cannot be stressed too strongly that any echo which just fails to cross the threshold will produce no synthetic echo (Figure 5.6). Thus, the probability of detection is heavily dependent on the setting of the threshold (see Section 5.2.4.4).

5.1.6.2 Picture brightness

Although the data for a synthetic picture must be read into the memory in real time, it can be read out more than once in any given pulse repetition period. As a result, a single fixed time base (usually equivalent to 12 miles) is used and the range scale selector is used to control the rate at which each line is read into the memory. This approach makes it possible to produce an inherently brighter picture which is suitable for viewing in a wide range of ambient light conditions. The problems of viewing a real time display with a hood or visor are well known. The dim characteristic of such a picture, particularly on the shorter range scales, arises to a great extent from the very short time available in which to draw (or write) each line. In the

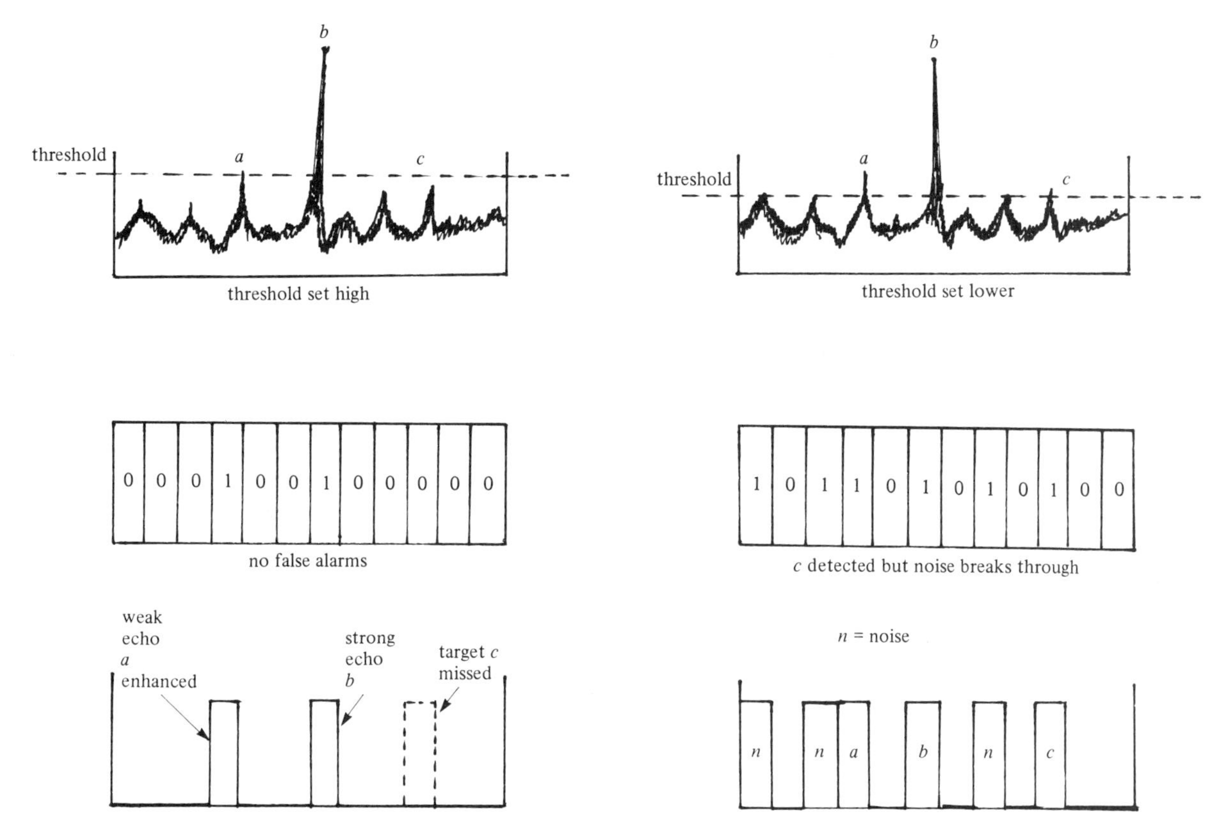

Figure 5.6 Target detection and threshold setting

case of a synthetic picture, on range scales less than 12 miles, the line is written more slowly and this makes a contribution to overall brightness. On the range scales greater than 12 miles the effect is reversed, but this is easily offset by the fact that, in any case, each line can be written more than once.

Despite the obvious increase in brightness provided by storage, in conditions of direct sunlight a rotating trace display may well require some form of partial shading, as the actual level of brightness achieved varies from set to set. If the picture is displayed on a raster or television-type display (see Section 5.1.7), then true daylight viewing will be more closely approached.

5.1.6.3 Sound levels

From an engineering point of view, the idea of a fixed time base is attractive for a variety of reasons, one of which is that it lends itself to the use of fixed deflection coils. In older displays, the electromagnetic coils which produced the spot deflection for each line were physically rotated about the neck of the cathode ray tube. A more attractive method is to use fixed coils to produce a rotating magnetic field. This is generally not practical at the very high deflection speeds required to produce a real time display on very short range scales. To the navigator, the most immediate indication of a fixed coil display is likely to be its quietness when compared with the familiar noise associated with mechanical rotating coils. Fixed coil displays greatly facilitate the production of screen graphics.

5.1.6.4 Levels of echo brightness

In an analogue display, echoes may have any amplitude over a continuous range from receiver noise to peak white. Presently most synthetic displays generate the echoes artificially at one, two or three levels only. (There are some indications that this number may well be increased, but it is by no means certain; one system offers sixteen levels and converts back to analogue after processing.) Clearly some amplitude intelligence has been lost in the process of storage, but the usefulness of such intelligence is a matter of debate. Some mariners do draw conclusions from the amplitude of displayed echoes, but the validity of such conclusions is open to question when, for example, one considers that a yacht with a radar reflector may return an echo whose strength compares favourably with that from a medium sized ship. Further, although the range of echo strengths is continuous, one can speculate as to how many different levels of brightness can be displayed by the phosphor. This is probably not more than eight.

Irrespective of the debate as to the intelligence that can (or cannot) be deduced from a knowledge of the strength of the returned echo, it is beyond dispute that an understanding of this loss of a continuous range of echo strengths is essential when judging the correct setting of the gain and tuning controls if observing a synthetic display (see Sections 5.2.4.4 and 5.2.4.5).

5.1.6.5 Coastline identification

It has long been recognized that the ability of civil marine radar to reproduce faithfully the finer detail of a charted coastline is very limited because the size and shape of each individual echo is heavily dependent on horizontal beam width and pulse length. This characteristic is likely to be exacerbated by the radial elongation of artificially generated echoes.

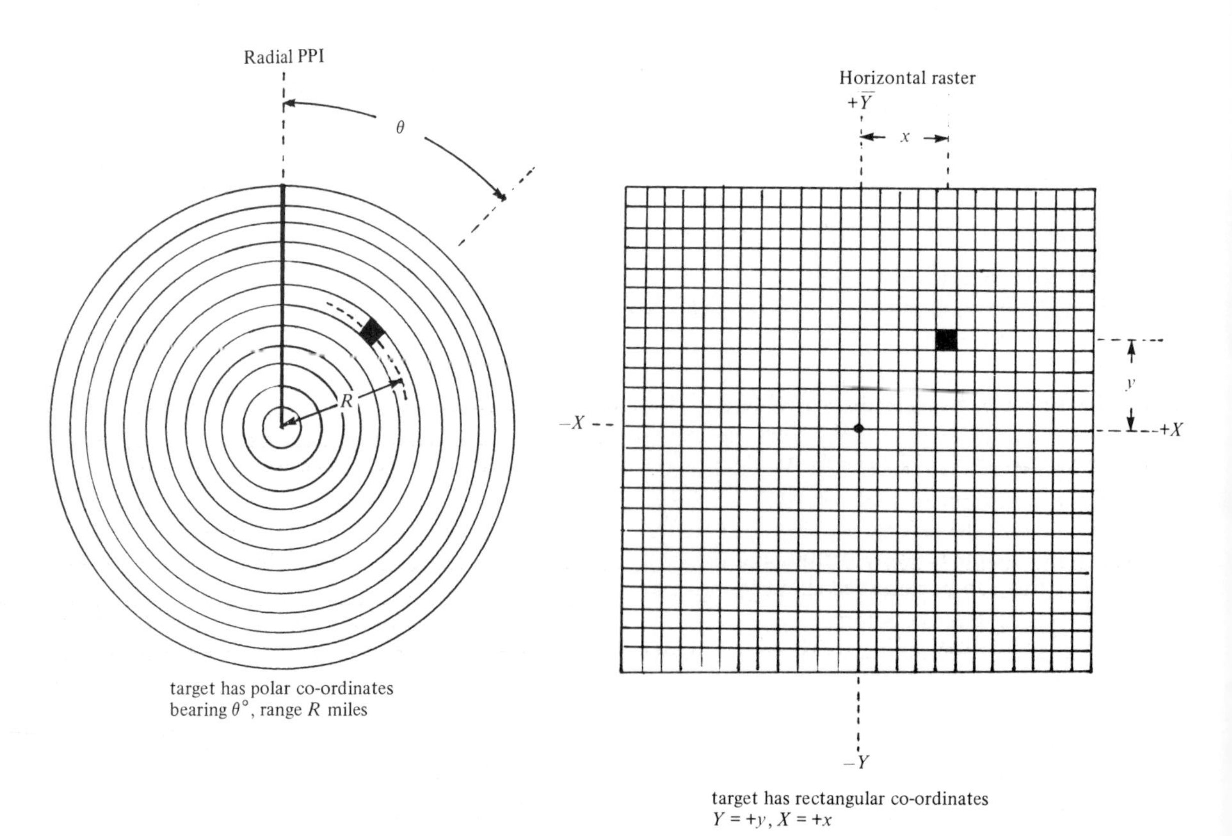

Figure 5.7 The principle of scan conversion

5.1.7 *Raster displays*

Most civil marine radars fitted in the early and mid 1980s produced the stored picture on a conventional radial scan display. Since then there has been a steady trend towards the display of the picture on a television screen. Such a display is commonly referred to as 'rasterscan'. Since 1985 many rasterscan displays have been 'type approved' by the UK Department of Transport and their fitting on board ships is becoming commonplace.

A raster is the name given to the pattern of horizontal lines which form a television picture. In order to present the radar picture on such a pattern of lines it is necessary to change the format of the stored data. It has been shown that the data is read in and stored in terms of bearing and range from the heading marker and origin respectively. By processing the data in a computer, the data can be reformatted in terms of northings and eastings, i.e. in rectangular co-ordinates on a Cartesian-type graticule. The calculation is similar to that performed by a navigator when using a traverse table to convert course (or bearing) and distance to d.Lat and departure. The technique is referred to as scan conversion (Figure 5.7).

Much of what has been said about synthetic pictures applies equally well to raster displays. However, its daylight viewing property is significantly better because the picture can be refreshed much more frequently. With a rotating trace display, while it is possible to brighten any given line by writing it more than once, such refreshing must be done within the pulse repetition period. Thereafter, that particular line cannot be refreshed until one antenna rotation (or scan) period has elapsed. By contrast, the writing of the horizontal line pattern of a raster is virtually independent of antenna rotation period, and the television-type picture can be refreshed about 90 times as often as its rotating trace counterpart.

The advent of rasterscan opens up the possibility of displaying other navigational data on or adjacent to the radar picture, and provides the opportunity to consider the use of colour. It is important that if colour is exploited it is used to convey intelligence and not just to make the picture pretty. Considerable study will be required to ensure this.

For a detailed treatment of rasterscan principles, see *Radar and ARPA Manual*.

5.2 Setting-up procedure and adjustment of controls

Design parameters such as transmitter power, antenna gain and receiver sensitivity determine the maximum performance that can be achieved by any given marine radar system. On any particular occasion the performance may be limited by weather and atmospheric conditions. It is axiomatic that, in order to achieve such performance, the controls must be correctly adjusted both in the initial setting-up routine and in the subsequent execution of specific operational tasks. This section deals particularly with the settings and procedures necessary to ensure optimum performance, by which is meant achieving the maximum possible probability of detection of targets. Some controls also affect the accuracy of the basic data that can be obtained from the display. These are mentioned where appropriate, but a more detailed consideration of accuracy is given in Section 5.10.

5.2.1 *Preliminary procedure*

Before switching the equipment on, the following preliminary checks should be carried out:

1 Ensure that the antenna is clear.
2 Check that the power switch, which makes the ship's mains available to the installation, is on.
3 Set the following front panel controls to zero effect:
   (a) Brilliance: this precludes the possibility of the screen being flooded with light when the display becomes operational. In any event, this is the first control which will be adjusted when setting up and it is convenient to start from zero effect.
   (b) Gain, contrast, range rings brilliance, variable range marker brilliance, sea clutter and rain

clutter (any automatic clutter controls should be switched off): all of the controls mentioned here have some effect on the overall brightness of the picture and, if they are not set to zero effect, they make it very difficult to correctly judge the setting of the brilliance control.

5.2.2 Switching on

It is convenient to consider that this operation carries out three distinct functions:

1 The supply of power to the antenna
2 The supply of power to the transmitter, receiver and display
3 The switching of the system from the standby condition to the operate condition

In some systems one may find an individual switch associated with each function. In others, it is common to find that a single switch combines functions 1 and 2. In a further variation a single multi-position switch may be employed. The application of power initiates a warming-up period which is likely to exceed one and a half minutes but which must not be more than four minutes. During this time the equipment is maintained in the standby condition and in some systems the antenna rotation is inhibited. While the set is warming up, the observer can make the following selections:

Orientation and presentation This selection is a matter of personal choice and navigational circumstances. The features which might influence such a selection are considered in Sections 5.3 and 5.4.
Range scale It is usually best to set up on a medium range scale. The likelihood of echoes and sea clutter which have not been affected by limiting (see Section 5.2.4.1) make it best suited for judging the quality of the picture when it is subsequently obtained.
Pulse length This should be appropriate to the range scale selected, e.g. medium range scale would suggest medium pulse length. In some cases the appropriate pulse length will be selected automatically with range scale, while in others the observer must make a decision.

5.2.3 Preparing the display

This involves the setting of controls directly associated with the cathode ray tube in order to prepare the tube to display a marine radar picture. If any of these controls are preset, it is still important to check that the setting is having the desired effect (they can go wrong!).

5.2.3.1 Setting the brilliance

The brilliance control adjusts the brightness of the spot of light which paints the picture. It should be set so that the rotating trace is barely visible. If the 'no-signal' brilliance is too low, weak echoes may not be displayed, while if it is too high, returned echoes may be difficult or impossible to see owing to lack of contrast.

5.2.3.2 Focus

This control adjusts the size of the spot, thus ensuring a sharp picture. To set the focus control, switch on the range rings and adjust the control until the rings appear as sharp as possible (choose a ring half-way between the centre and edge of the screen, as it is difficult to achieve perfect focus over the whole screen). Poor focus is evidenced by the ring appearing thick and 'woolly'. In modern marine equipment the focus is invariably preset and no user control is provided.

5.2.3.3 Centring

The origin of the trace must coincide with the centre of the graduated scale which surrounds the tube and the centre of the perspex cursor (see Section 5.11.5, Notice M.1158). The latter may be used for obtaining bearings. Centring is also essential prior to aligning the heading marker on the graduated scale.

Note Avoid parallax when centring by positioning the eye directly over the centre of the screen. This can be achieved by using the double engraved cursor lines (in transit) or the hood/visor.

5.2.3.4 Heading marker

All bearings are referred to this line. The picture can be oriented in one of three preferred ways:

Ship's head up In this case the heading marker must be aligned to 000° on the display bearing scale.
True north up In this case the heading marker must be aligned to that graduation of the display bearing scale which represents the ship's course, at the instant the vessel is 'right on'.
Course up In this case, the heading marker must be aligned to 000° on the display bearing scale at the instant the vessel is 'right on'.

In the second and third cases the display will be azimuth stabilized by the input of heading changes to a gyro repeater. After aligning the heading marker, observe it briefly to check that it is following variations in the ship's heading.

The features which are associated with various orientations are considered in Section 5.3.

If the heading marker is not correctly aligned on the display, all bearings read from the graduated scale around the PPI will be in error. This alignment should not be confused with that of the alignment of the heading marker with the ship's fore-and-aft line. The latter operation is concerned with ensuring that the heading marker contacts close and produce the heading marker signal at the instant the radar beam crosses the fore-and-aft line in the forward direction. This alignment cannot be checked without reference to both visual and radar observations. (See Section 5.11.5, Notice M.1158, paragraph 3.25.)

5.2.4 Obtaining the optimum picture

An optimum picture is obtained by adjusting the receiver controls to maximize the detection of weak echoes. The long established criteria on which such adjustment is based derive from the characteristics of the traditional analogue picture. Such criteria cannot invariably be applied directly to a synthetic picture. For this reason the adjustment of the controls will, in the first instance, be described in terms of an analogue picture. Subsequently, the case of a synthetic picture will be considered.

5.2.4.1 Setting the gain control (analogue displays)

The gain control adjusts the amount by which incoming signals are amplified. It should be advanced until a low level, close-grained speckling can just be seen all over the screen. This speckling is termed receiver noise. In general, noise is the technical term used to describe unwanted, random electronic signals. Minute noise signals are generated at all stages in the receiver, but those generated in the first stage receive most amplification. Thus, any target signal arriving from the antenna at the input of the receiver will not be detected unless it is stronger than the noise at that stage. The level of displayed noise is therefore of critical importance.

If the gain is set too low, weak echoes close to noise level will not be displayed. If the gain is set too high, weak echoes may be lost, not because they are not displayed, but because they will not be obvious owing to lack of contrast with the background (see Figure 5.8).

Further, excessive noise may also make it difficult to detect even strong targets. Because of the limited dynamic range of amplitudes which can be handled,

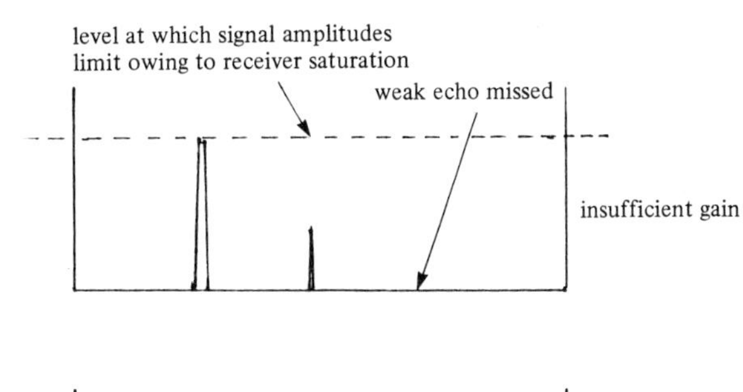

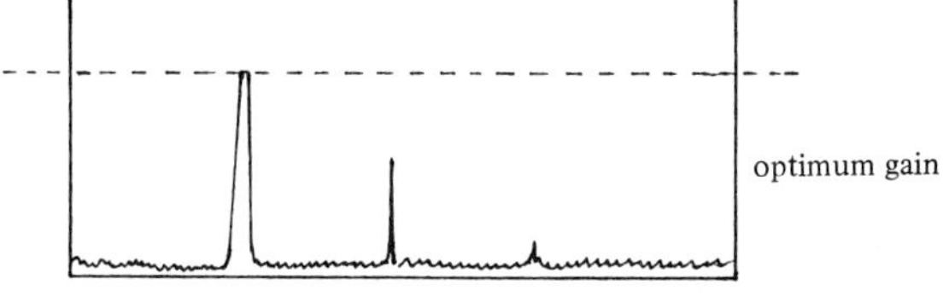

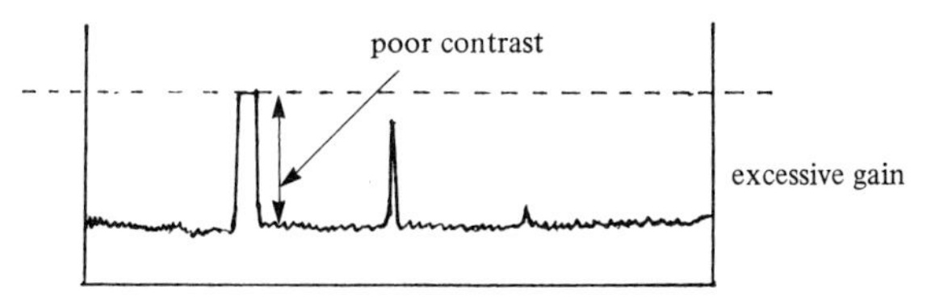

Figure 5.8 Gain control settings

strong signals may saturate the receiver. Consequently, an increase in gain may result in an increase in noise amplitude but make no improvement in the amplitude of the signals which have been limited. The result is a decrease in the contrast between the strong signals and the background noise.

5.2.4.2 Setting the tuning control (analogue displays)

The function of the tuning control is to adjust the frequency of the receiver to coincide with that of the transmitter, in which case the reception of echoes will be maximized. One of the most sensitive criteria on which the setting can be judged is the response of a weak land target (Figure 5.9). Strong echoes are avoided because they may limit, while land echoes are chosen to ensure a constant response against which to judge the setting of the control. By adjusting the tuning control to achieve the maximum displayed amplitude response from such a target, the correct setting of the tuning control may be easily and reliably found.

Alternative approaches involve adjusting the control to achieve:

1 The maximum radial extent of sea clutter echoes
2 The maximum area of rain clutter echoes
3 The maximum radial extent of the receiver monitor signal (see Section 5.2.5).

Some displays offer features such as 'magic eyes', tuning meters, and tuning indicators. Such devices are not required by the Performance Standards for Navigational Radar Equipment (see Section 5.11.1) and should not be used in preference to the above criteria.

The tuning may drift as the equipment warms up and to a lesser extent with the passage of time. The setting of the control should therefore be checked frequently during the first 30 minutes after switching on and periodically thereafter.

5.2.4.3 Setting the contrast control (synthetic displays)

Before considering the adjustment of the gain and tuning associated with a synthetic display, it is necessary to refer to the contrast control although it does operate at a later stage than both gain and tuning. Traditionally, the term contrast has been used to describe, in general, the difference between two levels of brightness and in particular the difference in brightness between the echoes and the background noise on an analogue display.

In a synthetic display, contrast is the name given to the control which adjusts the brightness of all synthetic echoes (see Figure 5.5). If the control is set to zero, no synthetic echoes will be displayed. Conversely, if the control is set too high, individual echoes will 'bloom', i.e. become large and defocused.

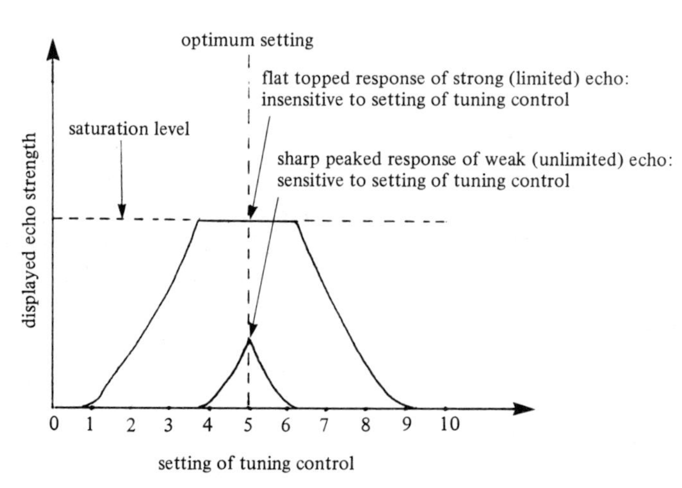

Figure 5.9 The tuning control setting

Between these extremes is a wide range of settings intended to allow the observer to adapt the overall brightness of the display to suit the ambient light conditions. (Compare this with the analogue case in which the brightness of the weaker echoes is limited by the extent to which the noise restricts the amount of gain that can be applied; see Section 5.2.4.1.) In some systems, normally single level systems, the contrast is preset by the manufacturer.

There is a 'chicken and egg' relationship between the contrast and gain/tuning. Some measure of contrast must be applied initially or there will be no visual signals on which to judge the setting of gain and tuning. Conversely it may not be possible to make a final decision as to the setting of contrast until after the gain and tuning have been adjusted, because the overall picture brightness is to some extent dependent on picture content. The correct strategy is to find an initial contrast setting which gives sufficient brightness to allow the subsequent setting of the gain and tuning controls. The operator's manual will normally give guidance but, if not, the setting can be found easily by experiment. If there is not enough time to experiment, set the contrast to the middle of its range. The setting is not crucial at this stage as the control may well be readjusted subsequent to the setting of gain and tuning. Further, the control only affects the brightness of the picture, not its content. The latter is dependent upon the setting of the gain and tuning.

The procedure is thus iterative and can be summarized as follows:

1 Select initial setting of contrast.
2 Set gain correctly (see Section 5.2.4.4).
3 Readjust contrast if necessary to suit ambient light conditions.
4 Set tuning correctly (see Section 5.2.4.5).
5 Readjust contrast if necessary.

5.2.4.4 Setting the gain control (synthetic displays)

The gain must be set so as to achieve the optimum relationship between the threshold level (which is preset) and the noise amplitude (which is controlled by the gain). Ideally we would wish the gain to be set such that the threshold is sufficiently low to detect all echoes yet sufficiently high to miss the noise. Such an ideal cannot be achieved. If the gain is sufficiently high for all detectable targets to cross the threshold, considerable noise will also cross the threshold. Conversely, if the gain is set so that no noise crosses the threshold, detectable targets might be missed (see Figure 5.6). Clearly a compromise setting must be found.

The long established criterion used in the analogue case is not easily applied to a single level synthetic picture. Any noise which crosses the threshold will be displayed with the same amplitude as a detectable echo. This is quite different from the analogue case, where there is a very obvious amplitude difference between most echoes and the noise background. However, with practice, the observer should be able to identify the noise by its random nature and the radial length of individual noise pulses. The peaks of the real noise are spiky in nature and thus the number of sucessive elements of a range word occupied by a single noise pulse is likely to be noticeably less than that occupied by a genuine target echo. This does not mean that a very short digital response is necessarily noise. A received echo which barely crosses the threshold may well activate only one range cell.

It is desirable that, in general, where a single level synthetic picture is provided, the observer has the facility to switch from analogue to synthetic picture on any range scale. If this is the case, there is no doubt that the analogue picture should be used to judge the setting of the gain in the traditional way. In some systems, two (or more) thresholds are used (Figure 5.10).

Generally, where two levels are offered, the observer is not able to view the analogue picture, or is restricted in such viewing to one range scale only (usually 12 miles, or the fixed timebase scale). There are signs that even where a single range scale analogue picture has been provided in the past, it may well be discontinued in the future. As before, if the analogue picture is available, then the gain setting should be judged with respect to real noise. If it is necessary to judge the setting of the gain with respect to the digital picture, a slightly different approach is more appropriate.

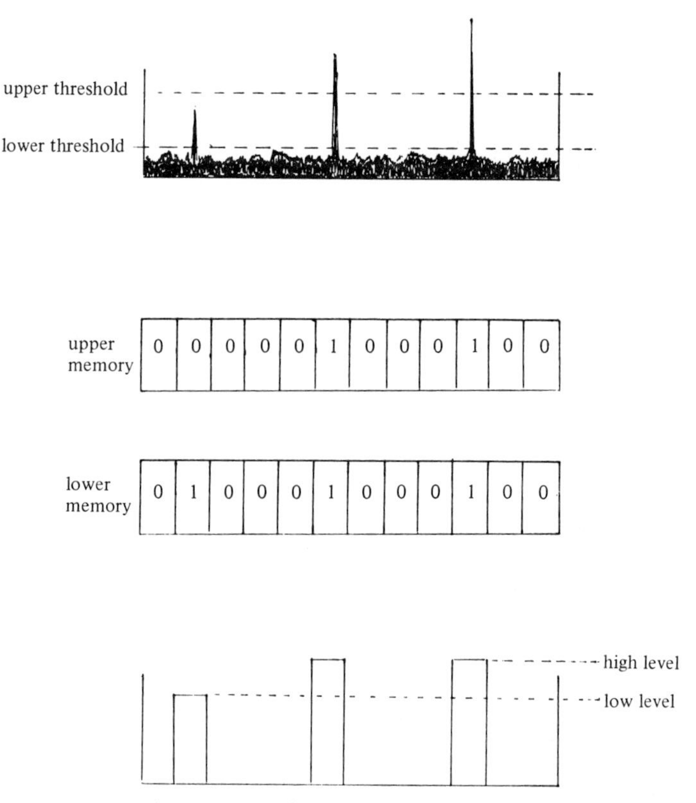

Figure 5.10 The principle of two video level synthetic picture

If two thresholds are used, the ideals referred to previously can be more closely approached and the setting of the control may relate more closely to the analogue case. This is particularly so if the thresholds are sufficiently close together to permit a setting of the gain such that the upper threshold is a little higher than the compromise value and the lower threshold a little lower. The maximum benefit is obtained when the ratio of high : low echo level is large (Figure 5.11).

The high level threshold position gives considerable immunity from unwanted noise pulses. The lower threshold position makes a positive contribution to the detection of echoes close to noise level. Clearly a fair number of noise pulses will cross this threshold but, because they are displayed at a fraction of high level amplitude, they do not destroy the contrast of the high level echoes. Further, although low level echoes will have the same amplitude as noise, their coherence, persistence and radial length will favour detection when compared with the random noise pattern. This mode of display makes it comparatively easy to judge the setting of the gain control with respect to a synthetic picture because it can be set so that the noise produces an even speckled background at low threshold.

In some systems, the ratio of high : low echo level may be smaller than illustrated in Figure 5.11, perhaps being closer to that of Figure 5.10. Ratios of 2 : 1 are not unknown. As the ratio increases, the setting of the gain control will present less of a problem.

5.2.4.5 Setting the tuning control (synthetic displays)

Traditionally the correct setting of the tuning control has been judged by reference to the displayed brightness of a weak land echo. It is not possible to apply

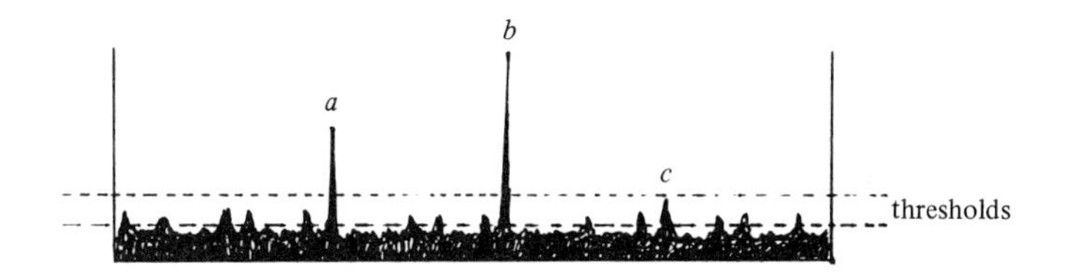

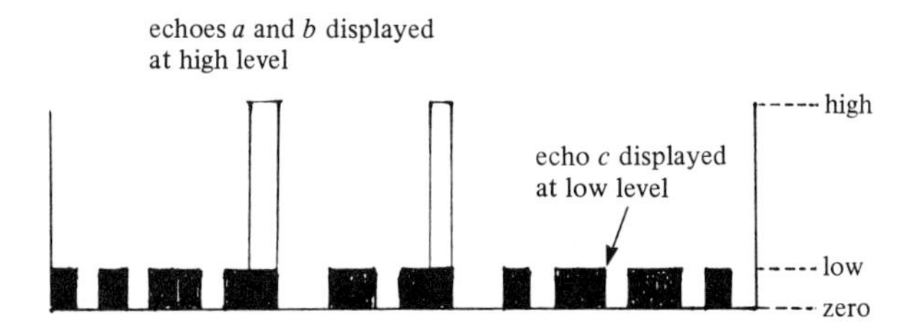

Figure 5.11 Two level picture with large high:low synthetic echo ratio

this technique to a simple synthetic picture because of the quantizing effect. This is illustrated in Figure 5.12. There may be little or no relationship between the amplitude of the synthetic echo and the analogue echo which produced it. The setting of the tuning control must not be judged with respect to the amplitude of the synthetic echoes. If an analogue picture is available then tuning should be carried out with reference to a real, weak land echo. If this is not possible, then the setting should be obtained by adjusting the control to achieve one of the following results:

1 The maximum radial extent of sea clutter echoes
2 The maximum area of rain clutter echoes
3 The maximum radial extent of the receiver monitor signal (see Section 5.2.5).

In the case of two (or more) level displays, it may be possible, with practice, to choose an isolated land target and tune for the maximum area displayed at low threshold. The tuning may drift as the system warms up and to a lesser extent with the passage of time. The setting of the control should therefore be checked frequently during the first 30 minutes after switching on and periodically thereafter.

5.2.4.6 Automatic tuning

In some systems there may be a form of automatic tuning.

Fully automatic tuning is correctly referred to as automatic frequency control (AFC) and is uncommon

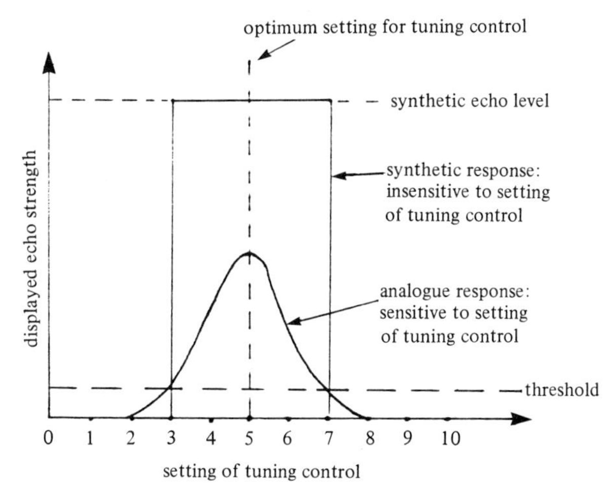

Figure 5.12 Tuning control setting – the effect of quantizing

in present-day civil marine radars. Such an arrangement, when functioning correctly, will tune the receiver from cold and maintain it continuously tuned to variations in transmitter frequency. This should not be confused with limited AFC or 'pull in' which is quite common. The latter will automatically compensate for small amounts of drift in the transmitter frequency but will not operate outside these small limits. Its successful operation is dependent on the observer tuning the receiver very close to the correct frequency in the first instance. Where automatic tuning is fitted, the observer should regularly check that the facility is operating correctly, either by supplementary use of the manual control or by reference to the receiver monitor signal (see Section 5.2.5).

5.2.5 Checking performance

Merely by viewing the displayed picture on a radar at sea, in general it is not possible to discern whether or not the set is giving optimum performance. Although the picture may look good, the observer cannot be sure that the equipment is operating at the level of performance that was intended by the manufacturer. In the open sea in calm, foggy weather, the absence of echoes, despite the presence of noise, could be due to a loss in performance rather than an empty ocean. Further, judging the setting of the tuning control would present a problem under these circumstances. Type tested sets are required by the Performance Standards for Navigational Radar Equipment (see Section 5.11.1) to have some means of checking performance and of checking that the equipment is correctly tuned in the absence of targets. The circuitry which provides such a facility is called a performance monitor, and when switched on produces a signal on the display, the radial length of which is a measure of the radar's performance (Figure 5.13).

The design of performance monitors has been approached in a variety of ways over the years. In some systems the overall performance (i.e. antenna, transmitter and receiver) is monitored by one device and the result displayed as a single signal. In others, the transceiver (transmitter and receiver) is monitored by one device while the power radiated from the antenna is monitored by a second device and the results displayed as two separate signals. The way in which the performance monitor signal is displayed is purely arbitrary, and a wide variety of methods exists. For this reason, the only safe procedure is to read the maker's manual in order to establish what signal to observe and to appreciate precisely what it measures.

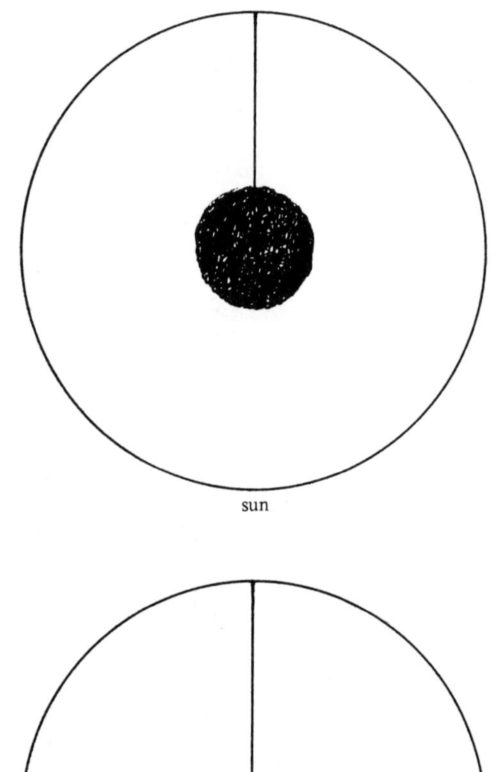

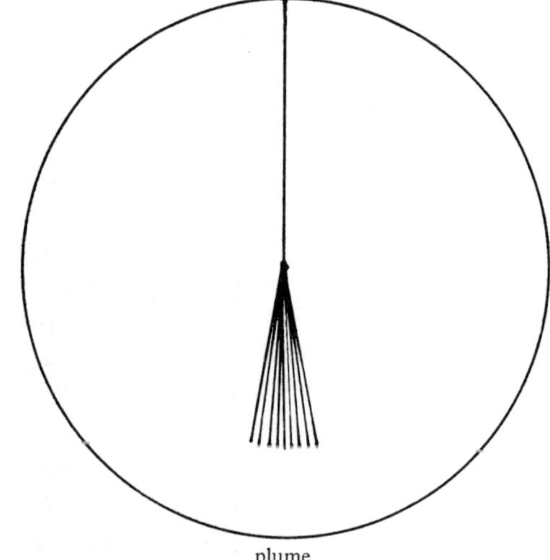

Figure 5.13 Examples of the display of performance monitor signals

The length or radius of the transceiver monitor signal is a function of the performance of the radar. It is measured (in miles and cables) by the manufacturer when the performance is known to be optimum. The following information should be shown on a tally close to the display:

1 The length of the signal consistent with optimum performance.
2 The length of the signal corresponding to a significant drop in performance.

It must be remembered that the length of the transceiver monitor signal is also dependent on the setting of the operational controls. When carrying out a check it is vital that controls are set as instructed in the maker's manual – otherwise the comparison is meaningless.

It cannot be stressed too strongly that the way in which the performance monitor signals are displayed, and the extent to which they monitor the performance, varies considerably from maker to maker – and even with different types made by the same company. Thus the important basic rule for carrying out a performance check is:

Consult the makers' manual and follow exactly the instructions given therein.

A performance check should be carried out as soon as practicable after setting up and thereafter at regular intervals. In Notice M.1158 (see Section 5.11.5) it is recommended that a check should be carried out before sailing and at least every four hours when a radar watch is being kept. This should be regarded as a minimum.

5.2.6 *Change of range scale and/or pulse length*

During normal operation of the radar system it will be necessary to change range scale from time to time. This may be for a variety of reasons, but attention is particularly drawn to the advice given in Notice M.1158, paragraph 3.6 (see Section 5.11.5).

A change of range may frequently be associated with a change in pulse length. Table 5.2 gives a comparison of short and long pulse length.

Table 5.2 A comparison of short and long pulse

	Short pulse	*Long pulse*
Energy content	Less energy, so use on short ranges	High. Good detection ranges. Use on long ranges and for poor response targets at short range
Minimum range	Good, so use on short ranges	Poor, so use on longer ranges where close targets are not so important
Range discrimination	Good	Poor
Effect on PPI paint	Short radial paint. Produces a well defined picture on short ranges where spot moves fast	Long radial dashes on short range owing to fast moving spot, *but* all right on long ranges with longer time base
Effect on sea clutter	Small extent and neat paint owing to (1) less energy (as targets are close, this is not so important) (2) shorter paint: better discrimination	Too much energy returned, so centre of screen masked. (1) Long paint: poor discrimination. (2) More energy: extended clutter area. (3) Severe masking: clutter and targets may be well above limit level
Use in rain	Possibly useful to detect targets within heavy rain (i.e. where rain echoes are up to limit level). There will be a general weakening of incoming echoes when short pulse is selected and a neater paint	Long pulse helps to combat attenuation caused by rain, and can thus be used to advantage for targets which lie beyond rain

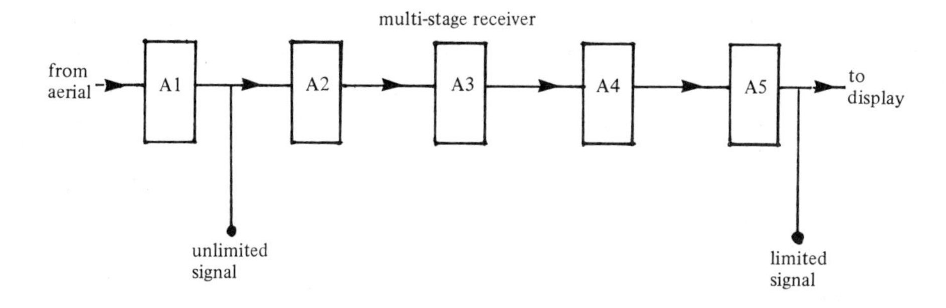

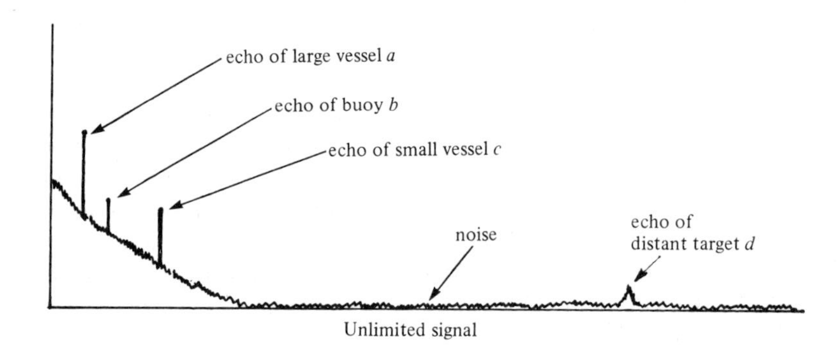

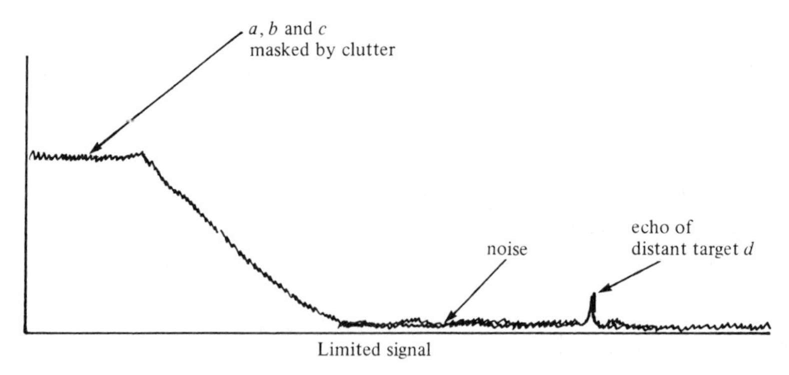

Figure 5.14 The masking effect of sea clutter returns

Particularly in older equipment, it may be found that the brilliance level changes with change of range scale and also with change of pulse length. In more modern sets, particularly those having a fixed time base, this is less likely to occur. However, it is prudent to establish whether or not such a change does take place. If it does, the correct procedure for ensuring that optimum picture is maintained is as follows:

1 Turn down the gain. This reduces the problems of afterglow but, more importantly, allows the brilliance level to be observed.
2 Change the range scale and/or pulse length.
3 Reset the brilliance control if necessary.
4 Restore the gain.

5.2.7 Sea clutter

Clutter is the name given to unwanted echoes which are returned by reflections from the sea waves around the ship (sea clutter) or from precipitation within detection range (rain clutter). Quite simply, echoes which are weaker than those returned by the clutter cannot be detected. This is an extremely important limitation of marine radar, and should be given particular consideration with regard to small vessels, small icebergs and similar floating objects, which may never be detected.

It does not follow that wanted echoes stronger than the clutter response will necessarily be detected with ease. Because of receiver limiting or saturation, responses much stronger than that of the clutter may be masked by the clutter signals. Skilled adjustment of the controls is necessary to ensure that targets just above clutter level do not go undetected.

5.2.7.1 The sea clutter control

Figure 5.14 shows one line of a radar picture, firstly as it would appear on an oscilloscope connected at an early point in the receiver (i.e. before any saturation has taken place), and secondly as it would appear just before being applied to the display. It can be seen that signals *a*, *b* and *c* have been masked by the presence of clutter. In order to detect *a*, *b* and *c*, the gain must be reduced to bring the clutter echoes below saturation level. If this is done by using the gain control, it will have the effect of removing echo *d* which is not in the clutter area. The latter is avoided by using the sea clutter control. This is sometimes referred to as swept gain or sensitivity time control (STC). Both of these terms describe its operation. It suppresses the gain at the beginning of each line by an amount selected by the observer (Figure 5.15). As the line is drawn, it restores the gain at a rate which is preset at the time of installation.

Clearly the ideal is for the observer to match the suppression curve to the clutter signals. It follows that if it is correct for one line of the picture it is likely to be incorrect for most of the others because the sea clutter response is not symmetrical through 360° of azimuth. It varies in sympathy with antenna direction and may vary randomly owing to the distribution of sea waves produced by shallow water and land effects. For this reason there is no single correct setting for the manual clutter control. The correct use of the control is to perform regular searching operations (see Section 5.2.7.2).

In modern radars which employ a logarithmic receiver it will be found that clutter is less troublesome and that the sea clutter control is more efficacious.

5.2.7.2 Use of the sea clutter control

The searching procedure can be summarized as follows:

1 Turn the manual sea clutter control to maximum effect and wait for the afterglow to fade.
2 Remove the suppression a step at a time, watching for the echoes to appear, until the clutter is again present.
3 Remember that sea clutter paints will tend to be random while those of targets will tend to be steady.
4 Repeat the search at frequent intervals. The frequency of such searches should in general be appropriate to the prevailing conditions, but in particular consideration should be given to the speed of the observing vessel, the visibility and the type of floating object which is likely to be encountered.

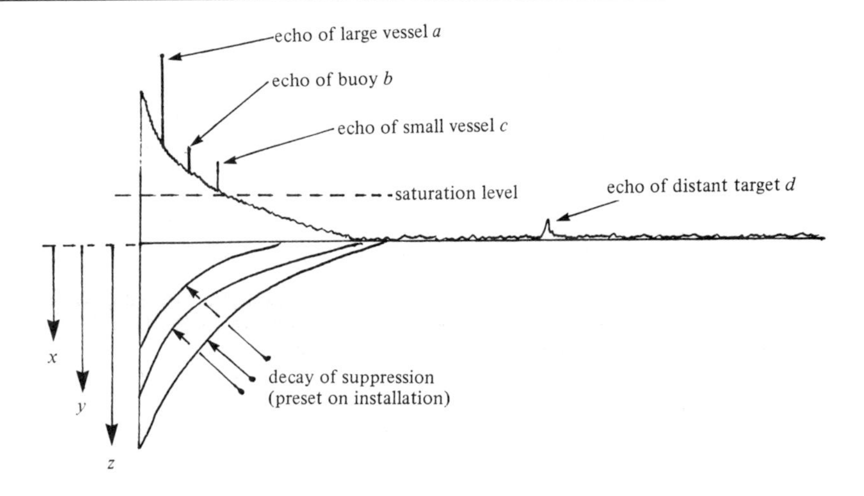

x,y,z represent varying degrees of initial suppression set by sea clutter control

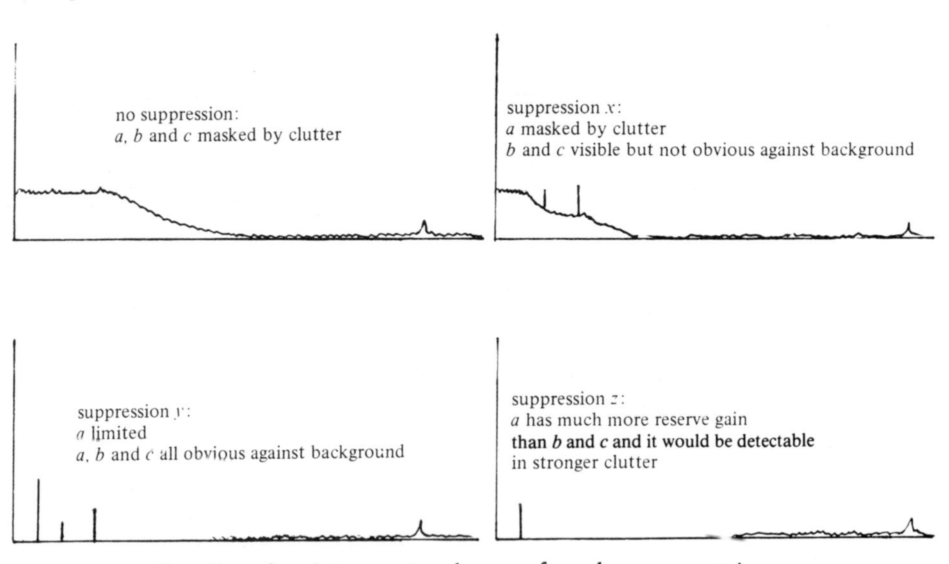

Figure 5.15 The effect of applying varying degrees of sea clutter suppression

5 Between searches, set the control so that just a few wave echoes are painting. It is worth remembering that if saturation is a problem it may be eased or removed by the selection of short pulse or using 10 cm (S band) radar transmission.

Successful use of the sea clutter control is dependent on an understanding of the theory of the phenomenon and practice in the performance of the searching technique. The latter takes time but is one of the most important skills associated with marine radar operation. Notice M.1158 (see Section 5.11.5) draws attention to the general need for clear weather practice. Sea clutter searching is a particular skill which merits such practice.

5.2.8 *Rain clutter*

The effect of rain on target detection in some ways is similar to that of sea returns but in other ways is quite different. Like sea clutter, rain can mask even very strong echoes by saturating the receiver and can preclude the detection of weaker targets within the rain area if their returns are weaker than that of the rain. On the other hand, precipitation echoes can occur anywhere on the screen and may well change their position quite rapidly. Further, the masking problem is made more serious by the fact that the radar pulse is attenuated on its two-way journey through the rain. It should be remembered that such attenuation will also affect the responses from targets beyond the rain area (see Section 5.2.8.3 and Table 5.2).

5.2.8.1 The rain clutter control

The rain clutter circuitry seeks to deal with the saturation problem in a way which is quite different from the local short range or general suppression of gain mentioned so far. The control is sometimes labelled differentiator or fast time constant (FTC) and it adjusts circuitry which is designed to respond only to increases in signal strength.

Figure 5.16 shows that the circuitry seeks not necessarily to prevent saturation but to remove the trailing edges of echoes and hence show the echoes separately by preventing overlap. In most systems it is possible to vary the amount of trailing edge removed and this facilitates systematic searching for targets within the rain area. It is worth mentioning that the differentiating effect is particularly useful for improving discrimination on short range pictures.

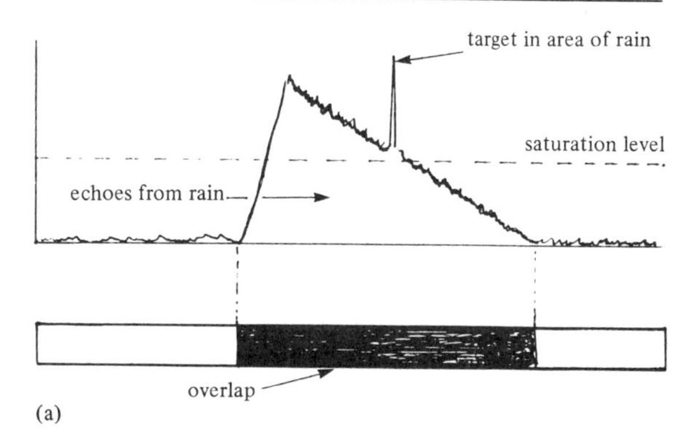

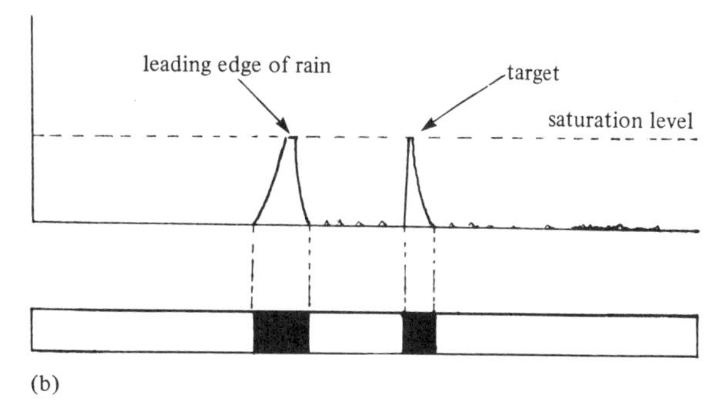

Figure 5.16 The effect of the rain clutter control
(a) No rain clutter control: displayed echoes of rain and target overlap radially
(b) Rain clutter control: there is radial (not necessarily amplitude) discrimination between leading edges of rain and target

5.2.8.2 Searching for targets in rain

It is most important to appreciate that although the rain clutter control is so called, it is not necessarily the most effective way of dealing with rain echoes. Many users hold the view that the use of the gain control in a searching fashion is much more effective. It might be argued that this view is to some extent subjective. However, the important point is that the navigator must realize that suppression and differentiation are two quite different attempts to solve the same problem. He must select whichever he finds most helpful. In some circumstances a combination of both techniques may be appropriate. In summary, the masking effect of rain may be dealt with as follows:

1 If the rain is close to the ship it may be possible to use the sea clutter control.
2 In general, the most effective technique is to use the gain control in a searching fashion.
3 If preferred, the rain clutter control may be used in a searching fashion, either alone or in combination with some suppression of the gain.
4 Where saturation is a problem, it may be eased or removed by selection of short pulse or 10 cm transmission (see Table 5.2).

5.2.8.3 Dealing with targets beyond rain

Because of the attenuating effect of rain, it may be difficult to detect weak targets beyond areas of rain. Such limitations can be minimized by using 10 cm transmission (S band radar) and long pulse. While searching for specific targets it may be useful to temporarily turn the gain above the normal setting or use an echo stretching facility, if available (see Section 5.2.12.1).

5.2.8.4 Other forms of precipitation

Broadly, what has been said about rain applies to a greater or lesser extent to all other forms of precipitation and to dust and sandstorms. In general, the strength of precipitation echoes and the extent of attenuation experienced depend on the size of the particles and the distance between them, when considered in relation to the transmitted wavelength.

5.2.9 Automatic clutter suppression

This facility is available on many modern radars which have a logarithmic receiver. Individual manufacturers name the control in various ways, but the technique employed is generally referred to as adaptive gain, and will not function correctly with a linear receiver.

The process is so called because the radar attempts to adapt the display gain continuously to the correct value. In order to achieve this it must deduce a suppression signal whose instantaneous value is appropriate to the clutter at that instant. It does this by using a device (known as an integrator) which follows the instantaneous mean level of the received signal. It is very carefully designed such that it responds at the same rate as the instantaneous variations of the average level of clutter. If its response is too slow it will not follow the average level accurately; if too fast it will tend to follow the target returns. The output of this device is inverted and hence used to apply the desired degree of suppression (Figure 5.17).

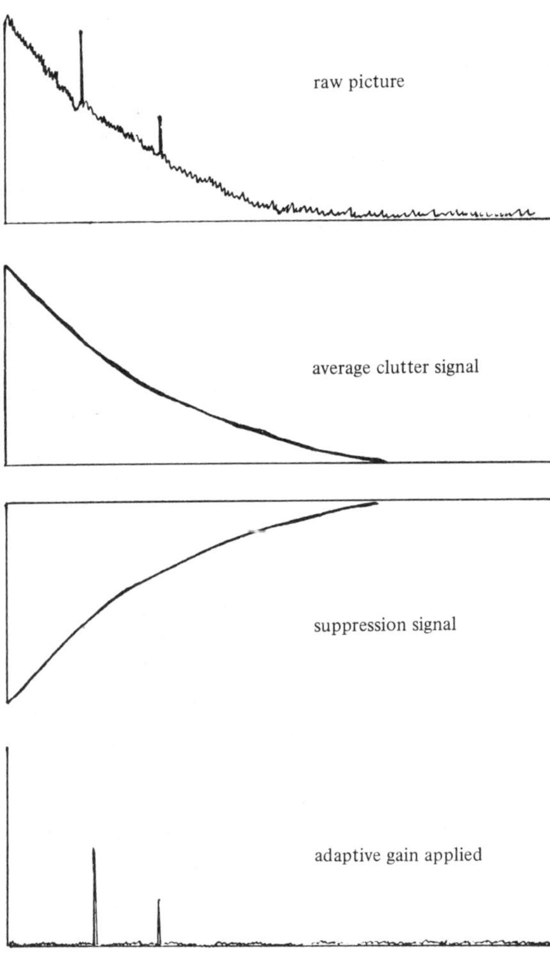

Figure 5.17 Automatic clutter suppression

In order to permit the adaptive gain signal to be generated, it is necessary to apply a small amount of differentiation (rain clutter) to the signal received at the display. The effect of this must be borne in mind and will be discussed in Section 5.2.9.1. A further technical feature is that the integrator cannot follow the instant change from zero clutter to maximum clutter at the start of the timebase. This is normally overcome by priming the integrator with the first mile (approximately) of signal from the previous timebase.

5.2.9.1 Practical application

Commonly, this control is integral with the manual clutter control. With the clutter control set to manual, the adaptive gain is inoperative and the manual sea and rain clutter controls function in the normal way. When adaptive gain is selected, the manual sea and rain clutter controls are rendered inoperative, a fixed amount of differentiation is applied and the gain is varied instantaneously and automatically according to the dictates of the suppression waveform. The manual gain remains operative and thus the operator controls the gain level which obtains in the absence of automatic suppression.

In some systems, independent adaptive controls are available for rain and sea clutter suppression. The former is effective over the full timebase while the latter is restricted to some preset range limit. In other systems, where a single control is offered, some manufacturers have placed a range extent limit on the operation while others have not.

The manual gain control should have been set correctly (half bright speckled background) with the adaptive gain off and clutter controls at zero effect. If the adaptive gain circuitry is functioning correctly, further adjustment of the gain should not, in general, be necessary. However, at any time that the observer suspects that excessive suppression is being applied, he may adjust the gain to produce a higher level of residual speckle in order to ensure maximum probability of detection of targets.

5.2.9.2 Effect on target detection

In general, the adaptive gain circuitry will have been set up by the manufacturers to cope with a large variety of sea and precipitation conditions. It cannot possibly achieve the ideal across the full range of sea conditions. In some conditions, particularly in severe clutter, it may well be found that skilled use of the manual sea clutter control will provide a higher probability of detection. It is particularly important to appreciate that if the adaptive gain circuitry is not set up correctly, or has drifted from its correct setting, targets may escape detection. The virtue of adaptive gain is that it will perform line by line, hour after hour, without tiring or losing concentration or being distracted. However, in certain circumstances it may miss targets close to clutter level. Thus the prudent use of adaptive gain is as a complement to manual clutter suppression and not as a total replacement.

If the vessel is in a river, narrow channel or enclosed dock area, the short range echoes may be extremely strong and will follow no mathematical law. The suppression produced by these effects may well remove wanted echoes. Manual operation is likely to be more effective in such cases.

5.2.9.3 Effect on coastlines

The adaptive gain sees land echoes in much the same way as it sees rain echoes. It will tend to display the leading edge and suppress that which lies behind it. The precise effect will vary depending on whether the beam is normal to or parallel with the coastline. For this reason, identification of coastal features is best performed with the adaptive gain switched off.

5.2.9.4 Effect on radar beacon and performance monitor signals

Both of these appear as a continuous echo train and will tend to be differentiated and suppressed by the operation of the adaptive gain. It is clearly important to switch to manual control to detect radar beacons or carry out performance checks.

5.2.10 The standby condition

The IMO radar performance specification (see Section 5.11.1) requires that type approved sets should have a standby mode from which the equipment can be brought to an operational condition within 15 seconds.

The detailed nature of the standby condition will vary with manufacturer, but in all cases transmission is inhibited. Thus, although the radar is virtually ready for immediate use, limited life components in the transmitter are not operational. A further, more general benefit is that the amount of radar-to-radar interference experienced by other vessels (who may have to operate) will be reduced over a potentially large geographical area. The life of components and hence the reliability of the radar set will be far less affected by continuous running than by frequent switching on and off, so that in periods of uncertain visibility it is better to leave the radar in full operation or on standby (see Section 5.11.5, Notice M.1158).

Rule 7(b) of the Collision Regulations requires that 'proper use shall be made of radar equipment if fitted and operational' (see Sections 22.1.3 and 22.2.3). There will be very many occasions on which 'proper use' will mean continuous running. However, there will be other occasions, for example in the open sea on a clear day with a fog forecast, where the standby mode would constitute proper use. Thus, compliance with the letter and spirit of the rule is ensured, as there is no unnecessary transmission in the meantime or excessive delay should the equipment be required quickly.

If the equipment is to be left on standby for a short period, the controls may be left in the optimum position, but the settings should be checked immediately on returning to the operational condition. If the equipment is likely to be on standby for a long period, it is probably better to set the controls to minimum effect and set up immediately on returning to the run condition.

5.2.11 True motion

Thus far, in considering the setting up and adjustment of controls, it has been assumed that the electronic origin of the display would be stationary at the geometrical centre of the screen. This will result in a 'relative motion presentation' in which targets will show their movement relative to the observing ship.

In a 'true motion presentation' the origin is driven across the screen in a direction and at a rate which corresponds with the true motion of the observing vessel. The input information which controls the movement of the origin may be measured with respect to the water or with respect to the ground.

If the input represents course and speed through the water, the presentation is said to be 'sea stabilized' and targets will show their movement through the water. This is the correct presentation for use in the extraction of collision avoidance data.

If the input information represents track and speed over the ground, the presentation is said to be 'ground stabilized' and targets will show their movement over the ground. Such movement is not appropriate to collision avoidance assessment in general and can be dangerously misleading if used inadvertently for this purpose.

The features and limitations of the various available presentations are discussed in detail in Section 5.4. In Sections 5.2.11.1 and 5.2.11.2 it is intended merely to state the practical procedure to be followed in setting up a true motion presentation.

5.2.11.1 Setting up a sea stabilized presentation

1 Check that the display is correctly orientated and azimuth stabilized.
2 Check that the display compass repeater (if fitted) is correctly aligned.
3 Select true motion presentation.
4 Select log or manual speed input. Check that the log is reading the correct water speed or that manual speed has been set to that value.
5 Adjust the reset controls so that the origin is placed to make best use of the screen, bearing in mind possible forthcoming manoeuvres.
6 Commence systematic plotting of the origin on the reflection plotter to ensure that it tracks in the correct direction and at the correct rate.

Remember that true motion is only as true as the course and speed information being supplied to it.

7 Be vigilant for loss of 'view ahead' as the origin

moves across the screen. Anticipate a suitable opportunity to reset the origin.

8 Reset the origin at an appropriate time and commence a new plot, to check the tracking of the origin.

5.2.11.2 Setting up a ground stabilized presentation

1 Correctly sea stabilize the display.
2 Plot an isolated ground-stationary target on the reflection plotter and deduce the set and rate of the tide. (The ground-stationary target will move at a rate equal to that of the tide but in the reciprocal direction.)
3 Where set and rate controls are provided, adjust them to the values deduced in step 2.
4 Continue systematic plotting of the ground-stationary target to ensure that it maintains its position. Readjust the tidal controls if necessary to remove residual movement. Remember, the tide may change with the passage of time and with the position of the vessel.
5 Continue systematic plotting of the origin, as this is the only really obvious indication that the display is ground stabilized.
6 In some systems, direct tidal input is not provided. The ground track is expressed in terms of 'course made good correction' and 'ground speed'. The course made good correction is defined as that angle between the course through the water and the ground track. It is measured in degrees port or starboard from the course through the water. The ground speed must be fed in manually. The observer is required to construct a vector triangle from which he deduces the course made good correction and ground speed. This method has the disadvantage that it is more time consuming in the deduction of the necessary quantities. Further, the values will change not only with position and the passage of time but also each time the observing vessel changes course or speed through the water. It does, however, have the advantage that it is easier to identify the sense of the adjustments necessary to remove the residual movement of the ground-stationary target. The residual movement can be easily resolved into fore-and-aft and athwartship components, the latter being removed by the course made good corrector and the former by manual speed.
7 Some systems exist in which the display can be ground stabilized by direct speed inputs from a dual axis, ground locked Doppler log.

It must be emphasized that a ground stabilized display shows the movement of targets over the ground and does not indicate their headings.

5.2.12 Miscellaneous controls

This section deals with a number of more specialized controls which will not be found on all civil marine radars.

5.2.12.1 Echo stretch

In a synthetic display, the radial length of the regenerated echo is normally fixed by the number of elements of memory activated by the detected response. Some systems offer an additional facility whereby the user can cause the echoes to be radially stretched in order to make them more obvious. The rules which govern the stretching vary somewhat with manufacturer, but in general the facility is limited to:

1 The longer range scales
2 Targets beyond a preset minimum range
3 Targets whose received echoes exceed a preset duration.

Such a facility may be particularly useful when one is first trying to detect distant land echoes. Once detection has been achieved and identification becomes a priority, the stretching effect will become counter-productive.

5.2.12.2 Interference rejection

Where picture data is stored, it is possible to filter out radar-to-radar interference. The logical rule on which the process operates is based on the assumption that the probability of interference appearing at the same range on two successive timebases is zero.

This is not unreasonable as the interference echoes have a completely random time relationship with the synchronous transmission, reception and timing cycle of the observer's radar system. Thus by comparing the current range word with the previous range word, it is possible to filter out interference from other radars. The signal so produced is referred to as correlated video. (The device will also remove the first line of the paint of each coherent target.)

Racon responses from some beacons located in the USA are rejected by this type of circuit. The problem only arises with the step-sweep type of racon, and occurs because in certain circumstances the radar does not receive two consecutive responses from the beacon while it is on a step embraced by the bandwidth of the receiver. A programme was scheduled to commence in March 1985 resulting in the modification of all US step-sweep racons over a period of about 9 months.

5.2.12.3 Second trace echo elimination

Some systems provide a facility whereby second trace echoes can be eliminated.

One method employed is to continually cause the PRF to vary sinusoidally about its nominal value, thus causing the period between pulses to 'breathe'. This has no effect on first trace echoes but, line by line, changes the range at which a second trace echo appears, thus breaking it up so that it does not appear as a coherent paint. This technique is known as jitter.

An alternative approach is to provide a front panel switch which allows the observer to temporarily introduce a fixed change in PRF. This again will not affect first trace echoes but will hopefully shift the receiver rest period to such an extent that the second trace echoes are not displayed.

5.2.13 Switching off

When finally switching the radar off, it is prudent to set the controls as you would wish to find them when initially switching on, i.e. as set out in Section 5.2.2.

5.3 Picture orientation

In marine radar literature, there is frequently an understandable descriptive overlap between the concepts of picture 'orientation' and picture 'presentation'. In this text, orientation will be defined as the choice of directional reference to be represented by the 000° (or 12 o'clock) graduation on the fixed bearing scale around the tube. Put more informally, it describes which way up the observer chooses to look at the picture.

Picture presentation will be used to specify the reference with respect to which echo motion will be displayed. Putting this more informally, it advises the observer if he should expect the echoes to show their movement with respect to his own ship, or with respect to the water, or with respect to the ground.

5.3.1 Ship's head up orientation (unstabilized)

In a ship's head up orientation the heading marker is aligned to 000° (12 o'clock) on the fixed bearing scale. Because no azimuth stabilization is applied, the heading marker remains in that position irrespective of changes in the heading of the observing ship. Thus in this orientation, the 000° graduation (12 o'clock) on the fixed bearing scale represents the observing vessel's instantaneous heading. The most obvious characteristic produced by the absence of azimuth stabilization is that any change of heading by the observing vessel results in an equal and opposite angular movement of all targets. This gives rise to the following significant practical problems:

1 Where a large alteration of course is made, any areas of land echoes are smeared across the screen. This makes them difficult to identify, and can obscure isolated fixed or floating targets in the afterglow.
2 As the observing vessel yaws about her chosen course, an angular 'wander' is superimposed on the movement of all other echoes. This makes it difficult to take bearings because the echoes are moving angularly (and perhaps rapidly when viewed from a

small vessel) and the heading must be read at the same time in order to deduce a true bearing. In some systems the latter limitation is offset by the fitting of a stabilized bearing ring outside the fixed bearing scale.

3 The angular wander due to yaw makes it very difficult to assess the direction of the movement of an echo from its afterglow trail. The angular smear is often more dramatic than the steady relative motion trail and is thus more likely to catch the eye. Assumptions based on such angular movement can be dangerously misleading (the Collision Regulations, rule 7(c); see Section 22.3.1). Collisions have occurred due to such mistaken assumptions.

4 The lack of azimuth stabilization detracts considerably from the efficacy of the reflection plotter, whose surface is either fixed or at best rotatable by hand. The observer is constrained either to mark the plotter only when the vessel is right on course or to rotate the surface (if this is possible) by an equal and opposite amount to the yaw. The former is irritating and time consuming, while the latter requires skill, practice and concentration. Neither is particularly attractive, but if the observer is required to reflection plot on an unstabilized display, failure to take account of the yaw can produce a dangerously misleading plot.

The only positive feature of a ship's head up unstabilized orientation is its correspondence with the view from the wheelhouse windows. This is to a great extent subjective, but some officers do find it awkward or uncomfortable to view a picture which is oriented north up (see Section 5.3.2), particularly on southerly courses. Many pilots also find ship's head up appropriate to particular areas of their work, such as narrow channels and lock entrances, where they feel the angular smearing is secondary when compared with the importance of port and starboard through the window corresponding with left and right on the radar screen. Figure 5.18 shows a comparison of orientations.

In conclusion, it is noteworthy that in almost all modern civil marine radars, the ship's head up orientation is unstabilized.

5.3.2 True north up orientation (stabilized)

In true north up orientation, the heading marker is aligned, at an instant when the vessel is right on course, to that graduation on the fixed bearing scale which corresponds with the ship's true course. As a result, the 000° (12 o'clock) graduation on the fixed bearing scale represents true north. By means of azimuth stabilization, changes in the vessel's instantaneous heading are reflected by a sympathetic angular movement of the heading marker, thus maintaining true north in alignment with the zero of the fixed bearing scale (see Figure 5.18). Such sympathetic movement of the heading marker prevents the angular wander of echoes so characteristic of an unstabilized display. This removes all the shortcomings of the ship's head up unstabilized orientation which were listed in Section 5.3.1.

This orientation corresponds with the chart, and many officers find this very agreeable or at least acceptable. However, as mentioned previously some users do find it awkward or uncomfortable, particularly on southerly courses and in some pilotage situations.

In conclusion, it must be stressed that azimuth stabilization is an essential feature of true north up orientation.

5.3.3 Course up orientation (stabilized)

In a course up orientation, the heading marker is aligned to 000° (12 o'clock) on the fixed bearing scale at an instant that the vessel is right on course. By azimuth stabilization, changes in the vessel's instantaneous heading are reflected by sympathetic angular movement of the heading marker, thus maintaining the ship's course (referred to as the reference course) in alignment with the zero of the fixed bearing scale. As in the case of the true north up orientation, the angular wander of echoes is removed and the same consequent benefits accrue.

The obvious virtue of this orientation is its ability to remove the shortcomings of the ship's head up unstabilized orientation while maintaining the heading marker in a substantially (though not exactly) head up position, provided that the vessel is not forced

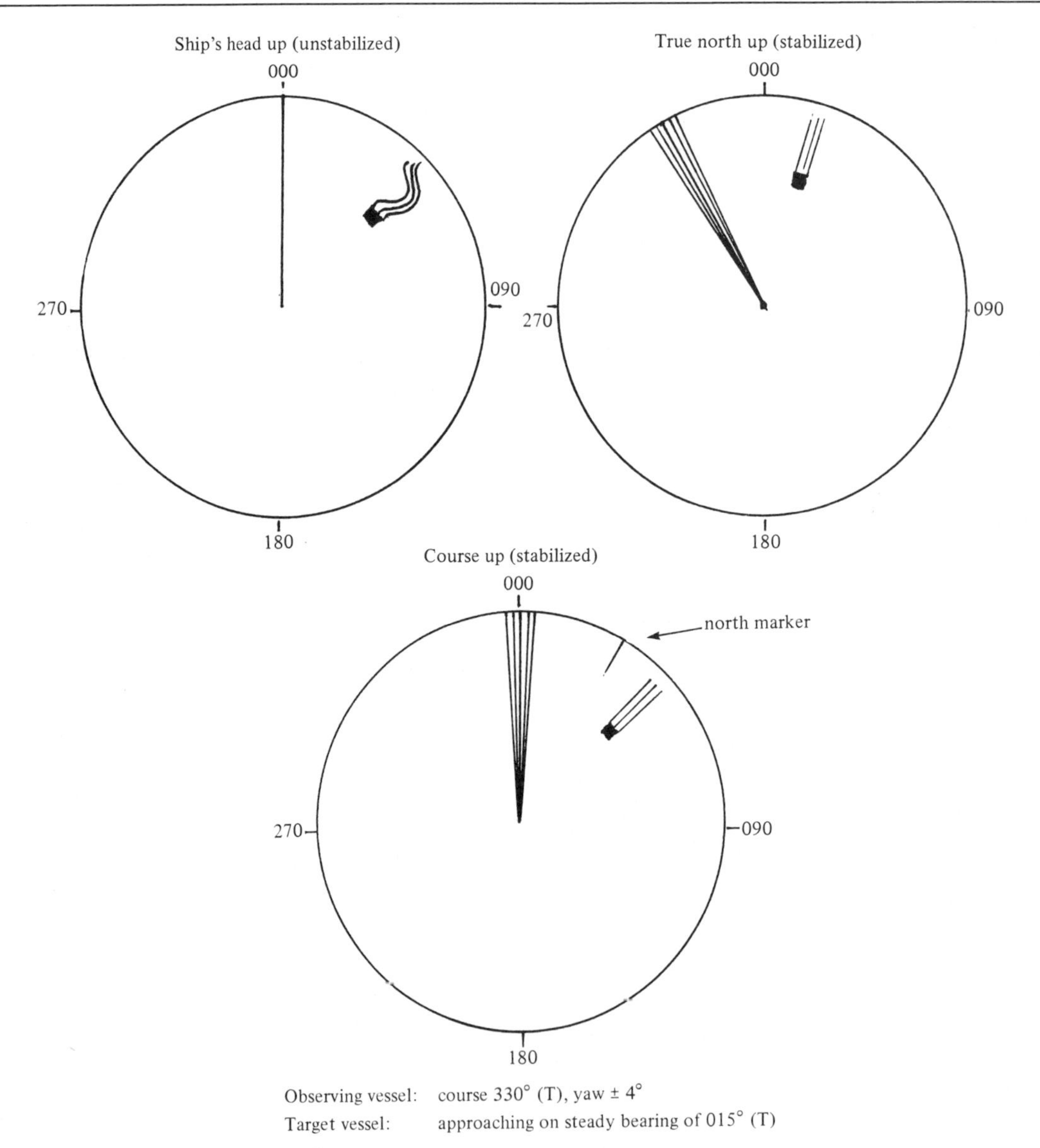

Figure 5.18 Orientations compared

to stray far from the reference course. When a major alteration becomes necessary, and the vessel is steadied on the new course, the orientation (although not meaningless) will have become less useful. The problem is that the orientation is 'previous course up' and the picture must be oriented to return the heading marker to 000° on the fixed bearing scale at an instant that the vessel is right on the new course (Figure 5.19). In most modern systems this can be done very simply at the press of a button. To assist the observer in interpreting the course up orientation, some systems display a north marker or a full synthetic bearing scale in computer graphics at the edge, but on the face of, the cathode ray tube.

Table 5.2 Picture orientations compared

Feature	*Orientation*		
	Ship's head up, unstabilized	*True north up, stabilized*	*Course up, stabilized*
Blurring when observing vessel yaws or alters course	Yes: can produce very serious masking	None	None
Measurement of bearings	Awkward and slow	Straightforward	Straightforward
Compatibility with reflection plotter	Very limited	Straightforward	Straightforward
Angular disruption of target trails when observing vessel yaws or alters course	Yes: can be dangerously misleading	None	None
Correspondence with wheelhouse window view	Perfect	Not obvious	Virtually perfect except after large course change
Correspondence with chart	Not obvious	Perfect	Not obvious

5.3.4 Data extraction – orientations compared

The extraction of data for navigation and collision avoidance will be discussed in detail in succeeding sections. However, in considering the influence of orientation on data extraction it is expedient to note that all techniques are based on the requirement to measure the range and bearing of identified targets and hence to track the movement of such targets. Considered against these twin requirements, the ship's head up unstabilized orientation has nothing to offer other than its subjective appeal, because by its very nature it regularly disrupts the steady state condition conducive to measurement of bearing and tracking of echo movement (see Figure 5.18).

True north up and course up orientations do not exhibit this angular disruption and hence are equally superior in fulfilling the stated requirements. Fortunately they are complementary in that while one is north up, the other is orientated in such a way as not to alienate the user who has a ship's head up preference (see Figure 5.18).

5.3.5 Double stabilization

When civil marine radars were first fitted, ship's head up was the only orientation available. As gyro compasses became more common, true north up became available as an alternative. Despite the obvious advantages of stabilization, many users held to ship's head up because of their subjective preference for an orientation which accorded with the view from the wheelhouse window. As time passed, the switching configuration was such in many (though not all) systems that ship's head up and azimuth stabilization could not be selected simultaneously. The fairly obvious logic of this is that, by definition, for ship's head up you wish the heading marker to remain at 000° (12 o'clock) and thus you must not introduce azimuth stabilization. The unfortunate effect of this was that in general the user had to choose between the benefits of azimuth stabilization and his subjective preference for which way up he liked to view the picture. (In some systems it was possible, with the appropriate knowledge, to switch in the azimuth stabilization irrespective of orientation and hence some users discovered how to produce their own 'unofficial' course up orientation. The general marketing of a course up facility as described in Section 5.3.3 became common in the 1970s.)

In the late 1950s one manufacturer conceived a way

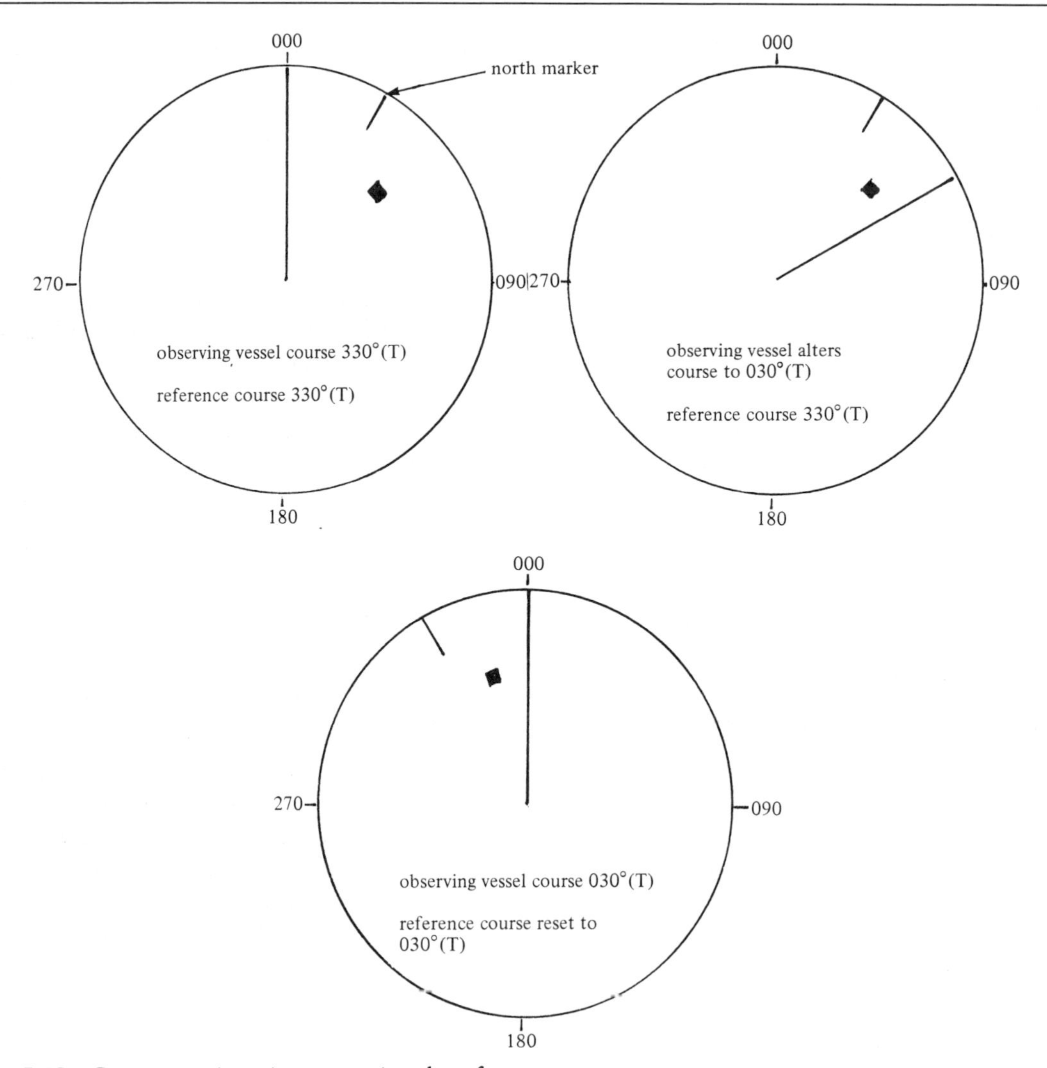

Figure 5.19 Course up orientation – resetting the reference course

of resolving the conflict whereby it appeared that if ship's head up was preferred, the advantages of azimuth stabilization were precluded. The conflict was resolved by double azimuth stabilization. The picture was azimuth stabilized to remove angular wander, and the tube (including the reflection plotter) was azimuth stabilized so that it was physically turned in such a way that the heading marker was always aligned with the 000° graduation on the fixed bearing scale within which the tube rotated. This arrangement of double stabilization ingeniously combined the advantage of an orientation which corresponded with the wheelhouse window view and the advantages of azimuth stabilization. Such an orientation could truly be described as ship's head up stabilized. The system was very complex because of the electromechanical arrangements necessary to relay picture signals, control voltages and power supplies to a continually rotating cathode ray tube. This particular method (which was called BONUS–BOws or North Up Stabilized) is now of historical interest as the equipment has not been produced for many years.

In the early 1970s another manufacturer utilized double stabilization with the aid of an electronic image-retaining panel. The latter behaves rather like the phosphor of a cathode ray tube, with the essential difference that the images can be retained almost indefinitely or removed instantly by electronic control. An azimuth stabilized picture was generated on a cathode ray tube and exposed to the image-retaining panel. The image-retaining panel was televised by a closed circuit television camera against the background of, amongst many other features, a fixed bearing scale and an azimuth stabilized bearing scale. The image-retaining panel was azimuth stabilized so that it was continuously rotated to maintain the heading marker coincident with the 000° graduation on the fixed bearing scale. The camera output was displayed on a 625 line television screen with the heading marker in the 000° (12 o'clock) position.

The system (known as Situation Display Radar, SDR) could, like BONUS, be truly described as ship's head up stabilized, because the target trails produced on the image-retaining panel were not affected by the yawing of the observing vessel. However, no reflection plotter could be fitted and it was not practical because of parallax to plot on the face of the display. Even if it had been, all the plots in any one sequence would require to be made when the ship was on the same heading. Instantaneous changes in the ship's heading, while producing no angular wander on the image-retaining panel, would produce angular movement with respect to the television screen. There is a very large number of situation display radars at sea, but it will become progressively more of historical interest because the equipment is no longer produced.

Some modern synthetic pictures can, to a greater or lesser extent, eliminate the smearing of echoes in the unstabilized condition, by writing the stored picture on a very short persistence PPI (or on a television screen using a raster). The elimination of such smearing has led the orientation to be described, on occasions, as ship's head up stabilized, but such a description is open to debate. Some might consider it to be consistent with the restricted definitions of azimuth stabilization applicable only to ARPA, but that in itself is questionable (see Section 5.11.3: annex 1, definition of terms to be used only in connection with ARPA Performance Standards). Certainly it is difficult to see how it can be reconciled with the fact that changes in the observing vessel's heading will produce angular wander in the movement of echoes, which in conventional radar terms is the hallmark of an unstabilized orientation.

In the future it may well be that a rasterscan orientation will reproduce, by means of multiple memories, the SDR effect, i.e. ship's head up with azimuth stabilized echo trails.

5.4 Picture presentation

A radar picture presents the movement of echoes with respect to a chosen reference. The three preferred presentations are:

1 Relative motion presentation
2 True motion, sea stabilized presentation
3 True motion, ground stabilized presentation.

These will be discussed in turn to the extent necessary to appreciate the significance of the echo movement implicit in each presentation. The amount of intelligence which can be obtained by merely viewing such echo movement as indicated by the afterglow trails is very limited. For a full appreciation of an encounter with other vessels, systematic observation of detected echoes is essential. The latter is fully covered in Section 5.6.

5.4.1 Relative motion presentation

In a relative motion presentation, the electronic origin of the picture (which represents the observing vessel) is fixed and the movement of targets is shown with respect to the observing vessel (Figure 5.20). Three key facts essential to an understanding of the relative motion presentation are:

1 The echo of a target which is stationary in the water will move across the screen in a direction which is reciprocal to that of the observing vessel's heading, at a rate equal to the observing vessel's speed through the water.

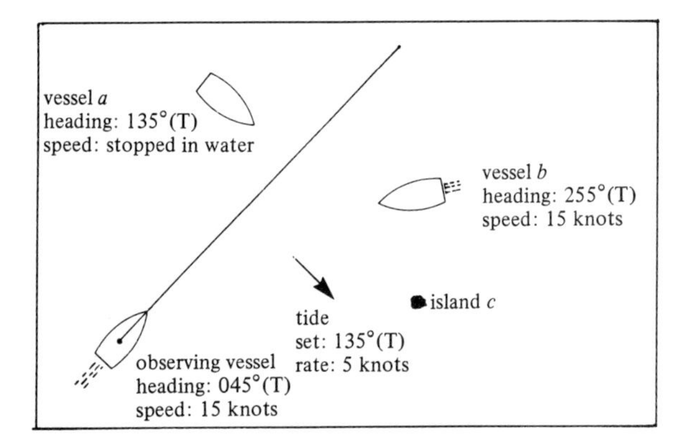

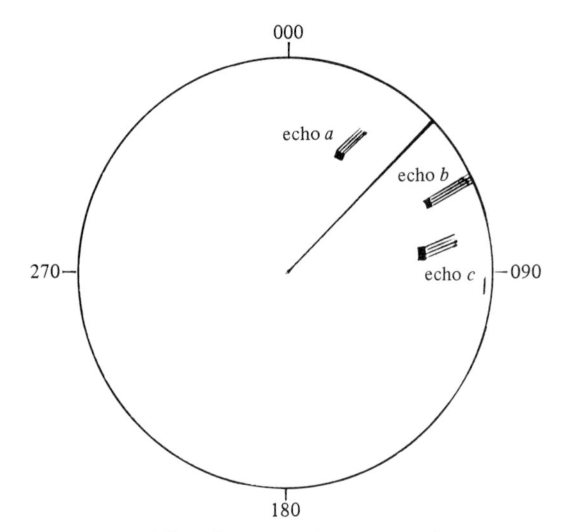

Figure 5.20 The relative motion presentation

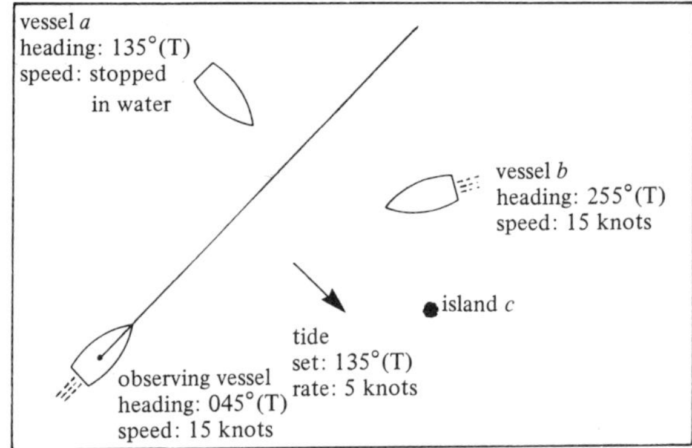

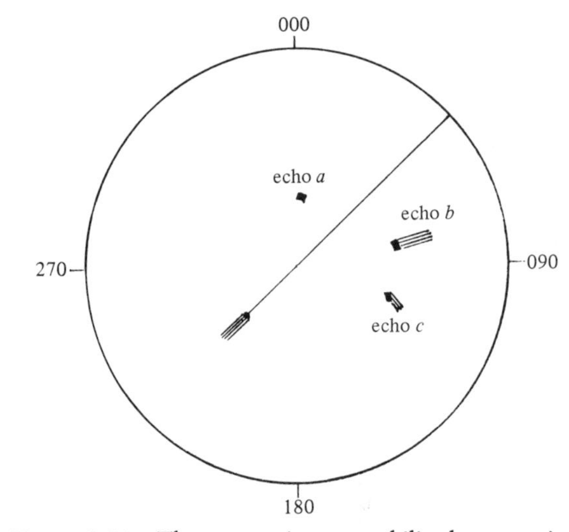

Figure 5.21 The true motion, sea stabilized presentation

2 The echo of a target which is moving through the water will move across the screen in a direction and at a rate which is the resultant (the vector sum) of the movement of the target through the water and the movement of the observing vessel through the water. It is thus impossible by merely looking at the display to reliably deduce the heading of such a vessel. Systematic observation and resolution of the vector triangle are necessary (see Section 5.6).
3 The echo of a land target, or a floating target at anchor, will move across the screen in a direction which is the reciprocal of the observing vessel's ground track at a speed equal to the speed of the observing vessel over the ground. (This characteristic is utilized in parallel indexing which is discussed in Section 5.7.)

Note In the absence of tide movements, 1 and 3 will be the same.

A true north up stabilized orientation has been chosen for the illustration in Figure 5.20, but course up stabilized orientation would have served equally well. (If a ship's head up unstabilized orientation was chosen, any yaw on the part of own ship would disrupt the echo movement.) The illustration also shows the display with the origin in the centre. This need not be the case. In some systems it is possible to off-centre the origin by about 60 per cent of the CRT radius. This is quite acceptable as the essential feature is that the origin be stationary. The advantage of the off-centred display is that it permits an extended view ahead, thus allowing the selection of a shorter range scale for any given range requirement, with the consequent benefit in scale size, scale speed of echo movement, and accuracy (see Sections 5.5, 5.6.2 and 5.10.1).

5.4.2 True motion, sea stabilized presentation

In a true motion, sea stabilized presentation the electronic origin of the picture is driven across the screen in a direction, and at a rate, that corresponds with the observing vessel's movement through the water (Figure 5.21). Three key facts are essential to the understanding of a true motion, sea stabilized presentation:

1 The echo of a target which is stationary in the water will remain stationary on the screen.
2 The echo of a target which is moving through the water will move across the screen in a direction and at a rate which corresponds with its motion through the water, i.e. it will indicate the true heading and speed of the target (if leeway can be neglected and tide is assumed uniform over the area).
3 The echo of a land target or a target at anchor will move across the screen in a direction which is the reciprocal of that of the tide, at a rate equal to that of the tide.

A true motion presentation cannot be achieved without azimuth stabilization. Figure 5.21 illustrates a true north up orientation, but course up could have been selected. The latter does however suffer from the disadvantage, that, in addition to origin resetting (see Section 5.2.11.1), reference course resetting is also necessary (see Section 5.3.3).

The realization of the foregoing features is entirely dependent on the accuracy of the tracking of the electronic origin. (Attention is drawn to the setting-up and monitoring procedure presented in Section 5.2.11.1 and to the error theory set out in Section 5.10.2.)

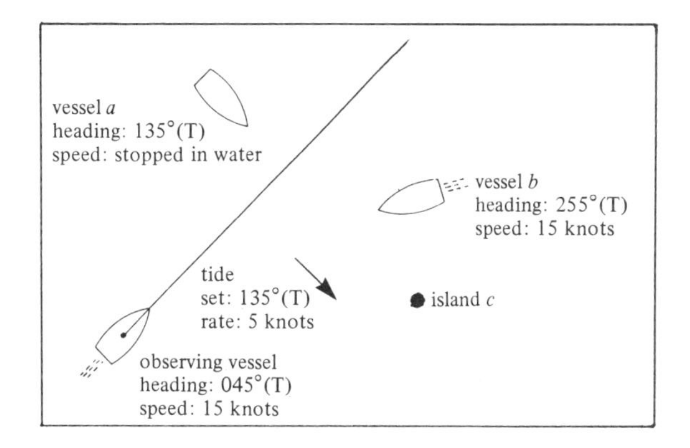

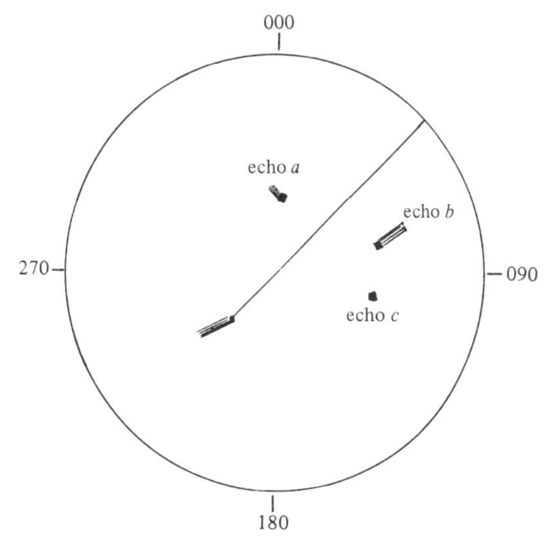

Figure 5.22 The true motion, ground stabilized presentation

5.4.3 True motion, ground stabilized presentation

In a true motion, ground stabilized presentation the electronic origin of the picture is driven across the screen in a direction, and at a rate, which corresponds with the observing vessel's track and speed over the ground (Figure 5.22). Three key facts essential to the understanding of a true motion, ground stabilized presentation are:

1 The echo of a target which is stationary in the water will move across the screen in a direction and at a rate corresponding to that of the tide.
2 The echo of a target which is moving through the water will move across the screen in a direction and

at a rate which corresponds with its movement over the ground. It is essential to appreciate that the echo movement *will not* indicate the heading of such a target.
3 The echo of a land target, or a target at anchor, will remain stationary on the screen.

A true north up orientation has been chosen for the illustration in Figure 5.22. A course up orientation can be chosen subject to the limitation of the need for resetting the reference course after an alteration of course.

No apology is made for restating the importance of appreciating that a ground stabilized presentation does not display the headings of moving targets, a knowledge of which is essential when applying the Collision Regulations.

5.4.4 Data extraction – presentations compared

When comparing orientations (see Section 5.3.4) it was noted that the various complex techniques used in collision avoidance and radar navigation are based on the ability to measure range and bearing of identified targets and hence to track the movement of such targets. The detailed influence of presentation on the use of radar for collision avoidance will be dealt with fully in Section 5.6 (and similarly the influence on radar navigation in Section 5.7), but in this section it is expedient to compare the presentation in terms of the following features:

1 Range and bearing measurement
2 Echo movement.

There is little to choose between the presentations in the matter of range and bearing measurement, apart from the fact that the mechanical cursor and fixed bearing scale cannot be used directly if the presentation involves off-centring. Range and bearing measurement are discussed fully in Section 5.5.

When comparing the presentations in terms of echo movement it is most meaningful to compare true motion, sea stabilized presentation with

1 Relative motion presentation; then
2 True motion, ground stabilized presentation.

(There is little practical value to be gained by making a direct comparison between relative motion and true motion, ground stabilized.)

5.4.4.1 Comparison of true motion (sea stabilized) and relative motion

1 In any encounter with other vessels in restricted visibility, it is necessary to establish firstly if a close-quarters situation is developing and, if so, secondly the heading (or aspect) of the other vessel(s) so that appropriate avoiding action may be chosen (see Section 22.2.5; and Notice M.1158, paragraph 3.5, in Section 5.11.5).

   The two presentations are truly complementary in the context of the above requirements. The relative motion presentation provides, by inspection, early indication of passing distance, while the true motion presentation similarly provides early indication of heading or aspect.
2 In a changing situation, the comparison is more favourable to true motion. If a target vessel manoeuvres, the change in relative motion is in general likely to be less obvious than the change in true motion, especially where the manoeuvre is an alteration of course. Further, while the relative motion will indicate a change in passing distance, the true motion will provide early and simple identification of the nature and magnitude of the target's manoeuvre. The latter information is likely to be of most immediate value in alerting the observer to the need to reappraise the situation.
3 The movement of targets on the relative motion screen is the resultant of two motions, the target's motion and the observing ship's motion. The Achilles' heel of relative motion is that, when the observing ship alters, it looks as if the rest of the world has altered too. No meaningful information other than range and bearing can be deduced until the observing vessel's manoeuvre is complete. Even then the question has to be asked: has the change in echo track, resulting from the observing vessel's manoeuvre, masked a manoeuvre by a target? In a true motion presentation, target manoeuvres can be not only noticed but positively identified irrespective of manoeuvres performed by the observing vessel. Thus again, in the changing

Vessel / Time	Observing	a	b
1000	Course 000 (T) Speed 15 knots	Course 090° (T) Speed 15 knots	Course 268° (T) Speed 7 knots
1012	Course and Speed maintained	Reduced Speed to 7 knots	Altered Course to 315° (T)
1024	Alters Course to 060° (T)	Course 090° (T) Speed 7 knots	Course 315° (T) Speed 7 knots
1036	Course 060° (T) Speed 15 knots	Course 090° (T) Speed 7 knots	Course 090° (T) Speed 7 knots

Echo *c* is a lightvessel. The tide set 090° (T) throughout

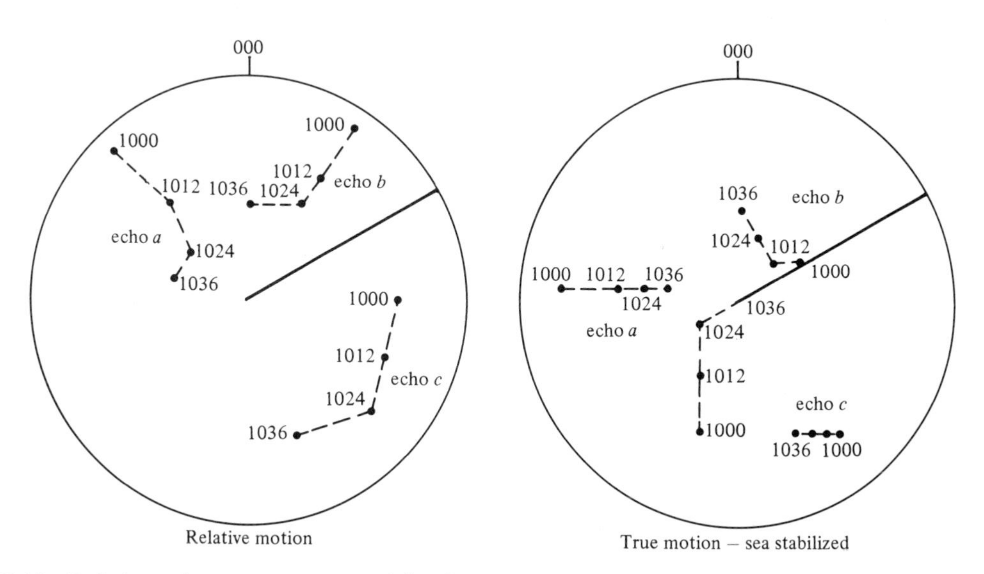

Figure 5.23 Relative and true motion (sea stabilized) presentations compared

situation, true motion compares favourably with relative motion.

4 In the navigational context, relative motion is ideally suited to parallel indexing because the latter exploits the basic characteristic of such a display, i.e. that the movement of land echoes is the reciprocal of the observing vessel's ground track (see Sections 5.7.2.2 and 5.9.3). True motion, sea stabilized, is not suited to parallel indexing. However, it does facilitate identification of stationary or near stationary echoes and can be used to establish the set and drift if a ground-stationary echo can be identified.

Figure 5.23 shows a comparison of relative motion and true motion, sea stabilized presentations.

It must be stressed that while mere inspection of the display will give early indications of target movement, conclusions must be based on systematic observations (see Section 5.6).

In making this comparison, it is not intended to fuel the argument that one presentation is superior to

Table 5.3 Summary of the features of relative motion and true motion, sea stabilized presentations

Feature	*Relative motion*	*True motion, sea stabilized*
Assessment of passing distance	Available by inspection from simple tracking on reflection plotter	Not directly available
Assessment of aspect (and speed)	Not directly available	Available by inspection from simple tracking on reflection plotter
Identification of target manoeuvre	Not directly available	Available by inspection from simple tracking on reflection plotter
Suitability for parallel indexing	Perfect	Not suitable
Continuity of target track during own vessel's manoeuvre	Disrupted	Maintained

the other. On the contrary it is intended to show that no one presentation has all the answers and that the two presentations are very much complementary and should be used as such in navigation control (see Table 5.3).

5.4.4.2 Comparison of true motion, sea stabilized and ground stabilized

1 A sea stabilized presentation offers direct indication of the heading of other moving vessels (subject to any error introduced by leeway), whereas ground stabilization gives no such indication. As the entire philosophy of collision avoidance is based on a knowledge of the headings of other vessels in an encounter, the sea stabilized presentation is the correct data source for collision avoidance. A ground stabilized presentation can be dangerously misleading if used mistakenly as a source of heading information. See Figure 5.24.
2 Both presentations require the input of course and speed measured with respect to the water. Such information should be readily and continuously available. In addition, ground stabilization requires the input of tidal information. The latter must be deduced and may well vary with position and with the passage of time. As a result, in practice it is difficult and time consuming to effect and accurately maintain the ground stabilization of a conventional radar display. Where computer assistance is available, this limitation can be removed (see Section 5.9.2).
3 Ground stabilization facilitates the identification of ground-stationary targets. This may well be useful, for example, when attempting to differentiate between anchored and moving vessels when approaching a crowded anchorage or perhaps in a channel. Other examples of situations in which this facility may prove useful will suggest themselves, particularly to those officers using radar on vessels employed in specialized tasks in the proximity of fixed and moving targets in a tideway. For more general navigation, a form of parallel index plotting can be carried out using a ground stabilized presentation (see Section 5.7.2.2), although it is very much second best to relative motion parallel indexing. However, in the absence of computer assistance it may be difficult or impossible in practice to avail oneself of the foregoing benefits, which in any event must be considered in relation to the loss of target heading information. Sometimes it is claimed that the movement of ground-stationary targets on a sea stabilized presentation is a limitation of that presentation. It can also be seen as a merit because it gives indications of the tidal set and drift experienced.
4 In summary, if conventional radar is considered, there is no doubt that in almost all circumstances sea stabilization is more appropriate, as ground stabilization has little if anything to offer. It gives no heading information, is difficult to effect, and has few practical applications.

Where computer assistance is available (see Section 5.9.2), ground stabilization has more to offer. It is

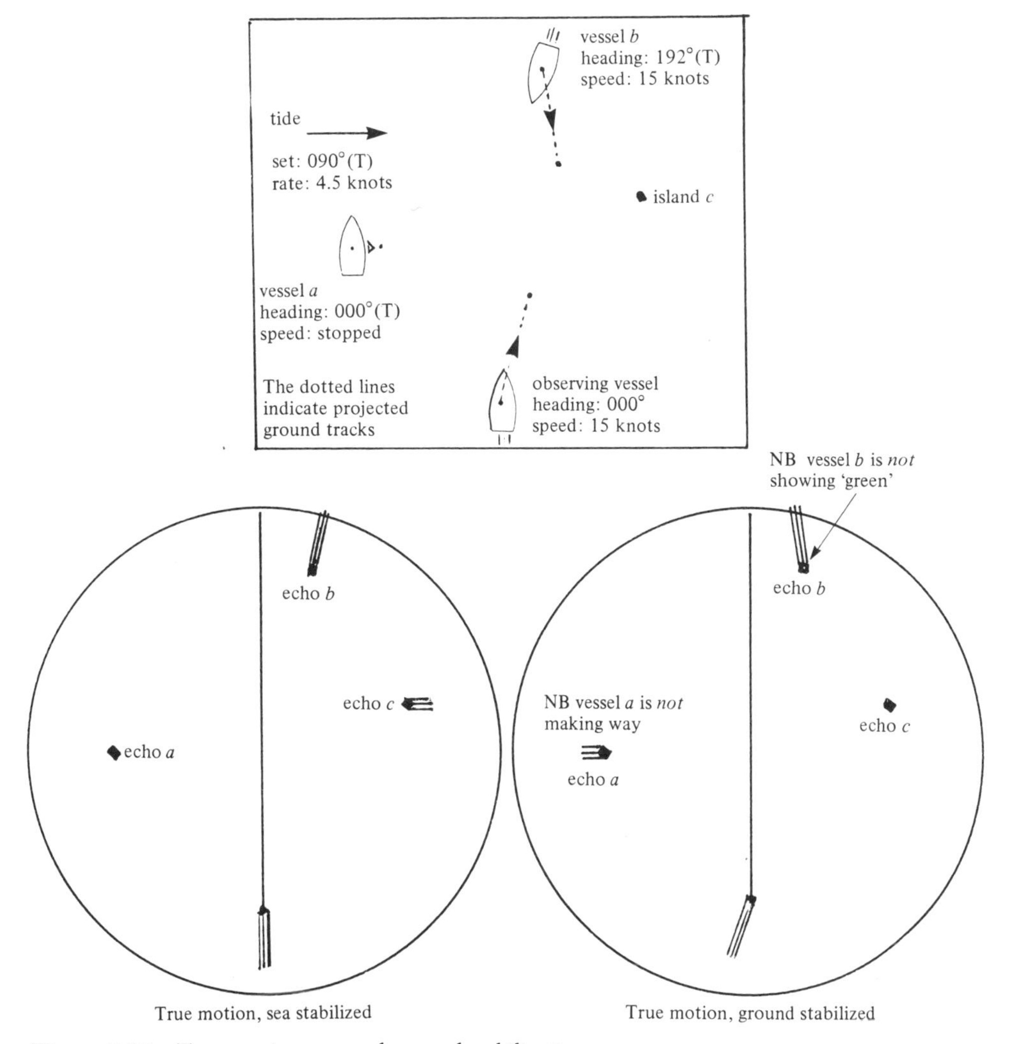

Figure 5.24 True motion, sea and ground stabilization

easily achieved, facilitates the benefits outlined in point 3, and is often complementary to ARPA mapping facilities (see Section 5.9.3.3). Nevertheless, its conflict with the role of the radar in collision avoidance may remain.

5.5 Extraction of data

The application of all radar techniques used in navigation control is based on the ability to measure range and bearing. In this section consideration will be given to the accurate extraction of such basic data. It is assumed that the reader is familiar with the range and bearing measurement facilities available on modern marine radars to the extent to which they are described in *Radar and ARPA Manual*.

5.5.1 Measurement of range

The IMO Performance Standards for Navigational Radar Equipment (see Section 5.11.1) set out the minimum accuracy of which a type tested set must be capable, and this is discussed in Section 5.10.1. However, there are a number of ways in which the observer must ensure that the potential accuracy is realized:

1 Adjust the brilliance of the rings, variable range marker (VRM) or electronic range and bearing line (ERBL) to obtain the finest possible line.
2 Measure the range to the nearer edge of the displayed echo.
3 Use the rings when the target is on or close to a ring.
4 Use the VRM (or ERBL) to interpolate between the rings.
5 Regularly check the VRM or ERBL against the rings.
6 Use the most open scale appropriate.

5.5.2 Measurement of bearing

As in the case of range measurement, the performance specification sets out minimum accuracy limits. However, again there are a number of practical precautions which the observer must take in order to ensure that the potential accuracy is realized:

1 It is important to check regularly that the heading marker accurately represents the ship's fore-and-aft line (see Notice M.1158, paragraph 3.25, in Section 5.11.5).
2 Ensure that centring has been carried out before using the mechanical cursor.
3 Ensure that the heading marker has been correctly aligned on the bearing scale around the screen.
4 Use an appropriate range scale with the target as near to the edge of the screen as possible.

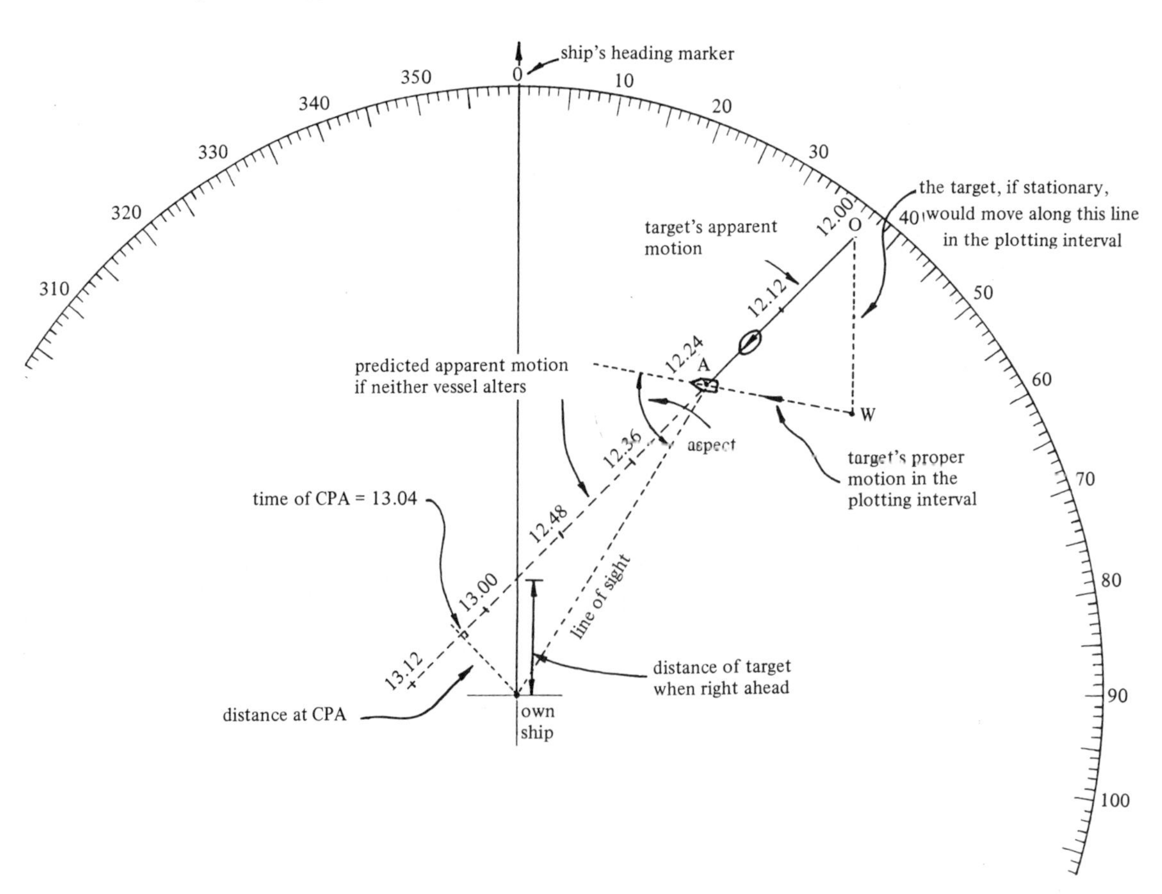

Figure 5.25 Radar plotting: the basic plotting triangle

5 Take care to avoid parallax when using the mechanical cursor or parallel lines associated with it.
6 If using a ship's head up unstabilized display, the ship's head must be read at the instant of taking bearings.
7 Check the electronic bearing line (EBL) or ERBL by superimposing it on the heading marker and noting the bearing.
8 For small isolated targets align the cursor with the centre of the target.
9 *Temporarily* reduce the gain if it will give a more clearly defined echo.

In general, it should be borne in mind that radar range position circles have a much higher inherent accuracy than is obtainable from position lines derived from radar bearings. Further, radar bearings are not as accurate as visual bearings (see *Radar and ARPA Manual*).

5.6 Radar plotting techniques

The shape of a radar echo will give no indication of aspect except at very close range (see *Radar and ARPA Manual*). Also, it is possible to deduce only very limited information by viewing the movement of echoes across the screen. For a full appreciation of an encounter with other vessels, systematic observation of the movement of echoes is essential. The term radar plotting has traditionally been used to decribe

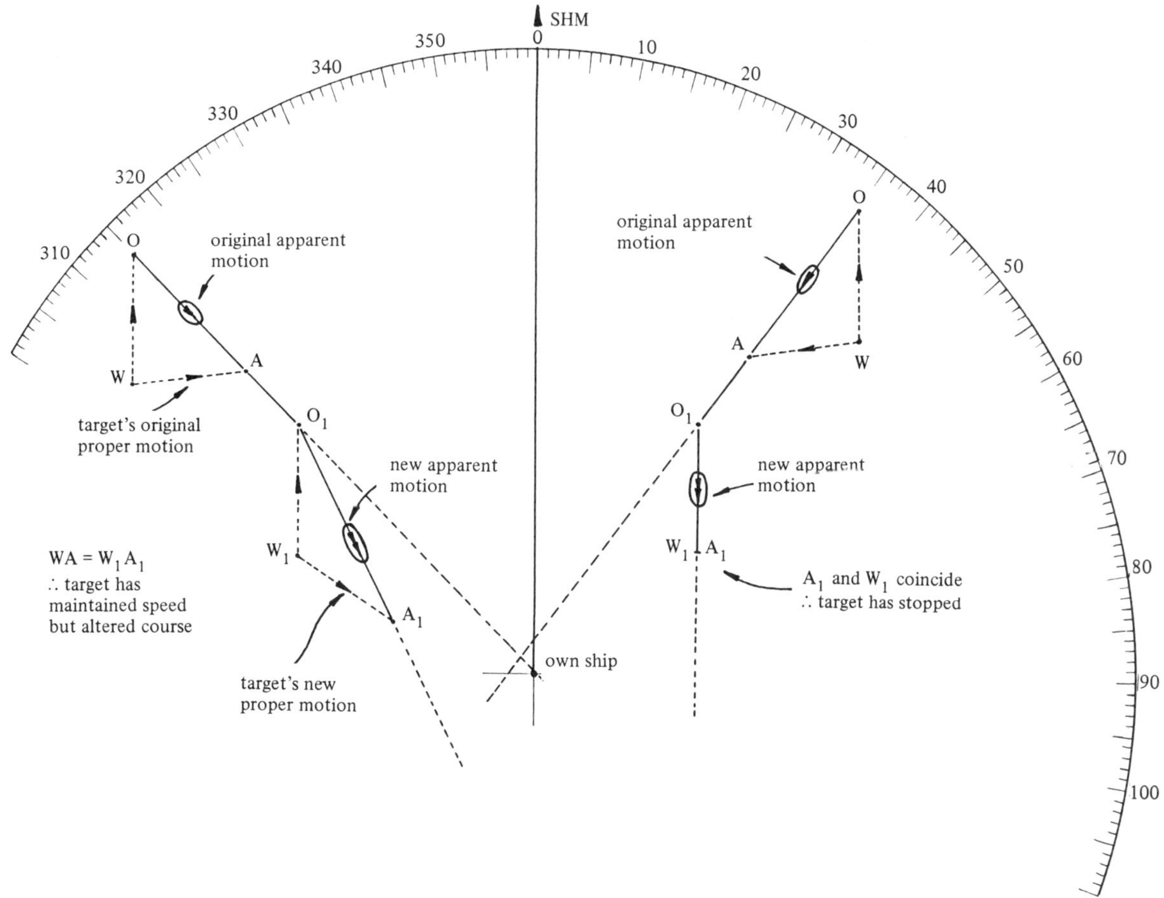

Figure 5.26 Radar plotting: the plot when targets alone manoeuvre

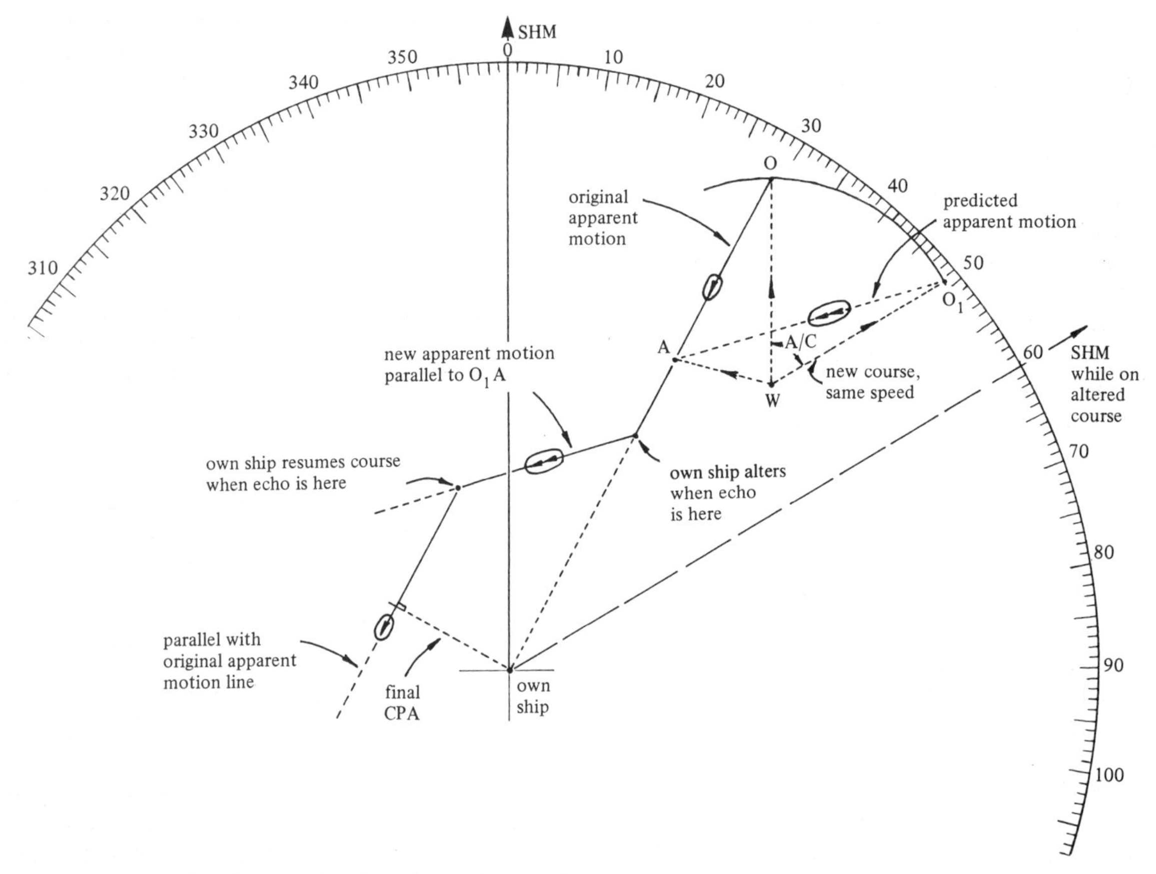

Figure 5.27 Radar plotting: the plot when only own ship manoeuvres

such systematic observation. Rule 7(b) of the Collision Regulations (see Section 22.3.1) leaves the mariner in no doubt that radar plotting (or equivalent systematic observation of detected objects) is an essential constituent of the proper use of radar. Although the regulations do not define the term radar plotting, consideration of the Steering and Sailing Rules leads one to the conclusion that all collision avoidance decisions must be based on a knowledge of the appropriate data and that, in the particular case of vessels not in sight of one another, steps must be taken by a radar equipped vessel to additionally derive any information which is lacking owing to the inability to see the other vessel(s). Such considerations effectively specify the information which must be extracted by radar plotting (see also Section 22.2.4).

Figures 5.25–5.27 show the fundamental radar plotting ideas. The information which can be extracted by radar plotting may be summarized as follows:

1 The nearest approach (or closest point of approach, CPA) and time to nearest approach (or time to closest point of approach, TCPA) of other vessels
2 The heading/aspect and speed of other vessels
3 The effect of any manoeuvre by the observing vessel on the CPA and TCPA of other vessels.

The availability of this information allows the mariner to:

1 Determine if a close-quarters situation is developing

2 Establish the nature of the encounter and hence select a seamanlike avoiding manoeuvre
3 Evaluate the effectiveness of any chosen manoeuvre.

The responsibility to extract such data will be discussed in detail in Section 22.2. In this section it is intended to consider the practical application of techniques which can be used to extract the data, and it is assumed that the reader is familiar with the theory underlying radar plotting (see *Radar and ARPA Manual*). All the techniques will produce the same information if carried to their conclusion, but the number of targets which can be handled in any given circumstances will vary with different plotting facilities.

5.6.1 *The paper plot*

The simplest form of plotting technique involves the use of a plotting sheet. This, although it may take various detailed forms, consists essentially of a pattern of concentric range circles, a peripheral bearing scale, and a scale of nautical miles. By taking a succession of ranges and bearings of a target at known time intervals, and laying them off from the centre of the range pattern, the direction and rate of the target's relative motion can be drawn and then extended to deduce the CPA and TCPA. It is essential to establish whether or not the target's relative motion is constant. Constancy will be indicated by the plots appearing (subject to the limits of plotting accuracy; see Section 5.10) in a straight line at equal spacings for equal time intervals (or unequal spacings for proportionally unequal time intervals). Only by taking timed observations and making this check can the plotter reliably establish whether or not the target has manoeuvred during the plotting interval. If the relative motion is not constant, then frequent timed observations must be made until a steady relative motion is obtained.

When a constant relative motion has been established, the target's true motion can be deduced by laying off the reciprocal of the observing vessel's course from the commencement of the steady relative motion, for a distance equal to that made good through the water by the plotting vessel in the time for which the steady relative motion was observed. The third side of the triangle so formed represents the target's course and distance travelled through the water for the chosen time interval. The plotter is then in a position to read off the target's aspect, course and speed.

If a manoeuvre is considered necessary, then a second triangle can be constructed using the target's true motion as a common side and a proposed course and speed for the plotting vessel. The third side of the triangle so formed represents the direction and rate of relative motion that will be achieved if the target maintains course and speed and the plotting vessel makes the proposed manoeuvre. This information enables the observer to decide if the proposed manoeuvre will achieve a safe passing distance.

Further, if the predicted apparent motion is transferred through the position of the target when the manoeuvre is complete, it facilitates detection of target manoeuvres and regular monitoring of the effectiveness of the action till the target vessel is finally past and clear.

If the forecast is fulfilled, then the time to resume course and/or speed is easily deduced. A line parallel to the original relative motion is drawn tangential to the passing distance which has to be achieved after resumption. The point at which this cuts the fulfilled forecast represents the position of the target at which resumption can take place. If the forecast has not been fulfilled, a target manoeuvre has occurred and the plotting procedure must be commenced afresh.

The technique described above is correctly termed a complete relative plot. It is possible to carry out a true plot on a sheet of paper and then resolve to obtain the relative motion (see *Radar and ARPA Manual*). The latter technique suffers badly from the need to plot the observing vessel before each and every observation of the target can be laid off, and care has to be taken to select the own ship position from which the CPA is measured. Either plot when complete will yield the same information but, when one is plotting manually on paper, the relative plot is quicker and simpler by comparison.

All paper plots are time consuming because of the need to measure ranges and bearings, transfer the information to the plotting sheet and perform the constructions and resolutions. As a result, the number of targets that can be handled by one plotter is very limited.

5.6.2 *Reflection plotting on a relative motion presentation*

In principle this is identical to the first technique described in Section 5.6.1, but in practice is very much quicker because there is no need to measure the data or lay it off elsewhere. Accuracy improves for the same reasons. It is a much more practical technique as the observer views the plot against the background of the entire radar picture and not in selective isolation or in different ambient light conditions. The plot will have to be restarted on long term change of range scale, but there is the compensation that, when reducing range scale, the plotting scale increases and on the shorter range scales large plotting triangles can be built up quickly. The provision of a reflection plotter will increase the number of targets that can be dealt with in any given situation. The limiting effect of an unstabilized orientation on the efficacy of a reflection plotter should be borne in mind (see Section 5.3.1).

5.6.3 *Reflection plotting on a true motion, sea stabilized presentation*

In theory this technique equates with the true plot mentioned in Section 5.6.1, but in practice is infinitely superior. It is essential to maintain a plot of own ship, but it requires negligible time and effort to track the movement of the origin on the reflection plotter.

This technique has the following significant merits, which compare favourably with reflection plotting on a relative motion presentation:

1 Target headings are obtained directly from the true plot.
2 Target manoeuvres are in general more obvious.
3 Target manoeuvres can be evaluated by inspection.
4 The continuity of the plot is unaffected by manoeuvres carried out by the observing vessel.

Against the foregoing merits must be set the limitation that the CPA is not available directly. The latter can be determined by constructing an OAW triangle for each target on the base of the existing WA (see Figures 5.25–5.27). However, it may be found more expedient to employ the zero speed control. The control can be used to temporarily stop the movement of the electronic origin and hence produce relative motion tracks for a period of sufficient length to make it possible to establish the CPA of chosen targets. Having established this information the observer can then restore the speed input on the basis that the CPAs will be fulfilled provided that the targets do not manoeuvre. The true motion presentation will give immediate indications should such manoeuvres occur.

Practised use of reflection plotting on a true motion presentation further increases the number of targets that can be handled. This is a view particularly held by officers operating continually in high density traffic, although some deep sea officers hold a subjective preference for the relative motion presentation.

5.6.4 *Reflection plotting on dual displays*

Although relative motion and true motion presentations are complements rather than competitors, comparison has been made between them because in some vessels the navigator will have to make a choice which will depend upon his own weighting of the factors discussed. In many vessels the choice will not be necessary as two displays will be available, at least one of which will have true motion facilities. In these circumstances there is no doubt that the correct use of such an installation is to operate one display with relative motion presentation and one display with true motion presentation. Such operation makes it possible to obtain CPA and TCPA directly from the relative motion plot while heading/aspect are available directly from the true motion plot.

All this information is available directly without resort to resolution of the vector triangle. This represents a major reduction in workload when compared with reflection plotting on either single presentation and considerably increases the number of targets that can be handled by one observer in any given circumstance.

The one item of information which is not directly available is the forecast CPA for a proposed manoeuvre.

There can be no doubt that the forecast information is beneficial. Experience has shown that many officers who, because of the time consuming construction, were loath to attempt a forecast on a relative motion plot, became highly enthusiastic about forecast data when it was made readily available to them by computer assistance (see Section 5.8.3.3). However, in the case of dual displays the situation is that current relative and true data is quickly, easily and continuously available whereas forecast data is not. One must question, in these circumstances, whether or not the time spent on constructing forecasts on the relative motion presentation (which will to some extent duplicate information already available directly from the true motion presentation) might not make a greater contribution to the safe resolution of an encounter if spent on methodical appraisal of the readily available data. One cannot expect universal agreement on this point, but it is intended to debate the issue in the following paragraphs and show that the correct use of the data directly available from the true and relative presentations will provide sufficient information on which a collision avoidance manoeuvre can be safely based.

Where only relative motion data is directly available, the need to produce a forecast can be amply justified by the fact that when the observing vessel manoeuvres, the relative tracks of all targets will change. Casual observation of these changes will not reliably reveal whether or not the change is solely due to the manoeuvre of the observing vessel. Such change may well mask a manoeuvre by one or more target vessels.

A most dangerous situation arises in restricted visibility if a target vessel manoeuvres in such a way as to cancel the effect of the observing vessel's manoeuvre. It is essential that the observing vessel obtains as early a warning as possible of such a manoeuvre. When reflection plotting is on a relative motion presentation, the forecast provides the earliest possible indication of any target manoeuvre, and it is essential to carefully check the effectiveness of the action until the other vessel is finally past and clear, as required by rule 8(d) of the Collision Regulations (see Section 22.3.1). Reflection plotting on dual displays makes directly available the data necessary for compliance with the foregoing requirements. The true motion plot will facilitate immediate detection and evaluation of target manoeuvres before, during and after the avoiding manoeuvre of the observing vessel, while the relative motion plot allows continual monitoring of the CPA.'

The forecast clearly allows the mariner to predict the CPA before making a manoeuvre, and his choice of manoeuvre may well be influenced by a provisional or iterative forecast. Again we come to the conclusion that the forecast is useful; but is it essential in all circumstances? In particular, where dual displays are available, can the mariner choose a substantial and safe avoiding manoeuvre on the basis of the data directly available?

An early skill learned in watchkeeping is how to judge the correct manoeuvre for an approaching ship whose passing distance is too close and whose heading is known. This skill is developed over years of watchkeeping such that the decision can be made where a number of ships are involved, some of which present a hazard and others of which will present a hazard if certain manoeuvres are selected. The culmination of this should be the master's ability to make such strategic judgement in a multi-ship encounter. Dual displays with reflection plots provide directly the data on which such judgement can be based. The true motion plot makes target heading (and speed) continuously available, facilitating a seamanlike choice of strategy. Immediately the observing vessel has completed her manoeuvre, the relative motion plot will begin to indicate the new passing distances and confirmation can be made that they are in fact safe.

5.6.5 Appraisal aids

Appraisal aid is the term used, in general, to describe an aid to plotting which, by some semi-automatic means, assists the observer in manually tracking echoes but does not perform automatic tracking. A wide variety of appraisal aids has been produced in the past and many are now only of historical interest. In modern equipment, there are two common types of appraisal aid. It is intended not to describe specific equipment in detail but merely to explain their general use as plotting techniques.

5.6.5.1 Plot marker systems

These systems in general make it possible to extract true and relative motion data simultaneously from the same display. Usually there are about five markers, each of which consists of a short radial line having a bright spot at the end remote from the electronic origin of the display. True motion, sea stabilized presentation is selected and the bright end of the marker is placed over a chosen target. When activated, the bright end of the marker maintains constant range and bearing from the electronic origin of the display. True motion plotting is carried out on the reflection plotter. At any instant, the target's heading/aspect and speed are available directly from the true plot, while the line joining the bright end of the marker to the present position of the target, if produced, will give the CPA. In the special case of zero CPA, the target will move down the radial line.

There are some variations on this basic idea. In one example the plot markers are bright spots without a radial line and a free electronic bearing line (EBL), having timed markers superimposed on it, is used to lay off the relative motion line of the chosen target. The origin of the EBL is located on the required plot marker and the line is rotated to intersect the appropriate target. (In this particular example, relative motion presentation can also be selected, in which case the plot marker traces out the track of a water-stationary target, and a forecast facility is also provided.)

In all such systems the basic principle is that a single radar presentation is used to produce true or relative data directly from a reflection plot, while the plot marker(s) enable the observer to quickly deduce the complementary data. Maximum benefit is likely to be obtained where true motion is the selected presentation, as its versatility in multi-ship encounters, where many manoeuvres may take place, can be fully exploited. Its inability to produce CPA data directly is offset by the use of the plot marker(s). In these circumstances, the data made directly available by this technique (subject to any limitations imposed by the number of plot markers) is the same as afforded by reflection plotting on dual displays. One must bear in mind that the marker(s) must be reset after any discontinuity in the relative motion of the target(s). Such discontinuity will occur for all targets if the observing ship manoeuvres.

5.6.5.2 Passive plotting memory systems

In such systems the observer plots targets by placing an electronic marker over the chosen target and executing an enter function. The time, and the co-ordinates of the marker (and hence of the target), are automatically stored in a microcomputer memory. Subsequent similar observations enable the microcomputer to deduce the past movement of the echo. This is displayed as an electronic track-ahead line with time markers and as a digital readout of direction and rate. When relative motion presentation is selected, relative data is computed, and true data is computed if true motion presentation is selected. A free EBL arrangement makes it possible to deduce the complementary data, one target at a time. The number of plots which can be stored at one time may be limited.

The technique provides a quick and easy method of recording systematic observations, but a single display is somewhat limited in the number of targets for which it is practical to extract the complementary data. It must be stressed that such a system is *not* an automatic radar plotting aid, as the operation is entirely based on the observer reading in (albeit indirectly) radar ranges and bearings at chosen intervals. No active tracking of the echo takes place within the microcomputer.

5.7 Navigational techniques

The use of radar for navigation falls naturally into three phases:

1 The landfall phase, in which the ship's position may well be in considerable doubt and targets may be unfamiliar and difficult to recognize
2 The phase involving routine coastal navigation, where there may be more general certainty of the vessel's position but where effective use of fixing

and progress monitoring techniques requires organization, skill, practice and a thorough awareness of the radar's capability

3 The pilotage phase, in which some specific characteristics of the installation may require critical exploitation.

It would be impossible to design a single radar which was perfectly suited to all three diverse applications. The observer must be conscious of the various ways in which the system can be optimized for any particular application. In the case of a single radar system, flexibility is limited to the intelligent selection of pulse length, PRF and range scale, together with appropriate use of clutter and gain controls. Where two radar systems are fitted, flexibility (and reliability) is greatly enhanced, particularly if the systems are fully interswitched (Table 5.4). It is expedient in a preamble to a treatment of radar navigation techniques to recall the complementary characteristics that might be considered.

5.7.1 *Landfall navigation*

Radar indication is useful in all cases of landfall and should always be used as a cross-check against other sources of information (see also Section 13.3). It is, however, particularly useful in conditions of restricted visibility.

When approaching the land, the immediate requirement is to detect the presence of the land and immediately thereafter commence the process of identification of the land echoes which have been detected. If the observer knows the general area in which to look for the land echoes, initial detection may be assisted by a slight but temporary increase in the gain. The slight loss of contrast in return for an increase in amplification can usually be tolerated when searching in a specific area. The echo stretching facility, if fitted, may well be useful in drawing the observer's attention to initial detection. However, once such detection has been achieved, the facility should be switched off when attempting to identify land targets as the stretching will distort the appearance and can prove a hindrance. Because of the effects of discrimination and target response, it is in the nature of radar detection that target identification will almost invariably be difficult at long range. This situation is further complicated by the fact that

Table 5.4 Interswitching: potential flexibility

Feature	*Characteristic*	*Complementary characteristic*
Transmitter power	High: good for long range detection and rain penetration	Low: reduces false echoes at short range
Wavelength	3 cm: good for definition	10 cm: good for reducing clutter response and for rain penetration
Pulse length	Long: good for long range detection and for rain penetration	Short: good for short range discrimination and clutter reduction
Horizontal beam width	Narrow: good for bearing discrimination	Less narrow: increased strikes per target
Antenna height	High: good for long range detection; also reduces shadow sectors	Low: good for minimum range and reduces extent of clutter
Antenna location	Centred: good for pilotage	Off centre: provides alternative shadow sector pattern
Range scale	Long: good for early warning	Short: good for clutter searches and discrimination
Presentation	Relative motion: good for CPA	True motion: good for aspect

objects which are visually conspicuous (e.g. a lighthouse) are frequently poor radar targets. In seeking to positively identify targets when making landfall, the following procedures and factors should be considered.

1 Consult the chart for any targets that may afford easy and positive identification. A coded radar beacon is probably the only sure source of such information, but its range may well be limited by the height of its transmitter. The positive identification of one target is not only important in offering the opportunity for a fix, but also allows the subsequent identification of other targets by association.
2 From the chart, consider all targets which might give detectable echoes on landfall and determine their theoretical maximum detection ranges from the knowledge of aerial and target height. Note the meteorological conditions, and have regard to how detection ranges in general will be affected. Subsequently consider the targets' reflecting characteristics (aspect, material, surface, shape and size) and hence judge the likely practical detection ranges. Look especially for good, isolated composite targets such as tank farms and small built-up areas.
   Information on some targets may be obtainable directly from the tables of reported radar ranges given in the Admiralty Sailing Directions (see Section 11.3.1). When good landfall targets have been established, it is helpful to record these for future reference.
3 When measuring ranges from a detected 'coastline', be alert to the possibility that the coastline proper may be below the radar horizon and the responses observed may be from higher ground which lies further inland. Failure to appreciate this leads to a dangerous error as the measured range suggests that the vessel is further to seaward than is in fact the case. Further bear in mind that, as the range decreases, the lower, nearer land will rise above the radar horizon and may give the misleading impression of an apparent movement of the land towards the observer (i.e. a strong onshore tidal set).
4 As the vessel's position changes, the aspect of land targets and the direction of land shadow areas may change, further complicating the task of relating the pattern of echoes to the chart.
5 To maximize the possibility of long range target identification, exploit any interswitching arrangement to obtain the best discrimination and receiver sensitivity that can be combined with transmitter power and aerial height.

In general conclusion, it is extremely important not to rely on fixes derived from detected land echoes until they have been positively identified. As mentioned before, it is in the nature of radar detection that such identification will almost invariably be difficult at long range. Strandings (and how many near strandings?) have occurred where vessels have carefully and methodically laid off, on the chart, successive ranges and bearings of a wrongly identified target. When an echo appears in the general area in which a particular point of land is expected by DR, there appears to be a great temptation to ascribe immediately and unquestionably the hoped-for identity to that target. This temptation must be resisted until cross-checks have established that the fix is consistent with other sources of information. Such sources should include, where possible:

1 The estimated position based on the DR position, with due allowance for any leeway and tidal effects
2 The depth as indicated by echo sounder
3 Positions or position line(s) obtained from other navigation systems.

It is not intended to suggest that such other information is either superior or inferior to that obtained from radar in any given circumstance. The important point is that disparity between any two should alert the mariner to carefully consider the disparity and to avoid discarding either piece of information without good reason.

5.7.2 Coastal navigation

Once landfall has been successfully achieved, the problem becomes one not so much of recognizing the general locality of the coastline, but of positively identifying an adequate number of coastal features so as to allow the increased frequency and accuracy of position fixing and progress monitoring which must accompany proximity to the land (see also Sections 13.2 and 20.1.4).

Positive identification is crucial. Radar ranges afford a very high degree of potential position line accuracy (see Section 5.10.1), but this can be completely wasted if the target is wrongly identified. The reader is reminded of the following factors:

1 Horizontal beam width distortion may mask small bays and inlets, cause islands close to land to appear as peninsulas, and angularly extend points of land.
2 Any stretch of coastline may change its appearance with the progress of the observing vessel along its track, owing to changes in land aspect and land shadow areas. (For this reason radar charts, produced as a collage of radar photographs, have proved valueless.)
3 The edge of the displayed coastline may not be the high water mark but could be higher land behind.

Any interswitching arrangements should be operated to achieve the best discrimination consistent with adequate detection range. In a single radar system a similar approach should be followed. The two important radar techniques which can be used to monitor a vessel's track in coastal navigation are position fixing and parallel indexing.

In position fixing, the vessel's position is determined at intervals and thus its progress along the required track is monitored. Where a departure from the desired track is detected, course is adjusted to return the vessel to the desired track.

Parallel indexing is a simple and effective way of continuously monitoring a vessel's progress by observing the movement of the echo of a radar conspicuous navigation mark with respect to track lines previously prepared on the reflection plotter.

In general, the two techniques should be seen as complements rather than alternatives. The relative priority ascribed to each technique in any given circumstance is a matter for individual judgement. If the vessel's track is fairly remote from hazards and landmarks it may be considered that parallel indexing is not practical and that regular radar fixing (cross-checked against other systems) will ensure adequate warning of any departure from the desired track and provide sufficient monitoring of the vessel's progress along the track.

When the track passes closer to hazards and becomes more complex, involving course alterations and possibly critical waypoints, parallel indexing attracts high priority because of its ability, irrespective of visibility, to give simple, obvious and immediate indication of departure from the desired track. It also gives, by inspection, progress along track, thus facilitating extraction of time to run to wheel-over or other critical points. Margins of safety give visual indication of the departure from track that can be tolerated should collision avoidance considerations necessitate a manoeuvre. It must be appreciated that the ease of monitoring afforded by parallel indexing is gained by preparing the work beforehand. The success of indexing is dependent on the skill, care and accuracy with which this preparation has been carried out. Although indexing offers immediate data in these circumstances, fixing should not be disregarded. The vessel's position should be plotted regularly on the chart to cross-check the indexing data and maintain a record of the vessel's progress. Radar positions should be cross-checked with position lines from other sources.

5.7.2.1 Position fixing

The following points are considered worthy of recapitulation (see also Sections 14.4 and 20.1):

1 Radar ranges have much higher potential accuracy than radar bearings (see Section 5.10.1.1), and the best radar fixes are obtained using at least three ranges of positively identified targets. Where visibility permits, a visual bearing and a radar range combine the best of both sources.
2 Radar bearings should not be relied upon to give an accuracy better than about ±2.5° (see Section 5.10.1.2) and therefore do not yield very accurate

fixes. Where bearings are used it is best to select small isolated targets. If it is found necessary to use headlands, bearings which lie along the axis of the headland are preferable to those tangent to the edge of the land.

3 Any radar position line should be seen as a source of navigational information whose consistency with information from other sources should be continually kept under review.

5.7.2.2 Parallel indexing

As with position fixing, the following points are considered worthy of recapitulation (see also Sections 14.6.3 and 20.1.4):

1 In preparation for indexing, the passage must be carefully planned to ensure that chosen tracks are safe and practical in every respect.
2 Care must be taken in the selection of indexing targets to ensure that those selected will be detectable and identifiable on an appropriate range scale at all relevant stages of the passage.
3 Attention must be given to accuracy and to the danger of clerical errors particularly in the matter of scale when transferring the information from the chart to the reflection plotter.
4 When indexing, the usual performance and accuracy checks must not be neglected (see Notice M.1158 in Section 5.11.5 at 3.20).
5 Parallel indexing can be used with an azimuth stabilized relative motion presentation or with a true motion presentation. The latter is very much a second best. Indexing exploits the natural characteristic of a relative motion presentation, while a conventional true motion display can only be used if ground stabilized by active and continuous input of tidal data. This is difficult to achieve and maintain and renders the presentation unsuitable for collision avoidance (see Section 5.4.4.2). It is not practical to draw up the index lines beforehand and monitoring cannot be commenced until ground stabilization has been achieved. If ground stabilization is lost, monitoring is disrupted and difficult to restart.

5.7.3 Pilotage

In pilotage, short range scales with good minimum range and discrimination are essential if radar information is to make a practical contribution to navigation control (see also Sections 13.4, 15.3 and 21.5). Some high definition radars are specifically designed with these characteristics to make them effective as river radars. In the absence of such a radar, the system should be optimized for the pilotage phases of a voyage. In an interswitched system, this can be achieved by selecting the more suitable siting, the narrower beam width, the shortest pulse length and suitable gain and clutter settings. It should be borne in mind that the rain clutter control can be used to good effect in improving range discrimination. Where a single radar is fitted, aerial siting and bearing discriminations are fixed but range discrimination may be improved by appropriate use of pulse length and rain clutter setting.

In general, in pilotage, detection is not a problem but the display of fine detail is essential.

Progress monitoring is ideally serviced in most pilotage situations by the use of parallel indexing techniques. In such situations, the need for position monitoring is continuous and parallel indexing is well able to achieve this provided that suitable preparation has been carried out prior to arrival in the pilotage area. It must be said, however, that the docking phase of pilotage is one for which current civil marine radar equipment is not particularly suitable.

5.8 Computer assistance

This chapter has dealt with radar plotting and radar navigation techniques in terms of conventional radar systems, i.e. those in which the techniques have to be performed manually. This treatment is appropriate to the majority of British ships if the entire tonnage range is considered. However, since the early 1970s equipment capable of extracting the data automatically has become progressively more available for civil marine applications. Since 1980 the incidence of such equipment has increased greatly because of United States legislation, IMO resolutions and subsequent national legislation brought forward by other countries. Of particular importance was the IMO fitting schedule which extended from 1984 to 1988 (see Section 10.3.1, paragraph (j)).

Such equipments employ a computer to store, process and display data and are now referred to as automatic radar plotting aids (ARPA) because the majority of their functions are devoted to the automatic tracking of targets. This section (and Section 5.9) will give an exposition of the use of ARPA as an aid to carrying out radar plotting and navigation techniques. It is important to appreciate that the ARPA cannot make collision avoidance decisions. It merely carries out tasks and calculations which otherwise would have to be performed manually by the navigator. Further, there is no reason to suspect that the accuracy of the solutions to such calculations are any better, or any worse, than would be obtained by an observer who devoted suitable time and care to the operation. The dramatic characteristic of an ARPA is its ability to perform the tasks automatically and indefinitely, without fatigue or distractions, while handling a workload which, in terms of both volume and speed of completion, is far in excess of that with which a manual operator could cope. The role of the ARPA should be seen as freeing the navigator from routine and tedious tasks, thus increasing the number of targets he can cope with and allowing him to devote more time to the decision-making process crucial to navigation control.

5.8.1 Setting up for basic tracking

All ARPAs offer a wide range of facilities and in some cases the vast number of controls may be daunting. However, it is important to appreciate that while some of the facilities are essential and used almost continually, many are of limited use and subjective appeal. It is intended here to consider the setting up of the display to obtain the basic collision avoidance data, and subsequently (in Section 5.9) to explore the wider range of additional facilities.

Setting-up procedures will vary in detail with manufacturer, but the following guiding principles are important and common to all systems:

1. Ensure that the radar system supplying the raw data to the ARPA tracker is correctly set up (see Section 5.2). Particular attention must be paid to pulse length selection, tuning, and *any* controls which affect receiver gain (e.g. sea clutter). Failure to tune the receiver correctly will invariably reduce the probability of targets being detected and tracked by the ARPA. The effect of the setting of receiver gain controls on tracking varies from one manufacturer to another. In some systems the signal level fed to the tracker is independent of the operator's gain controls, being continuously set by automatic circuitry. In other systems, the manual gain control determines the signal level fed to both the tracker and the CRT. When meeting an unfamiliar ARPA it is prudent before sailing to establish which, if any, of the gain control settings affect the input to the tracker. This can be done quite simply by acquiring a target, adjusting each relevant control in turn, and observing any loss of tracking (see Section 5.8.6.3).
2. Where a separate control is provided, switch on the computer and await the completion of any self-diagnostic program (this will normally take less than one minute). When the computer has become active, in general ARPAs will default to a 'square one' condition in which all but the basic facilities are switched off. This ensures that the user obtains only the functions he positively requests and is not confused by the need to switch off a large number of unwanted facilities. Care must be exercised where analogue controls occur as the computer may be unable to reset these.
3. Check that the correct heading and water speed information is being fed to the computer, and adjust if necessary. It is vital that this check is made, otherwise no reliance can be placed on the tracking. The importance of heading and speed inputs cannot be overstressed, and the navigator must always be watchful for errors in these fundamental data sources.
4. The IMO Performance Standards for ARPA (see

Section 5.11.3) require that means be provided to adjust independently the brilliance of the ARPA data and the radar data, including complete elimination of the ARPA data. When setting up, check that the ARPA data brilliance has been set to a mid-range value. When tracking has commenced it can be adjusted to suit the ambient light conditions.

5 Ensure that the vector time control has been set to a non-zero value, preferably in the range 6–12 minutes. This ensures that when tracking commences the vector length displayed will, in general, be obvious but not excessive. In due course the length can be readjusted to suit the traffic conditions.

6 Most systems will, on being switched on, default to either relative or true vectors depending on the design philosophy.

7 Set in suitable safe limits for CPA and TCPA (see also Section 5.8.6.1).

8 If unfamiliar with the system, locate the joystick (or tracker ball) plus the acquire and cancel control. No further controls are necessary to commence basic tracking.

5.8.2 *Basic tracking*

Acquisition is the term used to describe the selection of a target for tracking and the initiation of such tracking. It may be carried out manually or automatically. Manual acquisition is simple and quick, and is required by the IMO Standards (see Section 5.11.3). Automatic acquisition is optional, complex and characterized by the employment of widely differing philosophies. Automatic acquisition is by no means universally popular with users (see Section 5.9.1) and in this section consideration will be restricted to manual acquisition.

Manual acquisition is effected by using a joystick or tracker ball to position a graphics marker over the chosen target, and then activating the acquire control. If a target is detected within a small area (known as the gate) associated with the marker, it will be 'acquired' and then tracked automatically. From sequential ranges and bearings taken at intervals of about 3–6 seconds, and taking into account the heading and speed of the observing vessel, the computer will calculate the relative motion and the true motion of the target. It will display either the relative or the true motion (according to control selection) as a vector consisting of a bright graphics line emanating from the 'now' position of the target and indicating, by its direction and length, the target's predicted motion in the future.

The IMO Standards (see Section 5.11.3) set out the minimum accuracy that must be achieved. Unfortunately, the appropriate section of the Standards does not lend itself to simple summary because the potential accuracy depends on the geometry of the OAW triangle generated by the approach of the target and the time for which steady state tracking has been carried out. For the present it is sufficient to say that, within three minutes of steady state tracking, the ARPA must reach a level of accuracy at least as good as the best that could be achieved by manual plotting. The question of accuracy is discussed further in Section 5.10.4.

It is important to appreciate that the operation so far described is, in principle, exactly the same as that carried out by an observer manually constructing a plot.

5.8.3 *Extraction of basic collision avoidance data*

To be consistent with the general philosophy of radar plotting as set out in Section 5.6, the extraction of basic collision avoidance data by the use of ARPA will be considered in terms of that data which is necessary to:

1 Determine if a close-quarters situation is developing;
2 Assess the nature of the encounter(s);
3 Evaluate the effect of possible avoiding manoeuvres.

5.8.3.1 Development of a close-quarters situation

The CPA and TCPA of the tracked targets can be quickly and easily read off by inspection, if relative vectors are selected and the vector time control adjusted so that the end of the chosen target's vector represents its predicted position at CPA. The range at

CPA can be read off using the VRM (or against the background of the fixed rings) while the TCPA is obtained by direct reference to the vector time.

As an alternative, indication of the likelihood of a close-quarters situation can be obtained by selecting true vectors and running out the vector lengths progressively to show the predicted development of the encounter(s). The dynamic nature of this technique appeals to many users, but it must be borne in mind that any evaluation of CPA/TCPA is a matter of trial and error rather than the single direct measurement afforded by the use of relative vectors. It is important to remember that the CPA is *not* represented by the point at which the true vectors intersect, except in the special case of zero CPA.

5.8.3.2 Assessment of the nature of encounters

An indication of the heading/aspect of all tracked targets, of sufficient accuracy for tactical judgement, can be obtained by direct inspection of the true vectors provided they are extended to a suitable length. A reasonable indication of target speed can be obtained by visual assessment of the ratio of target true vector length to own ship vector length.

5.8.3.3 Evaluation of avoiding manoeuvres

This can be effected by using a trial manoeuvre facility which is required for compliance with the IMO Performance Standards for ARPA (see Section 5.11.3). Such a facility allows the user to feed in an intended course and speed and hence display the vectors as they would appear if the intention was carried out.

By selecting trial manoeuvre and relative vectors the navigator can, by inspection, predict the CPA(s) which will result if the chosen manoeuvre is performed and the targets maintain their previously calculated courses and speeds. Such information is one of the factors to be considered when evaluating the suitability of any proposed manoeuvre. It is not, however, the only factor, and due account must be taken of the need to make a manoeuvre which is obvious to other vessels (see Section 22.2.6). The ease with which the trial facility allows the navigator to establish the course to steer for a chosen passing distance may encourage the choice of a small alteration. This temptation must be avoided and it should be remembered that the other vessels in the encounter may well be using more rudimentary methods of data extraction. As a result they will be less able to detect target manoeuvres and may have a different perception of what constitutes a safe passing distance.

If true vectors are selected when in the trial mode, only the observing vessel's vector will be affected by the input of proposed manoeuvres (the targets' true vectors will be unaffected). This combination may be found useful in judging the suitability of a manoeuvre in terms of its general nature and sense, but assessment of the resultant CPA can only be obtained by trial and error (rather than by the single direct measurement afforded by the use of relative vectors). It is extremely important to remember that the CPA is *not* indicated by the point at which the true vectors intersect except in the special case of a zero CPA.

In its simplest form, the trial assumes that the manoeuvre will be performed immediately and instantaneously. More complex systems allow for a non-immediate manoeuvre by providing for the display of the effect of a proposed delay between selection of the trial values and performance of the manoeuvre. This is an extremely useful and practical extension of the basic idea.

A further extension of the facility which is sometimes offered is the representation of a non-instantaneous manoeuvre by making allowance for the observing vessel's manoeuvring characteristics (turning circle and speed change characteristic).

Bearing in mind the accuracy of the CPA that the ARPA can be relied upon to afford (see Section 5.10.4), one is driven to the conclusion that the radius of the turning circle is likely to be small by comparison, except in the case of very large vessels. There is much more justification for attempting to simulate the speed change characteristic, particularly for large vessels, as this can have a significant effect on the CPA finally achieved.

However, it should be noted that some systems which simulate manoeuvring characteristics merely use a simple linear reduction rather than the exponential-type decay characteristic of most vessels. Notwithstanding the provision of simulated manoeuvring characteristics, the navigator must remember that the accuracy of CPA afforded by the ARPA does not justify reliance on the predicted value being precisely achieved. Manoeuvres and passing distances must be sufficiently large to account for the specified accuracy limits.

In general conclusion, it must be remembered that the trial manoeuvre facility gives a purely mathematical extrapolation into the future based on data gathered in the recent past, and can take no account whatsoever of possible future manoeuvres by tracked targets.

5.8.4 *Alphanumeric data readout*

The IMO Standards require that, on the request of the observer, certain specific data for any chosen tracked target should be immediately available in alphanumeric form. The data can be summarized as follows (see Section 5.11.3 for the full text):

bearing	CPA	course
range	TCPA	speed

The request is made by placing the gate marker over the chosen target and activating some form of read control. The data is usually displayed on a special screen but in some cases is written on the PPI.

The ready availability of the data provides a convenient method of measuring and monitoring a target's range and bearing. It is also useful as a supplement to the information available by inspection from the vector display, especially in the case of a target which is giving rise to particular concern.

It is prudent to check that there is sensible agreement between the vector and alphanumeric data. However, as they both originate from the same database, agreement between the two is no absolute guarantee of accuracy. Additionally some systems offer readout of bow crossing range (BCR) and bow crossing time (BCT). It is extremely important to appreciate that the significance of BCR is heavily dependent on the relative direction of approach. It is not equivalent to CPA as an assessment of the development of a close-quarters situation, and can be dangerously misleading if treated as such.

5.8.5 *Track history*

This facility, like so many ARPA functions, appears under a wide variety of names (plots, trails, history, tracks, etc.) according to the choice of the manufacturer. The IMO Standards (Section 5.11.3) require the equipment to be able to display on request at least four equally time-spaced past positions of any targets being tracked over a period of at least 8 minutes. The principal benefit of track history is that it enables an observer to establish if and how a particular target has manoeuvred in the recent past, possibly while the observer was temporarily away from the ARPA display on other bridge duties. Not only is this knowledge useful in showing the observer what has happened, but it may well help him in forming an opinion of what the target is likely to do in the future. Manoeuvres by targets may be for navigational purposes, and it may be possible to gain an indication from the general overview that this is the case. Alternatively there may be an indication from the nature of the manoeuvre, when considered in relation to the general traffic pattern, that it has been undertaken for collision avoidance, and hence the observer can be alert for a resumption of previous conditions.

While relative history will only make it possible to detect that a manoeuvre has taken place, true history makes it possible to positively identify the nature and sense of the manoeuvre. Relative history has thus little to offer, while true history is essential to gain the benefits previously described. Many systems offer true history only for this very good reason. Some systems offer both, and the argument is frequently put forward that the vectors should be accompanied by history of the same type, otherwise the mariner will become confused. This argument is difficult to sustain as confusion can best be eliminated by recognizing that only true history is useful and by standardizing on its display.

5.8.6 *Operational warnings*

These warnings are designed to draw the observer'

attention to specific developments associated with the detection, tracking or approach of tracked targets. The IMO Standards (see Section 5.11.3) require the provision of three types of warnings, which can be briefly described as:

1 Guard alarm
2 Dangerous target alarm
3 Lost target alarm.

It is possible to activate or deactivate the operational warnings.

5.8.6.1 Guard alarm

The ARPA is required to have the capability to warn the observer with a visual and/or audible signal of any distinguishable target which closes to a range or transits a zone chosen by the observer. The target causing the alarm must be clearly indicated on the display.

The implementation of this 'early warning' requirement has been effected in many different ways by individual manufacturers. In its simplest form, the observer has the option of setting the radii of two concentric guard rings. If a detectable target crosses either ring with a decreasing range, the warning will be activated. A facility is normally provided to desensitize chosen arcs of the guard rings should they be likely to produce excessive alerts due to, for example, the presence of land echoes. In more complex forms the guard alarm has been integrated with an auto-acquisition facility, and this will be discussed in Section 5.9.1.

The guard alarm acts essentially as an aid to the early detection of approaching targets. Its role should be regarded as being supplementary to and *not* as a replacement for the need to regularly observe the display in carrying out long range scanning as required by rule 7(b) of the Collision Regulations (see Section 2.3.1), or as an alternative to keeping a proper visual look-out. It is important to remember that if a target does not produce a detectable echo at the range represented by the guard ring, the warning will not be activated. By judicious choice of the ranges at which the guard rings are set, it is possible to provide for the long range detection of stronger echoes with the outer ring while employing the inner ring to detect those targets missed at long range. The guard ring facility probably has its greatest value in the open sea, where there are few targets and lapses of concentration are more likely. In conditions of high traffic density, in the proximity of land and in the presence of extensive clutter, the high incidence of alerts may well prove distracting and counter-productive.

5.8.6.2 Dangerous target alarm

The ARPA is required to have the capability to warn the observer with a visual and/or audible signal of any tracked target which is predicted to close to within a minimum range *and* time chosen by the observer. The target causing the warning must be clearly indicated on the display.

All systems are substantially similar in the way this warning is provided. It is possible to set numerical values of CPA and TCPA which will activate the alarm if *both* are violated by a target being tracked. The choice of the values set as safe limits is a matter of personal judgement in any given circumstance. Excessive alerts can be distracting and counter-productive.

5.8.6.3 Lost target alarm

The ARPA is required to clearly indicate if a tracked target is lost (other than out of range), and the target's last tracked position must be clearly indicated on the display.

This facility is useful as it immediately draws the observer's attention to the tracker's failure to continue to detect an echo which has been under tracking. This might well occur while the navigator was temporarily absent from the display on other bridge duties.

5.8.7 Equipment warnings

To comply with the IMO Standards (see Section 5.11.3) the ARPA is required to provide suitable warnings of malfunction to enable the observer to monitor the proper operation of the system. The precise arrangements made for compliance with this requirement vary considerably with manufacturer.

For a full treatment of this topic the reader is referred to *Radar and ARPA Manual*. However, the following points are worthy of general consideration:

1 The maker's manual should be studied to establish the general philosophy of the warning system and hence to make the maximum use of any automatic or semi-automatic diagnostic facilities that may exist.
2 While self-testing and diagnostic routines may detect many malfunctions, there are some faults that cannot be so detected (e.g. the failure of elements in an electronic readout, which can for example cause an 8 to display as a 3).
3 The ARPA tracker has no way of knowing what numerical values of course and speed to expect as inputs. Thus gyro and log warnings can only be relied upon to sense the absence of inputs. The navigator must accept complete responsibility for continually checking the accuracy of such inputs.
4 Test programs made available by the manufacturer should be used periodically to assess the overall performance of the ARPA against a known solution. Such programs are useful for training or can be extended for that purpose. The use of these programs for practice prior to sailing (especially where officers are not familiar with the equipment) is highly recommended.

5.9 Additional ARPA facilities

In this section it is intended to consider the use of additional ARPA facilities which are not specifically required by the IMO Performance Standards.

5.9.1 Automatic acquisition

This is not required by the IMO Standards but is permitted subject to the provision of a minimum of twenty tracking channels (as compared with the minimum of ten where manual acquisition only is provided). The requirement for double the number of channels recognizes the fact that in general the computer is likely to be much less selective than an experienced navigator. Probably because of the optional nature of the facility, a wide range of automatic acquisition systems has been produced. At the simple end of the spectrum, targets are selected for acquisition as they are detected by one of two guard rings as described in Section 5.8.6.1; at the other end of the spectrum, the equipment continually searches for targets in an extensive area whose boundaries are set by the user. Some of the latter equipment will maintain surveillance of up to 50 of the targets detected within the search area, then use a sophisticated program to distribute priorities to the targets, and finally decide on a 'top twenty' which are allocated to the display tracking channels. It is not intended here to explore the mechanics and philosophy of various automatic acquisition techniques. In this text it is intended merely to consider, in general, the contribution that automatic acquisition can make to the use of ARPA in navigation control.

It is commonly remarked that automatic acquisition reduces the observer's workload. This is true to some extent, but cannot be considered in isolation. Because the auto system may put vectors on targets which the navigator would not have personally selected, more vectors may be displayed than are really necessary, and in heavy traffic there may be doubts about priority. As a result, the auto system may contribute to the workload by requiring the navigator to acknowledge alarms, cancel targets and reallocate vectors.

Undoubtedly a finite portion of the navigator's time will have to be devoted to acquiring targets if manual acquisition is employed. However, manual acquisition is so quick and simple in all systems that it is questionable whether this time is really significant. It could be further argued that in fact it is time well spent, because the act of selecting and acquiring a target manually makes a greater impact and contributes more to the mariner's early appraisal of the situation than when the discovery is made by the computer.

Two major problems confront the designer of any automatic acquisition system:

1 How to distinguish ships from land, clutter and spurious echoes. In some systems manual intervention in the form of the insertion of exclusion zone

and the operation of signal suppression is necessary, whereas in others automatic methods are used.

2 How to allocate priorities if more than twenty targets are present. In some systems the priority is simply the twenty nearest targets, while in more complex systems sophisticated programs carry out prioritization on the basis of many criteria. In all cases the criteria *must* be made available to the user.

If the navigator elects to use automatic acquisition, it is essential that he makes himself aware of how the design philosophy deals with the foregoing problems. The arrangements can be quite complex and vary greatly with manufacturer. For this reason, when sailing with an unfamiliar ARPA, it is probably best to use manual acquisition at least until the features of the automatic acquisition have been thoroughly studied with the assistance of the user's handbook.

When and if to use automatic acquisition is, in the final analysis, a matter of subjective judgement. Clearly automatic acquisition can make a contribution to collision avoidance by virtue of its ability to acquire targets when the navigator has left the display to attend to other bridge duties. This characteristic is likely to be most useful in the open sea, where the number of targets is small and the danger of loss of concentration looms. (If used, it must be seen as a supplement, and not an alternative, to the keeping of a proper look-out.) In conditions of high traffic density, in the proximity of land and where there is a high incidence of clutter, automatic acquisition is likely to prove to be of limited, if any, value and in fact its use may be distracting and counter-productive.

5.9.2 Automatic ground stabilization

The difficulties of achieving ground stabilization on a raw radar display were outlined in Sections 5.2.11.2 and 5.4.4.2. With computer assistance, ground stabilization can be achieved easily and automatically. An isolated land target is selected as a reference and is tracked by one of the ARPA tracking channels. This makes it possible for the computer to calculate the track of the observing vessel over the ground and hence to maintain the movement of the electronic origin of the display in sympathy with it.

When using this facility the observer should be particularly watchful for other targets which approach the reference target and in particular for those which pass between the observing vessel and the reference target. The result of target swop in these circumstances will have dramatic effects on the entire presentation. If such an eventuality appears likely, it may be expedient to move the reference to another target.

As a general rule, ARPA software (i.e. computer programs) has been designed such that the same stabilization is applied to presentation *and* true vectors, i.e. both are either sea stabilized or both are ground stabilized. (There is currently one exception to this rule.) Thus in general, where automatic ground stabilization is selected, true vectors will indicate the ground track of targets and *not* their headings (Figure 5.28). Failure to appreciate this can render the presentation dangerously misleading if it is mistakenly used in the planning of collision avoidance strategy. One might expect the danger of observers being misled in this respect to be less than in the case of a raw radar display (see Section 5.4.3) because of the obvious angular separation of own ship vector and own ship heading marker. Experience has shown that this is by no means invariably the case.

The useful feature of automatic ground stabilization is that it makes true motion parallel indexing a practical proposition (which is scarcely, if at all, the case with a raw radar display) and makes it possible to have electronic maps (see Section 5.9.3.3) which remain stationary on the screen. Some users find the latter feature particularly attractive from the point of view of navigation. However, it must be stressed that the penalty paid is that the presentation may not afford target *heading* information and may therefore, in principle, be unsuitable for collision avoidance.

5.9.3 Navigation lines

This is the general term used to describe the facility available with most ARPAs whereby it is possible to draw and maintain, on the PPI, electronic lines of varying orientation, length and position.

A number of levels of navigation line packages are available on individual ARPAs, but it is not intended

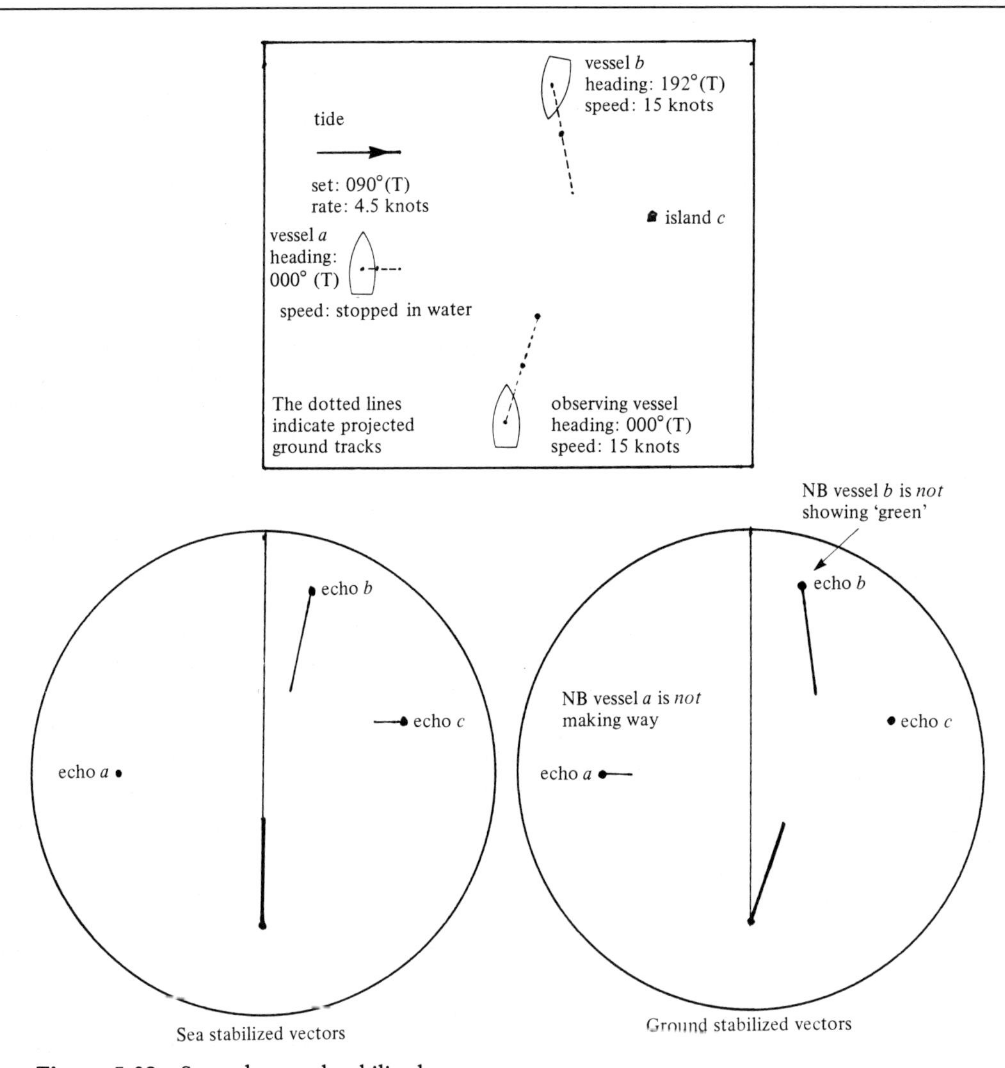

Figure 5.28 Sea and ground stabilized vectors

here to consider the individual details of these. In this text it is proposed merely to consider, in general, the contribution that such lines can make to the use of ARPA in navigation control.

5.9.3.1 Parallel indexing – relative motion presentation

The electronic navigation lines can be used to build up the familiar pattern of index lines and margins of safety normally required for the implementation of this technique (see also Sections 5.7.2.2, 14.6.3 and 20.1.4). The employment of navigation lines is a significant improvement on the use of chinagraph lines drawn on a reflection plotter. The scale of the lines is automatically changed to follow changes in the selected range scale. This obviates the need to draw a family of lines in anticipation of the use of several range scales in any given phase of a passage. It reduces the workload and the general clutter of lines with their potential for confusion (the blunder of mistaken range scale is probably one of the oldest and potentially most dangerous in the use of radar).

Provided a storage option has been purchased, the lines can be prepared well in advance and stored in memory until they are required. Facilities exist to store quite large numbers of sets of lines, making it possible to have quickly and easily available the index lines for many phases of a regular voyage.

The navigation line facility can be used simultaneously with relative vectors and sea stabilized true vectors, thus effectively combining the navigation and anti-collision functions of the radar. With the traditional reflection plotter methods, the sheer volume of index lines and plotting triangles can result in extreme congestion on a single display.

5.9.3.2 Parallel indexing – true motion, ground stabilized presentation

The index lines for this technique can be produced using a navigation lines facility (see also Sections 5.7.2.2, 14.6.3 and 20.1.4). For successful application of the technique, the automatic ground stabilization facility must be employed. No apology is made for restating the fact that under these circumstances the true vectors may *not* indicate the headings of targets, thus rendering the presentation unsuitable for collision avoidance. The above arrangements may thus fail to effectively combine the navigation and collision avoidance functions of the radar.

5.9.3.3 Maps

Maps composed of lines and points can be drawn and stored using electronic navigation lines, the more complex arrangements being commonly referred to as mapping facilities rather than navigation lines (see also Section 20.1.4). Such maps can range from a reference point and a few lines delineating the outline of a sandbank or traffic separation scheme to very detailed patterns comprising perhaps as many as 1000 elements and depicting all major features of navigational interest in one or more passages.

Theoretically the maps can be displayed on a relative motion presentation, or on a true motion presentation (either sea or ground stabilized). In some of the original systems the choice was restricted at the design stage and in some cases only one option was available to the navigator.

In older systems it may be found that maps can only be used safely with the ground stabilized true motion presentation, probably because it is reasonably straightforward to use automatic ground stabilization to maintain a reference target and hence a map in a fixed position on the screen. The penalty paid in almost all such systems is that the true vectors will also be ground stabilized and unsuitable for collision avoidance strategy. (One ARPA system developed in the early 1980s did offer the facility to provide a ground stabilized picture with sea stabilized vectors.) In the case of the other two presentations the problem is more complex because, to be of any practical use, the map must maintain its registration on a ground-stationary target which is moving across the screen. However, some manufacturers have made a positive feature of the ability to lock maps to a ground-stationary target which is moving across the screen. Such systems have the considerable merit of making the use of maps practical while maintaining the ability to produce sea stabilized true vectors.

In conclusion the ground stabilized, true motion presentation offers the subjective attraction of a stationary map but may well not satisfy the requirements of collision avoidance. The relative motion and true motion sea stabilized presentations which offer the facility to lock maps to a ground stationary target while maintaining sea stabilized vectors effectively serve the twin requirements of navigation and collision avoidance. Before making use of data from any map display it is *essential* to check that the map *has* maintained its registration with the fixed ground stationary features.

5.9.4 Potential collision points

When two ships are in the same area of sea it is always possible for them to collide. From the basic plotting triangle it is possible to determine the course to steer (if speed is maintained) to achieve interception. The point at which this course intersects the target's course line produced is defined as the potential point of collision (PPC) or alternatively as the potential collision point (PCP). It represents the point of simultaneous arrival of both vessels if the target stands on and the observing vessel steers the intercept course.

Such calculations have not formed part of standard Merchant Navy collision avoidance practice, not least because of the complexity and volume of work involved, except in the case of vessels performing specialized tasks (fleet auxiliaries etc.), which are in any event concerned with use of the data for interception and rendezvous rather than conventional collision avoidance.

With the advent of ARPA, manufacturers realized that such calculations could be carried out quickly, by computer, for all tracked targets. Some felt that the display of the PPCs might assist the mariner in making the correct collision avoidance decisions. It is intended here to consider the possible use of PPCs as an aid to collision avoidance on the assumption that the reader is familiar with the theory of the location, movement and interpretation of PPCs.

The argument for displaying the PPCs is that they assist collision avoidance strategy by showing the navigator, at a glance, those courses which are completely unacceptable. This, in fact, is the only function of PPCs. They do not, as is sometimes implied, indicate the safe courses. The PPCs do not give any indication of miss distance (other than in the zero case) and any attempt by the mariner to extrapolate either side of the point will be fraught with danger. A safe course is one which, amongst other things, results in a safe passing distance (rule 8(d) of the Collision Regulations; see Section 22.3.1) and assessment of the latter implies a knowledge of miss distance.

If PPCs are used, the following additional factors should receive due consideration:

1 Any target may have zero, one or two PPCs.
2 If no track line associates each tracked target with its PPC(s), care must be exercised in visually assessing this association.
3 On a relative motion presentation, the PPCs will not be stationary and can in certain circumstances show quite dramatic movement. (Movement will also occur on a ground stabilized, true motion presentation but in general it will be very much less obvious.)
4 On a relative motion presentation, where two targets are standing into a mutual close-quarters situation, their PPCs may well be widely separated on the observing vessel's display.
5 The location of the points is only valid for the observing ship's present speed.

5.9.5 Predicted areas of danger

The shortcoming of the PPC is its failure to afford indication of miss distance. The logical development is to construct, around the PPC, a plane figure associated with a chosen passing distance. The area within the figure is to be avoided to achieve at least the chosen passing distance, and is referred to as a predicted area of danger (PAD). The technique was patented by Sperry Marine Systems and is exclusive to the ARPA produced by that company.

In the early systems the area of danger was delineated by an ellipse, but in subsequent equipment hexagons have been used. The PAD approach is an extremely elegant solution to the problem of how best to present collision avoidance data. Its great virtue is that trial manoeuvre data (for a given speed) are implicit in the presentation. All theoretically possible courses (for a given speed and CPA) are displayed at a glance, which greatly assists in planning strategy. However, it would be a mistake to oversimplify the interpretation of the displayed data. The user must have a thorough understanding of the principles underlying the presentation, with particular reference to the location, movement, shape and change of shape of the PADs.

Such an understanding is of particular importance because in contrast with a vector presentation, which is an automated version of a manual technique performed by mariners for many years, the concept of PADs is completely novel. Based on such an understanding, the following points are worthy of consideration if using PADs in planning collision avoidance strategy:

1 Any target may have zero, one or two PADs.
2 The PPC is not at the centre of the PAD (except by coincidence).
3 The area of danger does not necessarily change symmetrically with change in miss distance.
4 The area of danger does not necessarily change linearly with change in miss distance (e.g. if the

heading marker strays a 'small' amount inside the PAD, it does not follow that the miss distance will be reduced by a 'small' amount).

5 The PAD may change its position, size and shape with the passage of time even when the target does not manoeuvre. Some of these changes may be quite dramatic at particular stages in an encounter (e.g. the PAD of a faster ship will collapse dramatically as it passes the observing ship). The successful user of the system must be able to link such a dramatic change with its origin in seamanship.
6 In showing all the mathematical possibilities at a glance, the presentation may reveal solutions which the mariner might overlook with a less sophisticated presentation of data. It is important to remember that the solutions are purely mathematical; the mariner has complete responsibility for deciding which solutions are seamanlike, and must guard against any temptation to make alterations whose nature, sense or size contravene the Collision Regulations.
7 The PADs are only valid for the present input speed.

5.9.6 Sectors of danger and sectors of preference

These comprise a further graphical method for indicating a possible range of headings consistent with a chosen miss distance. Sectors of danger (SODs) indicate the range of headings which will bring a target within the miss distance. Sectors of preference (SOPs) indicate those headings which will achieve at least the chosen passing distance. The use of the word 'sector' is not in complete agreement with the geometrical definition.

5.10 Error sources and accuracy

In this section it is intended to summarize the inherent accuracy considerations of which the navigator should be aware when making use of radar in navigation control. The inherent accuracy requirements for civil marine radars are set out in the IMO Navigational Radar Performance Standards and the IMO Standards for ARPA. References will be made to these documents, which are reproduced in Sections 5.11.1 and 5.11.3.

5.10.1 Accuracy of basic data

The basic data obtainable from a marine radar system is that of range and bearing measurement. Range and bearing accuracy is fundamental to the successful implementation of all radar techniques used in navigation control.

5.10.1.1 Range accuracy

The IMO Navigational Radar Performance Standards require that the fixed range rings and the variable range marker enable the range of an object to be measured with an error not exceeding 1.5 per cent of the maximum range of the scale in use or 70 metres, whichever is the greater. This represents a high level of inherent system accuracy, but it is important that this is realized by a good measurement technique as set out in Section 5.5.1.

5.10.1.2 Bearing accuracy

The Standards require that any means provided for obtaining bearings shall enable the bearing of a target whose echo appears at the edge of the display to be measured with an accuracy of ±1° or better. This requirement cannot be considered in isolation. All bearings are effectively measured from the heading marker, which is required by the Standards to have a maximum error not greater than ±1°. Where azimuth stabilization is provided, the accuracy of alignment with the compass transmission is required to be within 0.5°. If these error sources are aggregated, a bearing extracted from the display could be in error by 2.5° without the provisions of the Standards being contravened. As with range measurement, it must be appreciated that the foregoing represents inherent system accuracy, and good measurement techniques (as set out in Section 5.5.2) are essential if at least this accuracy is to be realized.

5.10.2 *True motion tracking accuracy*

The IMO Standards do not address this matter, but the maximum error in origin tracking permitted by the UK specification is 3° for direction and 5 per cent (or 0.25 knots, whichever is the greater) for speed. It is important to appreciate that these requirements relate to the agreement between the input signals and the resultant tracking. The navigator alone has total responsibility for continually ensuring that the inputs represent the *correct* values of course and speed (see Section 5.2.11.1).

Errors in origin tracking (inherent or otherwise) cannot be directly translated into errors in target tracks. The direction and magnitude of the error in a target's track depends on the relationship between the error in origin tracking and the course and speed of the target (Figure 5.29).

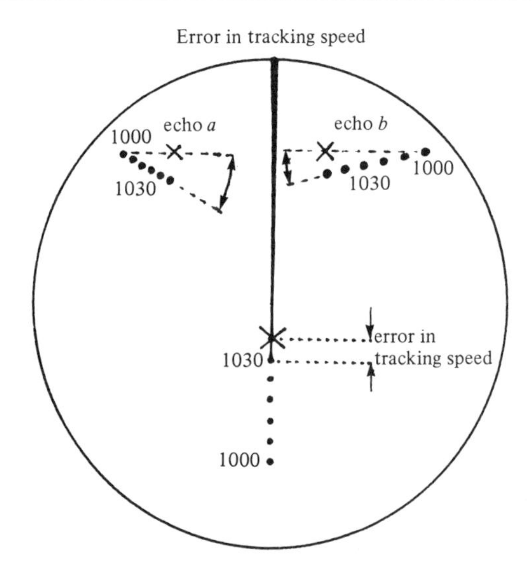

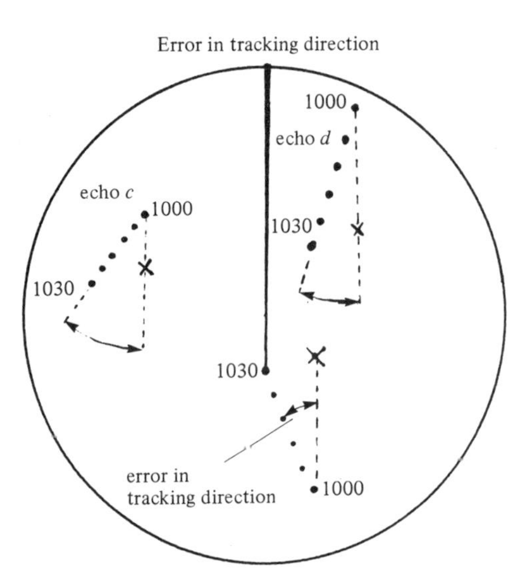

Figure 5.29 True motion tracking errors. X represents the correct positions for origin and targets at 1030. The error in the displayed headings at *a* and *c* is greater than is the case at *b* and *d*. Errors exist in the displayed speed of all targets.

5.10.3 *Radar plotting accuracy*

The inherent accuracy of the results which can be obtained from manual radar plotting depends on the inherent accuracy of the inputs to the calculation and the geometry of the particular plotting triangle. The inputs to the calculation are range, bearing, course, speed and time. Range, bearing and course accuracy were considered in Section 5.10.1. Speed, if derived from a log complying with the IMO Standards, should have an error not exceeding 5 per cent of the speed of the ship or 0.5 knots, whichever is the greater (see Section 2.4.1). Inherent errors (as opposed to mistakes) in timing should be negligible. It should be borne in mind that inherent errors do not include mistakes or inaccuracies introduced by the plotter in transferring the data from the source to the plot. In this text it is intended to consider the aggregate inherent accuracy which can be reasonably attributed to the deductions of a target's CPA, TCPA, course and speed. Over the years, the topic has been explored in a number of ways. The advent of the IMO Standards for ARPA has resulted in the production of authoritative values.

In deriving the accuracy requirement for automatic

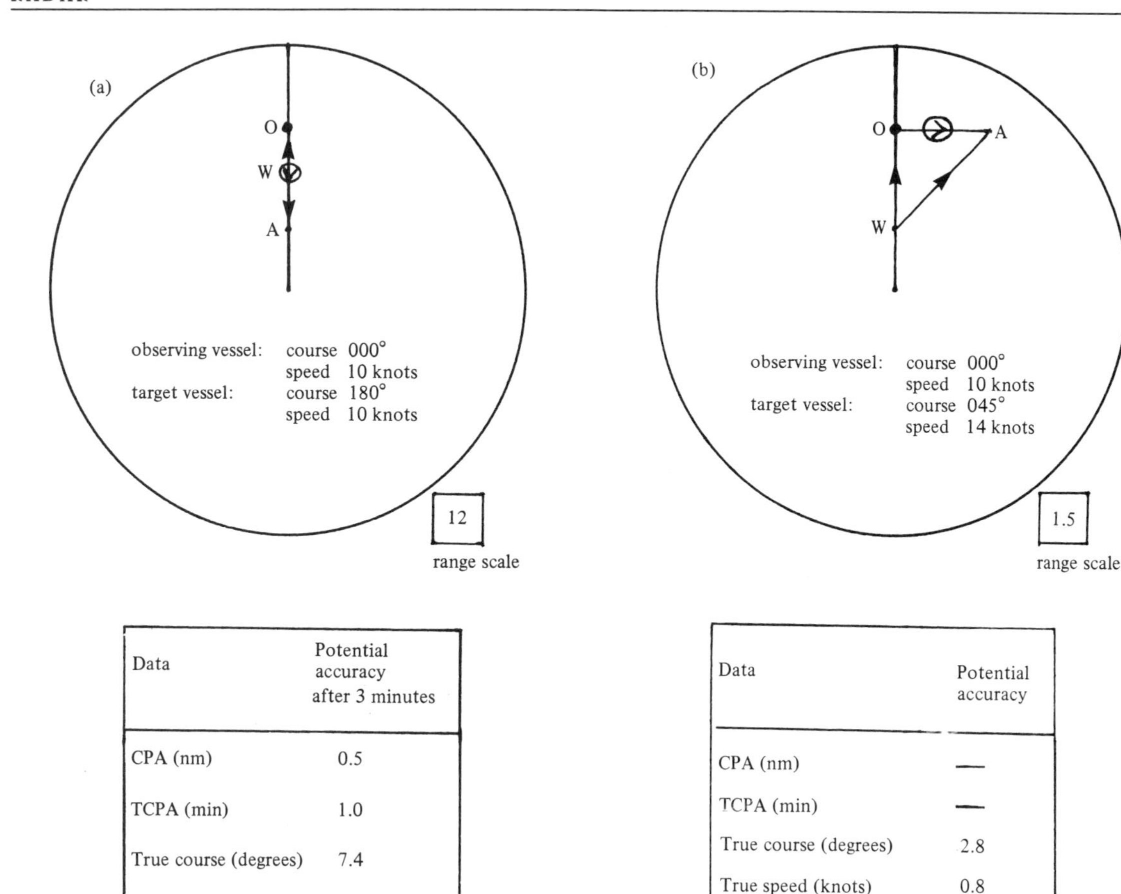

Data	Potential accuracy after 3 minutes
CPA (nm)	0.5
TCPA (min)	1.0
True course (degrees)	7.4
True speed (knots)	1.2

Data	Potential accuracy
CPA (nm)	—
TCPA (min)	—
True course (degrees)	2.8
True speed (knots)	0.8

Figure 5.30 *continued overleaf*

plotting, it was decided that the values should relate to the best possible manual plotting performance under environmental conditions of plus and minus 10° of roll. In arriving at these values, it was assumed that the sensor errors were appropriate to radar equipment complying with the IMO Radar Performance Standards. These values are thus useful as a guide to plotting accuracy. It must be remembered that although the ARPA is required to achieve these values within 3 minutes of steady state tracking, the manual plotter will require very much longer to achieve comparable accuracy.

Because of the dependence of potential accuracy on the geometry of the plotting triangle, the accuracies are specified for four separate scenarios, each illustrating a different geometry. The full details are set out in text and tabular form by the ARPA Standards, but for convenience they have been presented in pictorial form in Figure 5.30 (a)–(d).

Any treatment of errors is essentially statistical in its nature; it should be noted that the figures give 95 per cent probability values, i.e. results must be within the tolerance values on nineteen out of twenty occasions.

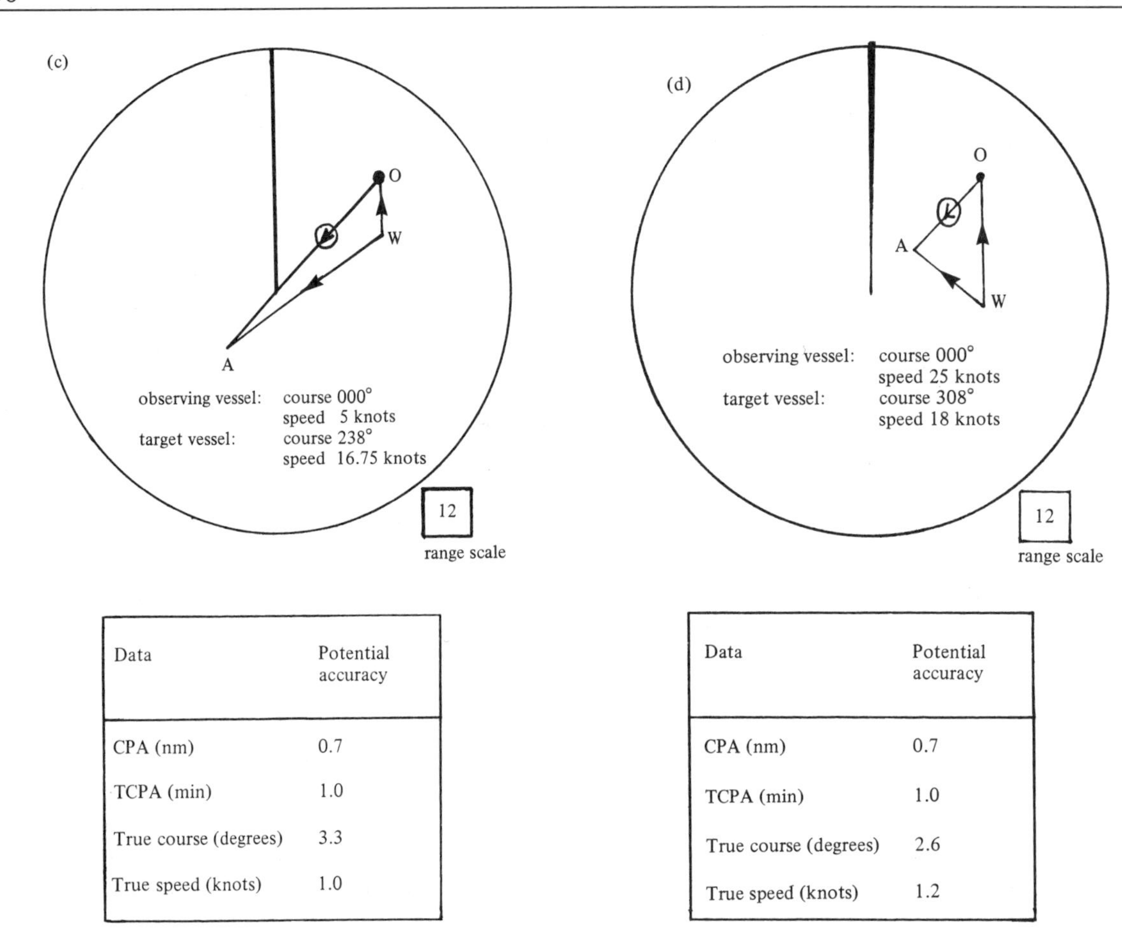

Data	Potential accuracy
CPA (nm)	0.7
TCPA (min)	1.0
True course (degrees)	3.3
True speed (knots)	1.0

Data	Potential accuracy
CPA (nm)	0.7
TCPA (min)	1.0
True course (degrees)	2.6
True speed (knots)	1.2

Figure 5.30 Plotting accuracy scenarios, (a) no. 1; (b) no. 2; (c) no. 3; (d) no. 4

5.10.4 Automatic plotting accuracy

The accuracy limits required after three minutes of steady state tracking have already been set out in Section 5.10.3. Further consideration of the IMO Standards for ARPA will show that intermediate values of accuracy are required to be achieved within 1 minute of steady state tracking. These values are associated with an early indication of the target's predicted motion, known as the target's motion trend. The Standards require that the ARPA's error contribution should remain small when compared with the input sensor errors for the scenarios on which the specification of 1 minute and 3 minute accuracies are based.

Excluding loss of the target, steady state tracking may be interrupted by:

1 A manoeuvre by the observing vessel
2 A manoeuvre by a target
3 An instantaneous error in course or speed input (e.g. a reduction in log speed input owing to a fault in the log circuitry).

Because individual ARPAs store and process the plotting data in different ways, it is not possible, without knowledge of the specific type of equipment, to state how basic calculated data such as CPA, TCPA, course and speed will be affected in the three-minute plotting period following an interruption to steady state tracking. Applying manual plotting principles one might for example assume that, in the event of a log fault developing, this fault would have no effect on the relative vectors in the subsequent plotting sequence. This is true in the case of some ARPAs but *not* in the case of others. The behaviour of both types of vectors during the smoothing period after interruption of steady state tracking varies according to the storage and tracking philosophy adopted by the designer. See *Radar and ARPA Manual* for a full treatment of tracking theory.

5.10.5 Position fixing accuracy

Position lines laid off on the chart will be subject to range or bearing accuracy as set out in Section 5.10.1. Where parallel indexing is carried out manually the same inherent accuracy will feature. Provided a suitable range scale is selected and the reflection plotter properly adjusted, there should be no loss in potential accuracy.

5.11 Extracts from official publications

Within Sections 5.11.1–5.11.5, the paragraph numbering of the original publications has been retained.

5.11.1 IMO Resolutions A222(VII), A278(VIII), A477(XII): Performance Standards for Navigational Radar Equipment

Installed before 1 September 1984

1 Introduction

1 The radar equipment required by Regulation 12 of Chapter V should provide an indication in relation to the ship of the position of other surface craft and obstructions and of buoys, shorelines and navigational marks in a manner which will assist in avoiding collision and in navigation.

2 It should comply with the following minimum requirements:

(a) Range performance

The operational requirement under normal propagation conditions, when the radar aerial is mounted at a height of 15 metres above sea level, is that the equipment should give a clear indication of:

(i) Coastlines:

At 20 nautical miles when the ground rises to 60 metres

At 7 nautical miles when the ground rises to 6 metres.

(ii) Surface objects:

At 7 nautical miles a ship of 5000 tons gross tonnage, whatever her aspect

At 3 nautical miles a small ship of length 10 metres

At 2 nautical miles an object such as a navigational buoy having an effective echoing area of approximately 10 square metres.

(b) Minimum range

The surface objects specified in paragraph 2(a)(ii) should be clearly displayed from a minimum range of 50 metres up to a range of 1 nautical mile, without adjustment of controls other than the range selector.

(c) Display

(i) The equipment should provide a relative plan display of not less than 180 mm effective diameter.

(ii) The equipment should be provided with at least five ranges, the smallest of which is not more than 1 nautical mile and the greatest of which is not less than 24 nautical miles. The scales should be preferably of 1:2 ratio. Additional ranges may be provided.

(iii) Positive indication should be given of the range of view displayed and the interval between range rings.

(d) Range measurement

(i) The primary means provided for range measurement should be fixed electronic range rings. There should be at least four range rings displayed on each of the ranges mentioned in paragraph 4.2, except that on ranges below 1 nautical mile range rings should be displayed at intervals of 0.25 nautical mile.

(ii) Fixed range rings should enable the range of an object, whose echo lies on a range ring, to be measured with an error not exceeding 1.5 per cent of the maximum range of the scale in use, or 70 metres, whichever is the greater.

(iii) Any additional means of measuring range should have an error not exceeding 2.5 per cent of the maximum range of the displayed scale in use, or 120 metres, whichever is the greater.

(e) Heading indicator

(i) The heading of the ship should be indicated by a line on the display with a maximum error not greater than ±1°. The thickness of the display heading line should not be greater than 0.5°.

(ii) Provision should be made to switch off the heading indicator by a device which cannot be left in the 'heading marker off' position.

(f) Bearing measurement

(i) Provision should be made to obtain quickly the bearing of any object whose echo appears on the display.

(ii) The means provided for obtaining bearings should enable the bearing of a target whose echo appears at the edge of the display to be measured with an accuracy of ±1° or better.

(g) Discrimination

(i) The equipment should display as separate indications, on the shortest range scale provided, two objects on the same azimuth separated by not more than 50 metres in range.

(ii) The equipment should display as separate indications two objects at the same range separated by not more than 2.5° in azimuth.

(iii) The equipment should be designed to avoid, as far as is practicable, the display of spurious echoes.

(h) Roll

The performance of the equipment should be such that when the ship is rolling ±10° the echoes of targets remain visible on the display.

(i) Scan

The scan should be continuous and automatic through 360° of azimuth. The target data rate should be at least 12 per minute. The equipment should operate satisfactorily in relative wind speeds of up to 100 knots.

(j) Azimuth stabilization

(i) Means should be provided to enable the display to be stabilized in azimuth by a transmitting compass. The accuracy of alignment with the compass transmission should be within 0.5° with a compass rotation rate of 2 r.p.m.

(ii) The equipment should operate satisfactorily for relative bearings when the compass control is inoperative or not fitted.

(k) Performance check

Means should be available, while the equipment is used operationally, to determine readily a significant drop in performance relative to a calibration standard established at the time of installation.

(l) Anti-clutter devices

Means should be provided to minimize the display of unwanted responses from precipitation and the sea.

(m) Operation

(i) The equipment should be capable of being switched on and operated from the main display position.

(ii) Operational controls should be accessible and easy to identify and use.

(iii) After switching on from cold, the equipment should become fully operational within 4 minutes.

(iv) A standby condition should be provided from which the equipment can be brought to a fully operational condition within 1 minute.

(n) Interference

After installation and adjustment on board, the bearing accuracy should be maintained without further adjustment irrespective of the variation of external magnetic fields.

(o) Sea or ground stabilization

Sea or ground stabilization, if provided, should not degrade the accuracy of the display below the requirements of these performance standards, and the view ahead on the display should not be unduly restricted by the use of this facility.

3 Siting of the aerial

The aerial system should be installed in such a manner that the efficiency of the display is not impaired by the close proximity of the aerial to other objects. In particular, blind sectors in the forward direction should be avoided.

Installed on or after 1 September 1984

1 Application

1.1 This Recommendation applies to all ships' radar equipment installed on or after 1 September 1984 in compliance with Regulation 12, Chapter V of the International Convention for the Safety of Life at Sea, 1974, as amended.

1.2 Radar equipment installed before 1 September 1984 should comply at least with the performance standards recommended in resolution A.222(VII).

2 All radar installations

All radar installations should comply with the following minimum requirements.

3 General

The radar equipment should provide an indication, in relation to the ship, of the position of other surface craft and obstructions and of buoys, shorelines and navigational marks in a manner which will assist in navigation and in avoiding collision.

3.1 Range performance

The operational requirement under normal propagation conditions, when the radar antenna is mounted at a height of 15 metres above sea level, is that the equipment should in the absence of clutter give a clear indication of:

.1 Coastlines:

At 20 nautical miles when the ground rises to 60 metres

At 7 nautical miles when the ground rises to 6 metres.

.2 Surface objects:

At 7 nautical miles a ship of 5000 tons gross tonnage, whatever her aspect

At 3 nautical miles a small ship of 10 metres in length

At 2 nautical miles an object such as a navigational buoy having an effective echoing area of approximately 10 square metres.

3.2 Minimum range

The surface objects specified in paragraph 3.1.2 should be clearly displayed from a minimum range of 50 metres up to a range of 1 nautical mile, without changing the setting of controls other than the range selector.

3.3 Display

3.3.1 The equipment should without external magnification provide a relative plan display in the head up unstabilized mode with an effective diameter of not less than:

.1 180 millimetres on ships of 500 tons gross tonnage and more but less than 1600 tons gross tonnage;
.2 250 millimetres on ships of 1600 tons gross tonnage and more but less than 10 000 tons gross tonnage;
.3 340 millimetres in the case of one display and 250 millimetres in the case of the other on ships of 10 000 tons gross tonnage and upwards.

Note Display diameters of 180, 250 and 340 millimetres correspond respectively to 9, 12 and 16 inch cathode ray tubes.

3.3.2 The equipment should provide one of the two following sets of range scales of display:

.1 1.5, 3, 6, 12 and 24 nautical miles and one range scale of not less than 0.5 and not greater than 0.8 nautical miles; or
.2 1, 2, 4, 8, 16 and 32 nautical miles.

3.3.3 Additional range scales may be provided.

3.3.4 The range scale displayed and the distance between range rings should be clearly indicated at all times.

3.4 Range measurement

3.4.1 Fixed electronic range rings should be provided for range measurements as follows:

.1 Where range scales are provided in accordance with paragraph 3.3.2.1, on the range scale of between 0.5 and 0.8 nautical miles at least two range rings should be provided and on each of the other range scales six range rings should be provided; or
.2 Where range scales are provided in accordance with paragraph 3.3.2.2, four range rings should be provided on each of the range scales.

3.4.2 A variable electronic range marker should be provided with a numeric readout of range.

3.4.3 The fixed range rings and the variable range marker should enable the range of an object to be measured with an error not exceeding 1.5 per cent of the maximum range of the scale in use, or 70 metres, whichever is the greater.

3.4.4 It should be possible to vary the brilliance of the fixed range rings and the variable range marker and to remove them completely from the display.

3.5 Heading indicator

3.5.1 The heading of the ship should be indicated by a line on the display with a maximum error not greater than ±1°. The thickness of the displayed heading line should not be greater than 0.5°.

3.5.2 Provision should be made to switch off the heading indicator by a device which cannot be left in the 'heading marker off' position.

3.6 Bearing measurement

3.6.1 Provision should be made to obtain quickly the bearing of any object whose echo appears on the display.

3.6.2 The means provided for obtaining bearing should enable the bearing of a target whose echo appears at the edge of the display to be measured with an accuracy of ±1° or better.

3.7 Discrimination

3.7.1 The equipment should be capable of displaying as separate indications on a range scale of 2 nautical miles or less, two small similar targets at a range of between 50 per cent and 100 per cent of the range scale in use, and on the same azimuth, separated by not more than 50 metres in range.

3.7.2 The equipment should be capable of displaying as separate indications two small similar targets both situated at the same range between 50 per cent and 100 per cent of the 1.5 or 2 mile range scales, and separated by not more than 2.5° in azimuth.

3.8 Roll or pitch

The performance of the equipment should be such

that when the ship is rolling or pitching up to ±10° the range performance requirements of paragraphs 4 and 5 continue to be met.

3.9 Scan

The scan should be clockwise, continuous and automatic through 360° of azimuth. The scan rate should be not less than 12 r.p.m. The equipment should operate satisfactorily in relative wind speeds of up to 100 knots.

3.10 Azimuth stabilization

3.10.1 Means should be provided to enable the display to be stabilized in azimuth by a transmitting compass. The equipment should be provided with a compass input to enable it to be stabilized in azimuth. The accuracy of alignment with the compass transmission should be within 0.5° with a compass rotation rate of 2 r.p.m.

3.10.2 The equipment should operate satisfactorily in the unstabilized mode when the compass control is inoperative.

3.11 Performance check

Means should be available, while the equipment is used operationally, to determine readily a significant drop in performance relative to a calibration standard established at the time of installation, and that the equipment is correctly tuned in the absence of targets.

3.12 Anti-clutter devices

Suitable means should be provided for the suppression of unwanted echoes from sea clutter, rain and other forms of precipitation, clouds and sandstorms. It should be possible to adjust manually and continuously the anti-clutter controls. Anti-clutter controls should be inoperative in the fully anti-clockwise positions. In addition, automatic anti-clutter controls may be provided; however, they must be capable of being switched off.

3.13 Operation

3.13.1 The equipment should be capable of being switched on and operated from the display position.

3.13.2 Operational controls should be accessible and easy to identify and use. Where symbols are used they should comply with the recommendations of the organization on symbols for controls on marine navigational radar equipment.

3.13.3 After switching on from cold the equipment should become fully operational within 4 minutes.

3.13.4 A standby condition should be provided from which the equipment can be brought to an operational condition within 15 seconds.

3.14 Interference

After installation and adjustment on board, the bearing accuracy as prescribed in these performance standards should be maintained without further adjustment irrespective of the movement of the ship in the earth's magnetic field.

3.15 Sea or ground stabilization (true motion display)

3.15.1 Where sea or ground stabilization is provided the accuracy and discrimination of the display should be at least equivalent to that required by these performance standards.

3.15.2 The motion of the trace origin should not, except under manual override conditions, continue to a point beyond 75 per cent of the radius of the display. Automatic resetting may be provided.

3.16 Antenna system

The antenna system should be installed in such a manner that the design efficiency of the radar system is not substantially impaired.

3.17 Operation with radar beacons

3.17.1 All radars operating in the 3 cm band should

be capable of operating in a horizontally polarized mode.

3.17.2 It should be possible to switch off those signal processing facilities which might prevent a radar beacon from being shown on the radar display.

4 Multiple radar installations

4.1 Where two radars are required to be carried they should be so installed that each radar can be operated individually and both can be operated simultaneously without being dependent upon one another. When an emergency source of electrical power is provided in accordance with the appropriate requirements of chapter II–1 of the 1974 SOLAS Convention, both radars should be capable of being operated from this source.

4.2 Where two radars are fitted, interswitching facilities may be provided to improve the flexibility and availability of the overall radar installation. They should be so installed that failure of either radar would not cause the supply of electrical energy to the other radar to be interrupted or adversely affected.

Symbols for controls on marine navigational radar equipment

1 List of controls to be identified by symbols

The following switches and variable controls are considered to be the minimum required to be marked by symbols:

Radar on – standby – off switch
Aerial rotation switch
Mode of presentation switch – north up or ship's head up
Heading marker alignment control or switch
Range selection switch
Pulse length selection switch – short or long pulse
Tuning control
Gain control
Anti-clutter rain control (differentiation)
Anti-clutter sea control
Scale illumination control or switch
Display brilliance control
Range rings brilliance control
Variable range marker control
Bearing marker control
Performance monitor switch – transmitted power monitor or transmit/receive monitor.

2 Code of practice

The following code of practice should be used when marking radar sets with recommended symbols:

2.1 The maximum dimension of a symbol should not be less than 9 mm.

2.2 The distance between the centres of two adjacent symbols should not be less than 1.4 times the size of the larger symbol.

2.3 Switch function symbols should not be linked by a line. A linked line infers controlled action.

2.4 Variable control function symbols should be linked by a line, preferably an arc. The direction of increase of controlled function should be indicated.

2.5 Symbols should be presented with a high contrast against their background.

2.6 The various elements of a symbol should have a fixed ratio one to another.

2.7 Multiple function of controls and switch positions may be indicated by a combined symbol.

2.8 Where concentric controls or switches are fitted, the outer of the symbols should refer to the larger diameter control.

3 Symbols

3.1 The symbols attached hereto should be used for controls on marine navigational radar equipment.

3.2 The circles shown around the following symbols are optional:

Symbol 4: aerial rotating
Symbol 9: short pulse
Symbol 10: long pulse
Symbol 17: scale illumination
Symbol 22: transmitted power monitor
Symbol 23: transmit/receive monitor.

Symbols for controls on marine navigational radar equipment

No.	Symbol	Description	Function
1		Off	To identify the 'off' position of the control or switch
2	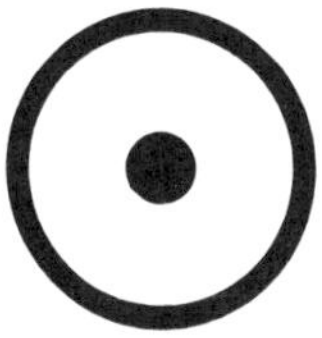	Radar on	To identify the 'radar on' position of the switch
3		Radar stand-by	To identify the 'radar stand-by' position of the switch
4		Aerial rotating	To identify the 'aerial rotating' position of the switch
5		North up presenta-tion	To identify the 'north up' position of the mode of presentation switch
6	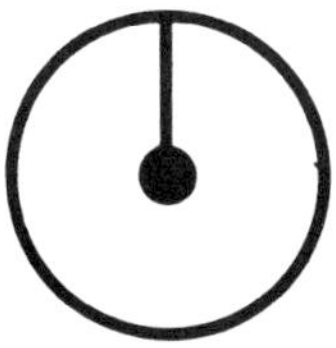	Ship's head up presenta-tion	To identify the 'ship's head up' position of the mode of presentation switch
7		Heading marker alignment	To identify the 'heading marker alignment' control switch
8		Range selector	To identify the range selection switch
9		Short pulse	To identify the 'short pulse' position of the pulse length selection switch
10		Long pulse	To identify the 'long pulse' position of the pulse length selection switch
11		Tuning	To identify the 'tuning' control
12		Gain	To identify the 'gain' control

No.	Symbol	Name	Purpose
13		Anti-clutter rain minimum	To identify the minimum position of the 'anti-clutter rain' control or switch
14	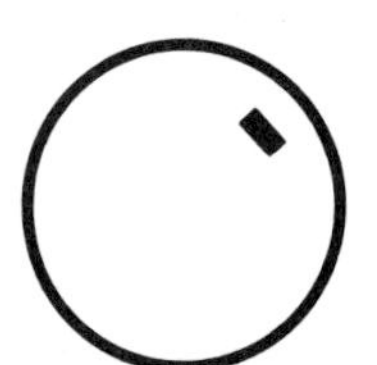	Anti-clutter rain maximum	To identify the maximum position of the 'anti-clutter rain' control or switch
15		Anti-clutter sea minimum	To identify the minimum position of the 'anti-clutter sea' control
16		Anti-clutter sea maximum	To identify the maximum position of the 'anti-clutter sea' control
17		Scale illumination	To identify the maximum position of the 'scale illumination' control or switch
18		Display brilliance	To identify the maximum position of the 'display brilliance' control
19		Range rings brilliance	To identify the maximum position of the 'range rings brilliance' control
20		Variable range marker	To identify the 'variable range marker' control
21		Bearing marker	To identify the 'bearing marker' control
22		Transmitted power monitor	To identify the on position of the 'transmitted power monitor' switch
23		Transmit/receive monitor	To identify the on position of the 'transmit/receive monitor' switch

5.11.2 UK Statutory Instrument 1984, no. 1203

Part III Radar installation

Radar performance standards and interswitching facilities

18(1) Every radar installation required to be provided shall comply with the performance standards adopted by the organization and shall, in addition, comply with the relevant performance specifications issued by the Department of Transport.

(2) Interswitching facilities:

(a) Where such a radar installation includes additional radar units and facilities for interswitching, at least one arrangement of units when used together shall comply with all the requirements of this part of these regulations.

(b) Where two radar installations are required to be provided on a ship, they shall be so installed that each radar installation can be operated individually and both can be operated simultaneously without being dependent upon one another.

Provision of plotting facilities

19 Facilities for plotting radar readings shall be provided on the navigating bridge of every ship required to be fitted with a radar installation. In ships of 1600 tons gross tonnage and upwards constructed on or after 1 September 1984 the plotting facilities shall be at least as effective as a reflection plotter.

Radar watch

20(1) While a ship which is required to be fitted with a radar installation is at sea and a radar watch is being kept, the radar installation shall be under the control of a qualified radar observer, who may be assisted by unqualified personnel.

(2) In every such ship a record shall be kept in the deck log book of the times at which radar watch is commenced and discontinued.

Serviceability and maintenance of radar installations

21(1) The performance of the radar installation shall be checked before the ship proceeds to sea and at least once every four hours whilst the ship is at sea and radar watch is being maintained.

(2) Every ship of 1600 tons or over required to be fitted with a radar installation which is going between the United Kingdom and locations in the unlimited trading area or between locations in the unlimited trading area shall be provided with at least one officer or member of the crew adequately qualified to carry out radar maintenance: provided that:

(a) If on an occasion on which a ship goes to sea, the officer or member of the crew adequately qualified to carry out radar maintenance is not carried because of illness, incapacity, or other unforeseen circumstance, but all reasonable steps were taken to secure the carriage on that occasion of a duly qualified officer or crew member, the provisions of this regulation which require such a ship on such a voyage to carry an officer or crew member adequately qualified to carry out radar maintenance shall not, subject to compliance with the conditions in subparagrah (b) below, apply to the ship during a period beginning with the day on which the ship goes to sea and ending either 28 days later or with the day on which the ship sails from its next port of call, whichever is the later.

(b) The conditions are that one such period shall not be followed immediately by any further period at sea during which the ship does not carry an officer or crew member adequately qualified to carry out radar maintenance and that the master, when going to sea on such an occasion shall:

  (i) Notify a proper officer of his intention not to carry a suitably qualified officer or crew member; and

  (ii) Make an entry of that notification in the ship's official log.

Qualifications of radar observers and radar maintenance personnel

22(1) For the purposes of these regulations, a person shall be deemed a 'qualified radar observer' if he holds:

(a) A valid Radar Observer's Certificate granted by the Secretary of State; or

(b) A valid certificate of attendance granted at the conclusion of a radar simulator course which has been approved by the Secretary of State; or

(c) A valid Electronic Navigation Systems Certificate granted by the Secretary of State; or
(d) A valid Navigation Control Certificate granted by the Secretary of State; or
(e) A certificate recognized by the Secretary of State as being equivalent to any of the certificates mentioned in (a), (b), (c) or (d).

(2) For the purposes of these regulations, an officer or crew member shall be deemed qualified to carry out radar maintenance if he holds:

(a) A Radar Maintenance Certificate granted by the Secretary of State; or
(b) An Electronic Navigational Equipment Maintenance Certificate granted by the Secretary of State; or
(c) A certificate recognized by the Secretary of State as being equivalent to either of the certificates mentioned in (a) or (b); or
(d) A certificate of proficiency to carry out maintenance on specified types of radar installations granted at the conclusion of a radar manufacturer's course which has been approved by the Secretary of State; or
(e) A special certificate to carry out maintenance on specified types of radar installations issued by the Secretary of State upon satisfactory written evidence that the applicant's employment, over a period of not less than 10 years between 25 May 1960 and 24 May 1980 has included the maintenance of marine radar installations.

Siting of radar installation
23(1) The antenna unit of the radar installation shall be sited so that satisfactory overall performance is achieved in relation to:

(a) The avoidance of shadow sectors;
(b) The avoidance of false echoes caused by reflections from the ship's structure; and
(c) The effect of antenna height on the amplitude and extent of sea clutter.

(2) The radar display shall be sited on the bridge from which the ship is normally navigated. The siting of one of the displays shall be such that:

(a) An observer, when viewing the display, faces forward and is readily able to maintain visual look-out;
(b) There is sufficient space for two observers to view the display simultaneously.

(3) The radar installation shall, where practicable, be mounted so as to prevent the performance and reliability of the installation being adversely affected by vibration and so that the installation will not, whilst in service, normally be subject to greater vibration than that specified in the General Requirements for Marine Navigational Equipment, 1982, issued by the Department of Transport.

Alignment of heading marker
24 The radar heading marker (and stern marker if fitted) shall be aligned to within 1° of the ship's fore-and-aft line as soon as practicable after the radar installation has been installed in the ship. Where interswitching facilities are provided, the heading marker shall be aligned with all arrangements of units. The marker shall be realigned as soon as practicable whenever it is found to be substantially inaccurate.

Measurement of shadow sectors
25 The angular width and bearing of any shadow sectors displayed by the radar installation shall be determined and recorded. The record shall be shown on a diagram adjacent to the radar display and be kept up to date following any change likely to affect shadow sectors.

Display sizes
26 A radar installation required to be provided which is installed on board a ship on or after 1 September 1984 shall provide a relative plan display having an effective diameter, without external magnification, of not less than:

(a) 180 millimetres on ships of 500 tons or over but less than 1600 tons;
(b) 250 millimetres on ships of 1600 tons or over but less than 10 000 tons;
(c) 340 millimetres in the case of one radar installation and 250 millimetres in the case of the other on ships of 10 000 tons or over.

5.11.3 IMO Resolution A422(XI): Performance Standards for Automatic Radar Plotting Aids (ARPA)

1 Introduction

1.1 Automatic radar plotting aids (ARPA) should, in order to improve the standard of collision avoidance at sea:

.1 Reduce the workload of observers by enabling them to automatically obtain information so that they can perform as well with multiple targets as they can by manually plotting a single target
.2 Provide continuous, accurate and rapid situation evaluation.

1.2 In addition to the general requirements for electronic navigational aids (resolution A.281(VIII)), the ARPA should comply with the following minimum performance standards.

2 Definitions

2.1 Definitions of terms used in these performance standards are given in annex 1 to this resolution.

3 Performance standards

3.1 Detection
3.1.1 Where a separate facility is provided for detection of targets, other than by the radar observer, it should have a performance not inferior to that which could be obtained by the use of the radar display.

3.2 Acquisition
3.2.1 Target acquisition may be manual or automatic. However, there should always be a facility to provide for manual acquisition and cancellation: ARPA with automatic acquisition should have a facility to suppress acquisition in certain areas. On any range scale where acquisition is suppressed over a certain area, the area of acquisition should be indicated on the display.

3.2.2 Automatic or manual acquisition should have a performance not inferior to that which could be obtained by the user of the radar display.

3.3 Tracking
3.3.1 The ARPA should be able to automatically track, process, simultaneously display and continuously update the information on at least:

.1 Twenty targets, if automatic acquisition is provided, whether automatically or manually acquired
.2 Ten targets, if only manual acquisition is provided.

3.3.2 If automatic acquisition is provided, a description of the criteria of selection of targets for tracking should be provided to the user. If the ARPA does not track all targets visible on the display, targets which are being tracked should be clearly indicated on the display. The reliability of tracking should not be less than that obtainable using manual recordings of successive target positions obtained from the radar display.

3.3.3 Provided the target is not subject to target swop, the ARPA should continue to track an acquired target which is clearly distinguishable on the display for five out of ten consecutive scans.

3.3.4 The possibility of tracking errors, including target swop, should be minimized by ARPA design. A qualitative description of the effects of error sources on the automatic tracking and corresponding errors should be provided to the user, including the effects of low signal-to-noise and low signal-to-clutter ratios caused by sea returns, rain, snow, low clouds and non-synchronous emissions.

3.3.5 The ARPA should be able to display on request at least four equally time-spaced past positions of any targets being tracked over a period of at least 8 minutes.

3.4 Display
3.4.1 The display may be a separate or integral part of the ship's radar. However, the ARPA display should include all the data required to be provided by a radar display in accordance with the performance standards for navigational radar equipment.

3.4.2 The design should be such that any malfunction of ARPA parts producing data additional to information to be produced by the radar as required by the performance standards for navigational equipment should not affect the integrity of the basic radar presentation.

3.4.3 The display on which ARPA information is presented should have an effective diameter of at least 340 mm.

3.4.4 The ARPA facilities should be available on at least the following range scales:

.1 12 or 16 miles
.2 3 or 4 miles.

3.4.5 There should be a positive indication of the range scale in use.

3.4.6 The ARPA should be capable of operating with a relative motion display with north up and either head up or course up azimuth stabilization. In addition, the ARPA may also provide for a true motion display. If true motion is provided, the operator should be able to select for his display either true or relative motion. There should be a positive indication of the display mode and orientation in use.

3.4.7 The course and speed information generated by the ARPA for acquired targets should be displayed in a vector or graphic form which clearly indicates the target's predicted motion. In this regard:

.1 An ARPA presenting predicted information in vector form only should have the option of both true and relative vectors.
.2 An ARPA which is capable of presenting target course and speed information in graphic form should also, on request, provide the target's true and/or relative vector.
.3 Vectors displayed should either be time adjustable or have a fixed time-scale.
.4 A positive indication of the time-scale of the vector in use should be given.

3.4.8 The ARPA information should not obscure radar information in such a manner as to degrade the process of detecting targets. The display of ARPA data should be under the control of the radar observer. It should be possible to cancel the display of unwanted ARPA data.

3.4.9 Means should be provided to adjust independently the brilliance of the ARPA data and radar data, including complete elimination of the ARPA data.

3.4.10 The method of presentation should ensure that the ARPA data is clearly visible in general to more than one observer in the conditions of light normally experienced on the bridge of a ship by day and by night. Screening may be provided to shade the display from sunlight but not to the extent that it will impair the observer's ability to maintain a proper look-out. Facilities to adjust the brightness should be provided.

3.4.11 Provisions should be made to obtain quickly the range and bearing of any object which appears on the ARPA display.

3.4.12 When a target appears on the radar display and, in the case of automatic acquisition, enters within the acquisition area chosen by the observer or, in the case of manual acquisition, has been acquired by the observer, the ARPA should present in a period of not more than 1 minute an indication of the target's motion trend, and display within 3 minutes the target's predicted motion in accordance with paragraphs 3.4.7, 3.6, 3.8.2 and 3.8.3.

3.4.13 After changing range scales on which the ARPA facilities are available or resetting the display, full plotting information should be displayed within a period not exceeding four scans.

3.5 Operational warnings

3.5.1 The ARPA should have the capability to warn the observer with a visual and/or audible signal of any distinguishable target which closes to a range or transits a zone chosen by the observer. The target causing the warning should be clearly indicated on the display.

3.5.2 The ARPA should have the capability to warn the observer with a visual and/or audible signal of any tracked target which is predicted to close to within a minimum range and time chosen by the observer. The target causing the warning should be clearly indicated on the display.

3.5.3 The ARPA should clearly indicate if a tracked target is lost, other than out of range, and the target's last tracked position should be clearly indicated on the display.

3.5.4 It should be possible to activate or deactivate the operational warnings.

3.6 Data requirements

3.6.1 At the request of the observer the following information should be immediately available from the ARPA in alphanumeric form in regard to any tracked target:

.1 Present range to the target
.2 Present bearing of the target
.3 Predicted target range at the closest point of approach (CPA)
.4 Predicted time to CPA (TCPA)
.5 Calculated true course of target
.6 Calculated true speed of target.

3.7 Trial manoeuvre

3.7.1 The ARPA should be capable of simulating the effect on all tracked targets of an own ship manoeuvre without interrupting the updating of target information. The simulation should be initiated by the depression either of a spring-loaded switch, or of a function key, with a positive identification on the display.

3.8 Accuracy

3.8.1 The ARPA should provide accuracies not less than those given in paragraphs 3.8.2 and 3.8.3 for the four scenarios defined in annex 2 to this resolution. With the sensor errors specified in annex 3, the values given relate to the best possible manual plotting performance under environmental conditions of ±10° of roll.

3.8.2 An ARPA should present within 1 minute of steady state tracking the relative motion trend of a target with the following accuracy values (95 per cent probability values):

Scenario	*Relative course (degrees)*	*Relative speed (knots)*	*CPA (nautical miles)*
1	11	2.8	1.6
2	7	0.6	—
3	14	2.2	1.8
4	15	1.5	2.0

3.8.3 An ARPA should present within 3 minutes of steady state tracking the motion of a target with the following accuracy values (95 per cent probability values):

Scenario	*Relative course (degrees)*	*Relative speed (knots)*	*CPA (nautical miles)*	*TCPA (min)*	*True course (degrees)*	*True Speed (knots)*
1	3.0	0.8	0.5	1.0	7.4	1.2
2	2.3	0.3	—	—	2.8	0.8
3	4.4	0.9	0.7	1.0	3.3	1.0
4	4.6	0.8	0.7	1.0	2.6	1.2

3.8.4 When a tracked target, or own ship, has completed a manoeuvre, the system should present in a period of not more than 1 minute an indication of the target's motion trend, and display within 3 minutes the target's predicted motion, in accordance with paragraphs 3.4.7, 3.6, 3.8.2 and 3.8.3.

3.8.5 The ARPA should be designed in such a manner that under the most favourable conditions of own ship motion the error contribution from the ARPA should remain insignificant compared with the errors associated with the input sensors, for the scenarios of annex 2.

3.9 Connections with other equipment

3.9.1 The ARPA should not degrade the performance of any equipment providing sensor inputs. The connection of the ARPA to any other equipment should not degrade the performance of that equipment.

3.10 Performance tests and warnings

3.10.1 The ARPA should provide suitable warnings of ARPA malfunction to enable the observer to monitor the proper operation of the system. Additionally, test programs should be available so that the overall performance of ARPA can be assessed periodically against a known solution.

3.11 Equipment used with ARPA

3.11.1 Log and speed indicators providing inputs to ARPA equipment should be capable of providing the ship's speed through the water.

Annex 1 Definitions of terms to be used only in connection with ARPA performance standards

Relative course — The direction of motion of a target related to own ship as deduced from a number of measurements of its range and bearing on the radar, expressed as an angular distance from north.

Relative speed — The speed of a target related to own ship, as deduced from a number of measurements of its range and bearing on the radar.

True course — The apparent heading of a target obtained by the vectorial combination of the target's relative motion and own ship's motion, expressed as an angular distance from north.

True speed — The speed of a target obtained by the vectorial combination of its relative motion and own ship's motion.

Note For the purpose of the definitions of true course and true speed there is no need to distinguish between sea and ground stabilization.

Bearing — The direction of one terrestrial point from another, expressed as an angular distance from north.

Relative motion display — The position of own ship on such a display remains fixed.

True motion display — The position of own ship on such a display moves in accordance with its own motion.

Azimuth stabilization — Own ship's compass information is fed to the display so that echoes of targets on the display will not be caused to smear by changes of own ship's heading.

North up — The line connecting the centre with the top of the display is north.

Head up — The line connecting the centre with the top of the display is own ship's heading.

Course up — An intended course can be set to the line connecting the centre with the top of the display.

Heading — The direction in which the bows of a ship are pointing, expressed as an angular distance from north.

Target's predicted motion — The indication on the display of a linear extrapolation into the future of a target's motion, based on measurements of the target's range and bearing on the radar in the recent past.

Target's motion trend — An early indication of the target's predicted motion.

Radar plotting — The whole process of target detection, tracking, calculation of parameters and display of information.

Detection — The recognition of the presence of a target.

Acquisition — The selection of those targets requiring a tracking procedure and the initiation of their tracking.

Tracking — The process of observing the sequential changes in the position of a target, to establish its motion.

Display — The plan position presentation of ARPA data with radar data.

Manual — Relating to an activity which a radar observer performs, possibly

	with assistance from a machine.
Automatic	Relating to an activity which is performed wholly by a machine.

Annex 2 Operational scenarios

For each of the following scenarios predictions are made at the target position defined after previously tracking for the appropriate time of one or three minutes:

Scenario 1

Own ship course	000°
Own ship speed	10 knots
Target range	8 nautical miles
Bearing of target	000°
Relative course of target	180°
Relative speed of target	20 knots

Scenario 2

Own ship course	000°
Own ship speed	10 knots
Target range	1 nautical mile
Bearing of target	000°
Relative course of target	090°
Relative speed of target	10 knots

Scenario 3

Own ship course	000°
Own ship speed	5 knots
Target range	8 nautical miles
Bearing of target	045°
Relative course of target	225°
Relative speed of target	20 knots

Scenario 4

Own ship course	000°
Own ship speed	25 knots
Target range	8 nautical miles
Bearing of target	045°
Relative course of target	225°
Relative speed of target	20 knots

Annex 3 Sensor errors

The accuracy figures quoted in paragraph 3.8 are based upon the following sensor errors and are appropriate to equipment complying with the performance standards for shipborne navigational equipment.

Note σ means standard deviation.

Radar
Target glint (scintillation) (for 200 m length target)
Along length of target σ = 30 metres (normal distribution)
Across beam of target σ = 1 metre (normal) distribution).

Roll-pitch bearing The bearing error will peak in each of the four quadrants around own ship for targets on relative bearings of 045°, 135°, 225° and 315° and will be zero at relative bearings of 0°, 90°, 180° and 270°. This error has a sinusoidal variation at twice the roll frequency.

For a 10° roll the mean error is 0.22° with a 0.22° peak sine wave superimposed.

Beam shape	Assumed normal distribution giving bearing error with σ = 0.05°
Pulse shape	Assumed normal distribution giving range error with σ = 20 metres
Antenna backlash	Assumed rectangular distribution giving bearing error ±0.5° maximum.

Quantization

Bearing	Rectangular distribution ±0.01° maximum
Range	Rectangular distribution ±0.01 nautical miles maximum.

Bearing encoder assumed to be running from a remote synchro giving bearing errors with a normal distribution σ = 0.03°.

Gyro compass
Calibration error 0.5°.
Normal distribution about this with σ = 0.12°.

Log
Calibration error 0.5 knots.
Normal distribution about this with 3σ = 0.2 knots.

5.11.4 UK Statutory Instrument 1984, no. 1203

Part VIII Automatic radar plotting aid installation

Automatic radar plotting aid performance standards

39 Every automatic radar plotting aid installation required to be provided shall comply with the performance standards adopted by the organization and shall, in addition, comply with the relevant performance specifications issued by the Department of Transport.

Siting and other requirements of automatic radar plotting aid installations

40(1) Where the automatic radar plotting aid installation is provided as an additional unit to a radar installation it shall be sited as close as is practicable to the display of the radar with which it is associated.
(2) Where the automatic radar plotting aid installation forms an integral part of a complete radar system, that radar system shall be regarded as one of the radar installations required by regulation 3(4)(b) and accordingly shall comply with the relevant requirements of part III of these regulations.
(3) The automatic radar plotting aid installation shall be interconnected with such other installations as is necessary to provide heading and speed information to the automatic radar plotting aid.

Use of an automatic radar plotting aid to assist in the radar watch

41 When at any time on or after 1 September 1985 a ship required to be fitted with an automatic radar plotting aid is at sea and a radar watch is being kept on the automatic radar plotting aid, the installation shall be under the control of a person qualified in the operational use of automatic radar plotting aids, who may be assisted by unqualified personnel.

Qualifications of observers using an automatic radar plotting aid to assist in keeping a radar watch

42 For the purpose of Regulation 41 of these regulations, a person shall be deemed to be qualified in the operational use of automatic radar plotting aids if he holds:

(a) A valid Electronic Navigation Systems Certificate granted by the Secretary of State; or
(b) A valid Navigation Control Certificate granted by the Secretary of State; or
(c) A valid Automatic Radar Plotting Aids Certificate granted by the Secretary of State; or
(d) A certificate recognized by the Secretary of State as being equivalent to any of the certificates mentioned in (a), (b) or (c).

5.11.5 UK DTp Merchant Shipping Notice M.1158: The Use of Radar and Electronic Aids to Navigation

3 The use of radar (including ARPA)

General

3.1 Collisions have been caused far too frequently by failure to make proper use of radar; by altering course on insufficient information; and by maintaining too high a speed, particularly when a close-quarters situation is developing or is likely to develop. It cannot be emphasised too strongly that navigation in restricted visibility is difficult and great care is needed even though all the information which can be obtained from radar observation is available. Where continuous radar watchkeeping and plotting cannot be maintained, even greater caution must be exercised.

Interpretation

3.2 It is essential for the observer to be aware of the current quality of performance of the radar set (which can be most easily ascertained by a performance monitor) and to take account of the possibility that small vessels, small icebergs and similar floating objects may escape detection.

3.3 Echoes may be obscured by sea or rain clutter. Adjustment of controls to suit the circumstances will help, but will not completely remove this possibility.

3.4 Masts and other obstructions may cause shadow sectors on the display.

Plotting

3.5 To estimate the degree of risk of collision with another vessel it is necessary to forecast her closest point of approach. Choice of appropriate avoiding action is facilitated by knowledge of the other vessel's course and speed, and one of the simplest methods of estimating these factors is by plotting. This involves knowledge of own ship's course and distance run during the plotting interval.

Choice of range scale

3.6 Although the choice of range scales for observation and plotting is dependent upon several factors such as traffic density, speed of the observing ship and the frequency of observation, it is not generally advisable to commence plotting on short range scales. In any case advance warning of the approach of other vessels, or changes in traffic density, should be obtained by occasional use of the longer range scales. This advice applies particularly when approaching areas of expected high traffic density when information obtained from the use of the longer range scales may be an important factor in deciding on a safe speed.

Appreciation

3.7 A single observation of the range and bearing of an echo can give no indication of the course and speed of a vessel in relation to one's own. To estimate this a succession of observations at known time intervals must be made.

3.8 Estimation of the other ship's course and speed is only valid up to the time of the last observation and the situation must be kept constantly under review, for the other vessel, which may or may not be on radar watch, may alter her course or speed. Such alteration in course or speed will take time to become apparent to the radar observer.

3.9 It should not be assumed that because the relative bearing is changing there is no risk of collision. Alteration of course by one's own ship will alter the relative bearing. A changing compass bearing is more to be relied upon. However, this has to be judged in relation to range, and even with a changing compass bearing a close-quarters situation with a risk of collision may develop.

3.10 Radar should be used to complement visual observation in clear weather to assist in the assessment of whether risk of collision exists or is likely to develop. It also provides accurate determination of range to enable action taken to avoid collision to be successful, bearing in mind the manoeuvring capabilities of own ship.

Clear weather practice

3.11 It is important that shipmasters and others using radar should gain and maintain experience in radar observation and appreciation by practice at sea in clear weather. In these conditions radar observations can be checked visually and misinterpretation of the radar display or false appreciation of the situation should not be potentially dangerous. Only by making and keeping themselves familiar with the process of systematic radar observation, and with the relationship between the radar information and the actual situation, will officers be able to deal rapidly and competently with the problems which will confront them in restricted visibility.

Operation

3.12 If weather conditions by day or night are such that visibility may deteriorate, the radar should be running, or on standby. (The latter permits operation in less than 1 minute, whilst it normally takes up to 4 minutes to operate from switching on.) At night, in areas where fogbanks or small craft or unlighted obstructions such as icebergs are likely to be encountered, the radar set should be left permanently running. This is particularly important when there is any danger of occasional fogbanks, so that other vessels can be detected before entering the fogbank.

3.13 The life of components, and hence the reliability of the radar set, will be far less affected by continuous running than by frequent switching on and off, so that in periods of uncertain visibility it is better to leave the radar either in full operation or on standby.

Radar watchkeeping

3.14 In restricted visibility the radar set should be permanently running and the display observed, the frequency of observation depending upon the prevailing circumstances, such as the speed of one's own ship and the type of craft or other floating object likely to be encountered.

The use of parallel index techniques as an aid to navigation by radar

3.15 General Investigations of casualties involving the grounding of ships, when radar was being used as an aid to navigation, have indicated that a factor contributing to the grounding was the lack of adequate monitoring of the ship's position during the period leading up to the casualty. Valuable assistance to position monitoring in relation to a predetermined navigation plan could have been given in such cases if the bridge personnel had used the techniques of parallel index plotting on the radar display. Such techniques should be practised in clear weather during straightforward passages, so that bridge personnel become thoroughly familiar with this technique before attempting it in confined difficult passages, or at night, or in restricted visibility.

3.16 The basic principle of parallel index plotting can be applied to either a stabilized relative motion display or a *ground* stabilized true motion display.

3.17 On a stabilized relative motion display the echo of a fixed object will move across the display in a direction which is the exact reciprocal of the *course made good* by own ship at a speed commensurate to that of own ship over the ground. A line drawn from the echo of the fixed object tangential to the variable range marker circle set to the desired passing distance will indicate the forecast track of the echo as own ship proceeds. If the bearing cursor is set parallel to this track it will indicate the course to make good for own ship. Any displacement of the echo from the forecast track will indicate a departure of own ship from the desired course over the ground.

3.18 On a ground stabilized true motion display, the echo of a fixed object will remain stationary on the display and the origin of the display (own ship) will move along the course made good by own ship at a speed commensurate to that of own ship over the ground. A line should be drawn from the echo of the fixed object tangential to the variable range marker circle set to the desired passing distance. If the electronic bearing marker is set parallel to this line it will indicate the course to be made good by own ship over the ground. Any departure of own ship from this course will be indicated by the drawn line not being tangential to the variable range marker circle. (The variable range marker circle should move along the line like a ball rolling along a straight edge.)

3.19 The engraved parallel lines on the face of the bearing cursor can be used as an aid to drawing the index lines on, say, a reflection plotter and to supplement the bearing cursor.

3.20 It should be borne in mind that parallel indexing is an aid to safe navigation and does not supersede the requirement for position fixing at regular intervals using all methods available to the navigator.

When using radar for position fixing and monitoring, check:

(a) The radar's overall performance
(b) The identity of the fixed object(s)
(c) Gyro error and accuracy of the heading marker alignment
(d) Accuracy of the variable range marker, bearing cursor and fixed range rings
(e) On true motion, that the display is correctly ground stabilized.

3.21 It must be remembered that parallel index lines drawn on the reflection plotter are applicable to one range scale only. In addition to all other precautions necessary for the safe use of the information presented by radar, particular care must therefore be taken when changing range scales.

3.22 Some ARPA equipments provide a facility to generate synthetic lines which may be used as aids to parallel indexing techniques. These can be particularly useful where changes of range scale make the use of reflection plotters inappropriate.

Regular operational checks

3.23 Users of radar are reminded that frequent checks of the radar performance should be made to ensure that it has not deteriorated.

3.24 The performance of the radar equipment should be checked before sailing and at least once every four hours whilst a radar watch is being maintained. This should be done by using the performance monitor where fitted.

3.25 It is recommended that checks of the heading alignment should be made periodically to ensure that correct alignment is maintained. The following procedures are recommended:

(a) *Centring the trace* Each time the radar is switched on, and at the commencement of each watch when the radar is used continuously and whenever bearings are to be measured, the observer should check that the trace is rotating about the centre of the display and should, if necessary, adjust it (the centre of the display is the centre of rotation of the bearing scale cursor).

(b) *Aligning the heading marker and radar antenna* Visually aligning the radar antenna along what appears to be the ship's fore-and-aft line is not a sufficiently accurate method of alignment. The following procedure is recommended for accurate alignment:

(i) Adjust accurately the centre of rotation of the trace. Switch off azimuth stabilization. Rotate PPI until the heading marker lies 0° on the bearing scale.

(ii) Select an object which is conspicuous but small visually and whose echo is small and distinct and lies as nearly as possible at the maximum range scale in use. Measure simultaneously the relative visual bearing of this object and the bearing on the PPI relative to the bearing scale. It is important that the visual bearing is taken from a position near the radar antenna in plan. Repeat these measurements twice at least and calculate the mean difference between bearings obtained visually and by radar.

(iii) If an error exists, rotate the PPI picture until the radar bearing is the same as the visual bearing.

(iv) If necessary adjust the heading marker contacts in the antenna unit to return the heading marker to 0° on the bearing scale.

(v) Take simultaneous visual and radar bearings as in (iii) above to check the accuracy of the alignment.

3.26 Alignment of the heading marker or correcting the alignment on a ship berthed in a dock or harbour, or using bearings of a target that has not been identified with certainty both by radar and visually, can introduce serious bearing errors. The procedure for alignment of heading marker should be carried out on clearly identified targets clear of a confusion of target echoes. The alignment should be checked at the earliest opportunity.

Automatic plotting aids (ARPA)

3.27 In addition to the advice given above, and the instructions given in the appropriate operating manual, users of ARPA should ensure that:

(a) The test programs (where fitted) are used to check the validity of the ARPA data.

(b) The performance of the radar associated with the ARPA is at its optimum.

(c) The heading and speed inputs to the ARPA are satisfactory. Correct speed input, where provided by manual setting of the appropriate ARPA contol(s) or by an external input, e.g. Doppler log, is vital for correct processing of ARPA data. Serious errors of output data can arise if heading and speed inputs to the ARPA are incorrect. In this context users should be aware of the possible hazards of using a ground stabilized mode of ARPA display when information on the movement of other ships is being used to assess a potential collision risk, particularly in areas where significant currents and/or tidal streams exist. On some ships, should the master gyro fail, the transmitting magnetic compass can operate the gyro repeaters and provide heading information to other equipment, including ARPAs and true motion radars. Users should bear in mind that in this mode the errors involved are magnetic and should be ascertained and applied accordingly.

(d) The use of audible operational warning signals to denote that a target has closed on a range, or transits a zone chosen by the observer, does not relieve the user from the duty to maintain a proper look-out by all available means. Such warning devices, when the ARPA is operating in an automatic acquisition mode, should be used with caution especially in the vicinity of small radar inconspicuous targets.

Users should familiarize themselves with the effects of error sources on the automatic tracking of targets by reference to the ARPA operating manual.

6 The Decca Navigator

6.1 System introduction

The Decca Navigator is a medium range electronic position fixing system based on hyperbolic principles. A chain normally comprises a master transmitting station and three slaves (Figure 6.1). Signals received aboard the vessel are compared in phase; the phase difference between master and slave is a measure of the vessel's position within a lane. Until recently, position was indicated by means of Decometer readings. These readings are plotted on a special chart, overlaid with a series of hyperbolic lattices. The lattice charts can be used normally for navigation as they contain all the necessary navigational data (unlike the plotting charts produced by some governments) and are kept corrected by means of Notices to Mariners.

It is important to remember that it is only the fractional or decimal pointer which is driven (as the vessel

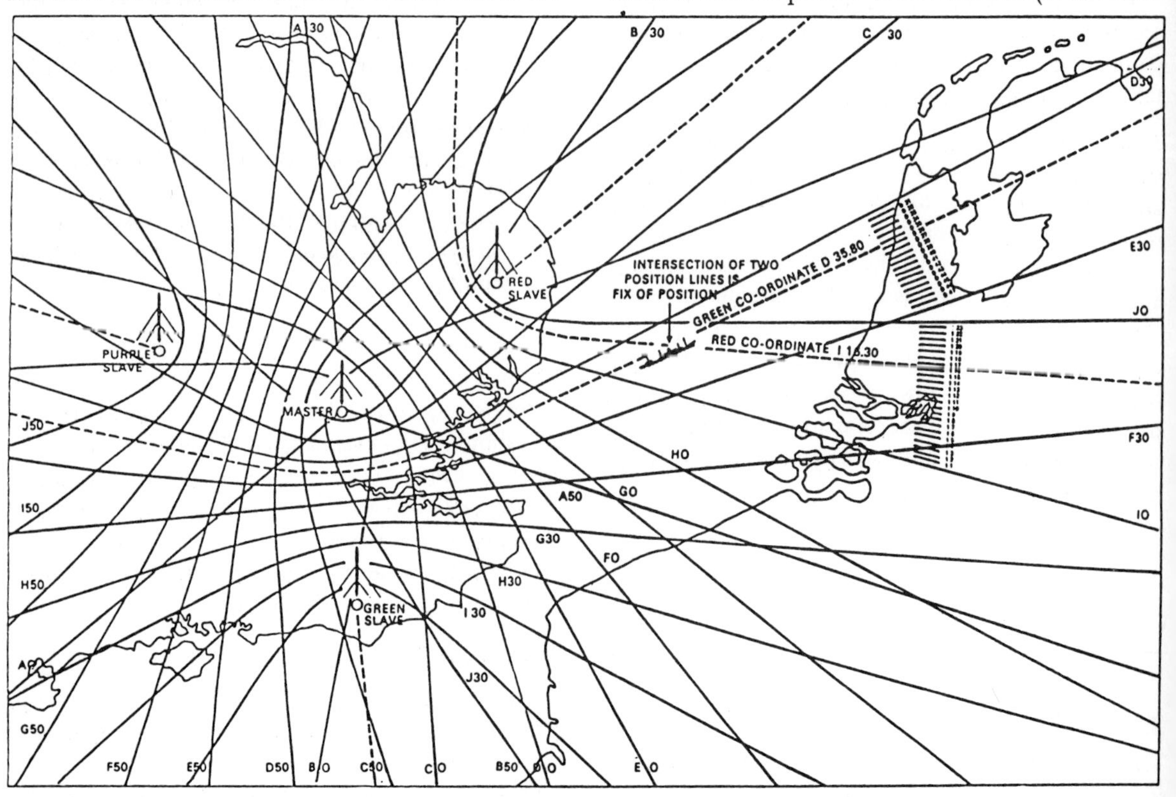

Figure 6.1 Typical Decca chain layout, showing fix plotted from two LOP readings

crosses the lanes) by the incoming signal. The lane counter is 'clocked' on (or off) one lane as each lane is crossed, and the zone is clocked on (or off) one zone as each zone is crossed. It is therefore essential to set up the receiver in some known position. The receiver then counts the number of lanes and zones crossed as the vessel travels along. Any error in the setting of the original lane and zone numbers will remain until the wrongly set Decometer is corrected.

The range of a chain is some 250 nm by night and some 400 nm by day from the master, although atmospheric conditions can both reduce and extend this range.

In recent times, Decca have moved away from lattices on overprinted charts and now incorporate a co-ordinate converter in the display. This converts the position derived from the incoming signal to latitude and longitude such that it can be plotted on an ordinary navigational chart.

Although receivers using the Decca transmissions are now being manufactured by numerous companies under a variety of trade names, the fundamental principles of operation are common to all.

Note Of the decometer type, only the Decca Navigator Mark 21 is still in use; other models (i.e. Marks 5 and 12) should by now have been withdrawn from service.

Information regarding the system, its operation and the services provided is contained in *THE DECCA NAVIGATOR – MARINE DATA SHEETS*. This volume is still required when using the Decca Navigator Marks 52 and 53 as it contains the data sheets giving the fixed and variable errors which must be allowed for when determining position. These receivers have their own user's manuals.

6.2 Setting-up procedure

6.2.1 *The Decca Navigator Mark 21*

This is the commercial model commonly available at sea today, and so the setting-up procedure will be the same for the vast majority of sets:

1 Decide from the chart which chain is to be used and set the number and letter on the chain selectors.
2 Using the best estimate of the vessel's position on the Decca lattice chart, note the zone letter and lane number, ready for setting up each Decometer.
3 Switch from off to lock 1 (Decometer pointers can be prevented from rotation by keeping the hold button pressed until the receiver has locked to the master signal). Successful locking is indicated by 'steady' and bright illumination of the lock lamp together with steady fraction pointers (hold having now been released).
4 Switch from lock 1 to ref. Decometer fraction pointers should move to indicate zero. If not, the zero controls should be used to set the respective fraction pointer to zero.
5 Switch to lock 2 and the condition described in 3 above should be restored.
6 Switch to op. The LI (lane indicator) should now work through the sequence master, red, green, purple, nothing, over a period of 20 seconds. After the break, the first reading to come up on the LI is the master, and the reading should lie within the tolerance range 23.8, 23.9, 00.0, 00.1, 00.2. If it does not, then press and release the LI zero button and confirm.
7 Set the Decometers with the zone letters and lane numbers already noted in 2 above. Cross-check the LI readings with the Decometer dials.

Note:

1 The Decometer reading must lie within the range: red 0 to 23.9; green 30 to 47.9; purple 50 to 79.9.
2 Because the fraction pointers change most rapidly, they should be read first, then the lane numbers, then the zone letter.
3 When fixing on the chart, identify the zone, then the lane and finally the fraction.

Periodic checks should be made to ensure:

1 That LI and Decometers agree.
2 That the first LI reading in each sequence is within 23.8 to 00.2.
3 That Decometer zeros are not in error (by switching to ref.).

The procedures for changing chain, interchain fixing and use at the fringes of the coverage area or where excessive skywave is present, are all well covered in the *Decca Operating Instructions*. It should be noted that under normal conditions the Decometer fraction pointers will provide best accuracy and should always be used. Under extreme skywave conditions, if the LI readings are within tolerance but disagree with the Decometer fractional readings, then the LI readings may provide the more accurate position.

6.2.2 The Decca Navigator mark 53

The mark 53 is a four channel integrating narrow band receiver incorporating both Normal and Lane Identification pattern positioning with full world-wide chain coverage. Long range and marginal area reception is facilitated by the provision of both electronic and dead reckoning rate aiding techniques.

Fixing accuracy is improved by a 'cross chain' fixing capability, which, in adjacent Decca chain areas, significantly improves the position line intercept.

The mark 53 has the ability to display vessel position in latitude/longitude format to 0.01 of a minute, in Decca co-ordinate format to 0.01 of a Decca lane or as distance and bearing from a defined way point location to a resolution of 0.1 nautical mile and one degree respectively.

One hundred way point locations can be inserted, either by defining their latitude/longitude, Decca co-ordinates or distance and bearing from the vessel, or another defined waypoint location.

'Craft position' waypoints can be inserted automatically by means of the 'Auto WP' control. This feature is of particular value to fishing vessels, survey vessels and in naval operations. True speed and course-made-good is calculated and displayed while up to nine route plans, based on a series of defined waypoints, can be stored to provide rhumb line steering guidance information and autopilot control.

Off-track distance, distance to go and ETA at waypoints can be displayed on demand. Comprehensive digital and analogue output interfaces allow the mark 53 to operate with integrated ships' management systems, colour video plotters, automatic electro-mechanical track plotters, remote displays, automatic pilot control systems and printers.

The mark 53 incorporates large, clear Liquid Crystal Displays for positional information read out. This is supplemented by a separate 40 character, alpha-numeric display which provides the user with setting-up and function prompts, reception status and other valuable navigation data.

The performance of the mark 53 is such that it will provide the optimum accuracy obtainable from the Decca Navigator System under all reception conditions and at all ranges.

(Racal-Decca Marine Navigation Limited)

6.3 Fixed and variable errors

As a result of extensive testing by Decca it has been found that, at certain fixed positions, observed Decometer readings:

1 Differ from those which are expected
2 Vary in a more or less predictable manner over periods of 24 hours and also with the time of year; these are variable errors.

This information is collated and published in the marine data sheets.

6.3.1 Fixed errors

These are published in the form of diagrams (see Figures 6.2, 6.3 and 6.4.) The values shown are in hundredths of a lane unit. Those which are circled should be subtracted from the reading as obtained from the Decometer. There is a separate diagram for each slave (colour).

With the Decca mark 52 and 53 there is a facility to feed in fixed errors. These may or may not be applied. If the correction for the fixed errors has been applied then an asterisk and the letter F will appear alongside the displayed reading, whether it be lat./long. or Decca lane numbers. Since fixed errors vary with position, they should be updated as the vessel proceeds. If corrections are being applied, then after an hour an alarm will sound to invite the operator to update the entered values.

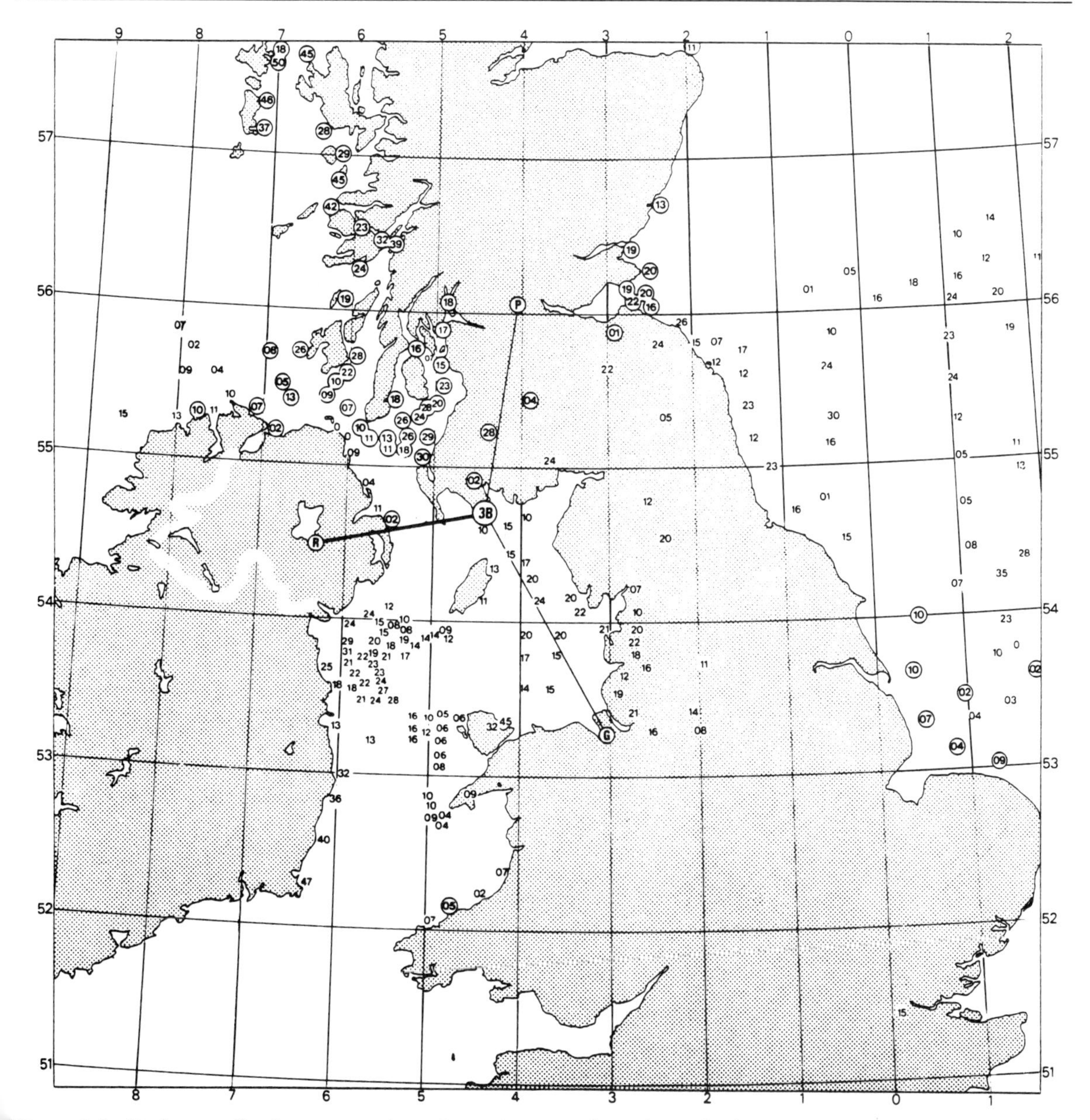

Figure 6.2 Red pattern fixed error corrections. Corrections to apply to observed red Decometer readings to overcome fixed errors. Values shown are in hundredths of a lane unit. Figures encircled should be subtracted; figures not encircled should be added

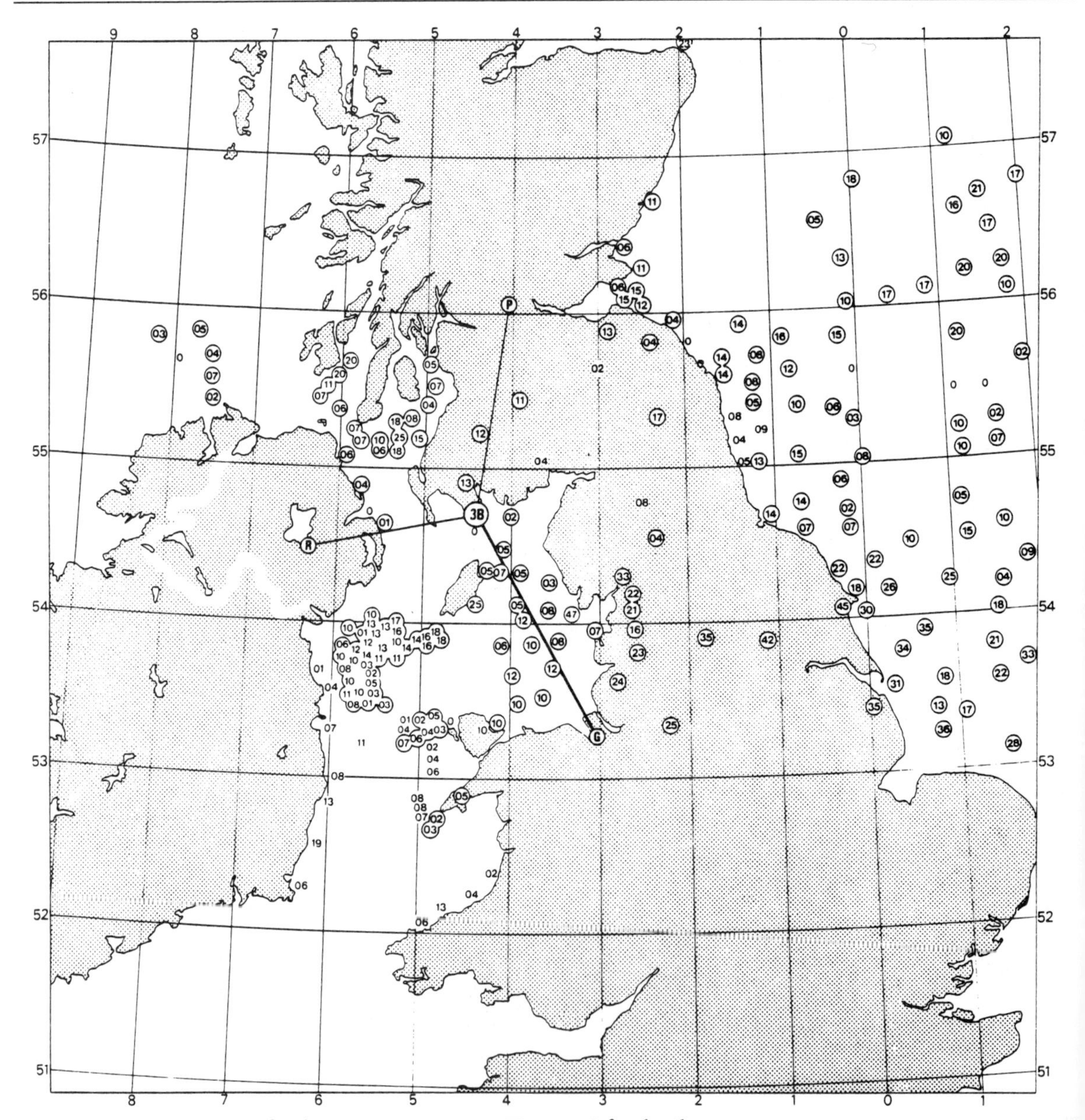

Figure 6.3 Green pattern fixed error corrections (see Figure 6.2 for details)

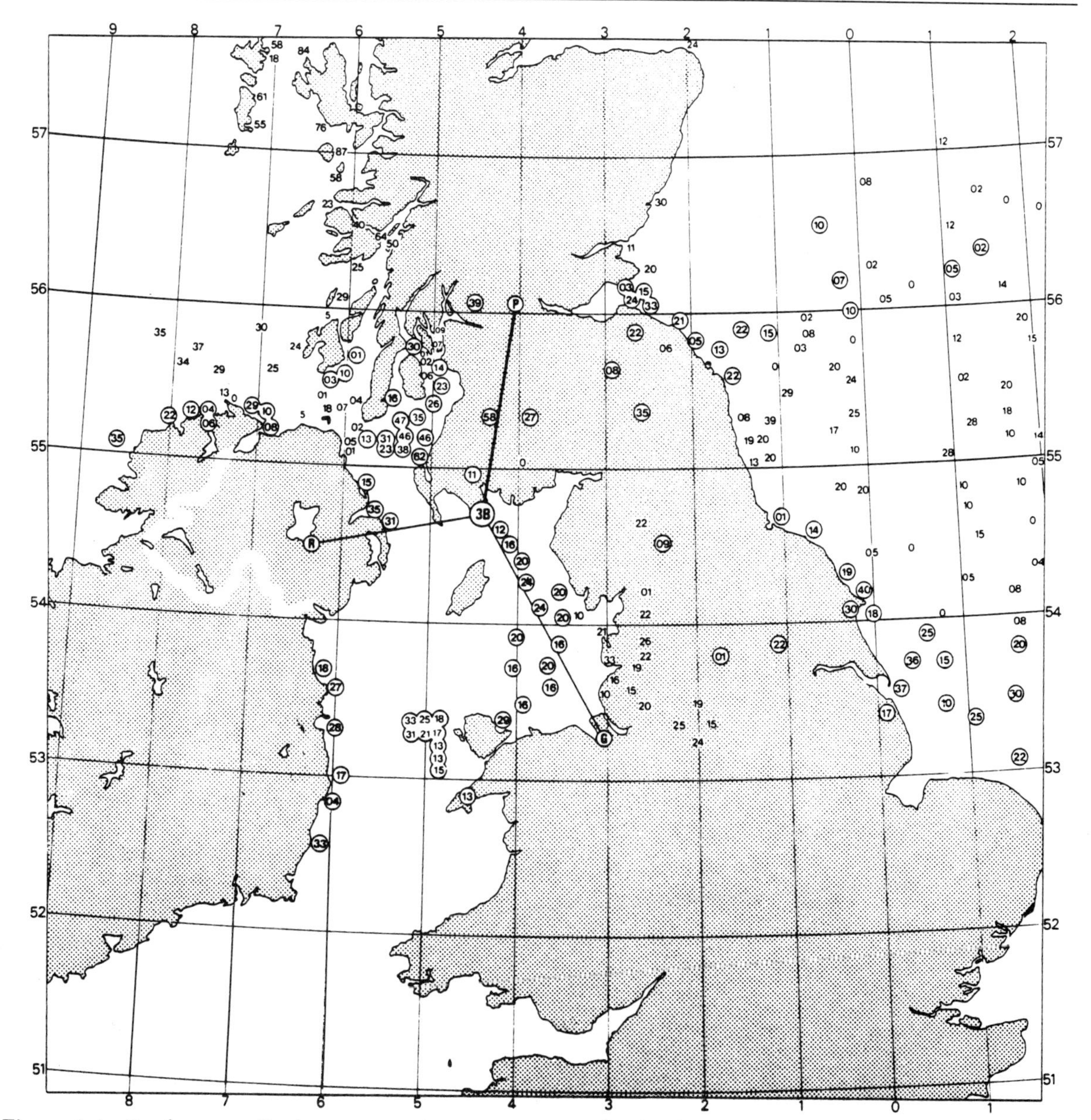

Figure 6.4 Purple pattern fixed error corrections (see Figure 6.2 for details)

6.3.2 *Variable errors*

These are published in the form of a reliability circle, centred on the plotted position, within which the ship's true position will lie. The information is obtained as follows:

1 Enter the time and season factor diagram (Figure 6.5) with month and time as arguments. Extract the Decca period (full daylight, halflight, etc.).
2 If the Decca period is full daylight, enter the full daylight coverage and accuracy diagram (Figure 6.6) with the approximate position. Read the variable error directly.
   If the Decca period is not full daylight, enter the predicted coverage and accuracy diagram (Figure 6.7) with the approximate position. Extract the contour letter (a, b, etc.).
3 Enter the random fixing error table (Table 6.1) with the Decca period and contour letter as arguments and extract the random fixing error.

This is the radius of a circle centred on the plotted position within which the ship can be expected to be with 68 per cent probability, i.e. on two out of three occasions.

The ship can be expected to be within twice the tabulated value of the position marked on the chart with 95 per cent probability, i.e. on nineteen out of twenty occasions.

Table 6.1 Random fixing errors at sea level in nautical miles: 68 per cent probability level

Decca Period	*Contour*					
	a	*b*	*c*	*d*	*e*	*f*
Half light	<0.10	<0.10	<0.10	0.13	0.25	0.50
Dawn/dusk	<0.10	<0.10	0.13	0.25	0.50	1.00
Summer night	<0.10	0.13	0.25	0.50	1.00	2.00
Winter night	0.10	0.18	0.37	0.75	1.50	3.00

The table gives the variable fixing errors not likely to be exceeded in more than one case out of three readings.

6.4 Fixing position

Having obtained the Decometer readings as advised in Section 6.2.1 and corrected for fixed errors as in Section 6.3.1, the lines of position should now be laid off on the chart. Using a Decca ruler for interpolation, a position line should be drawn in the vicinity of the vessel's position. If all three colours are available (i.e. red, green and purple lattices are printed on the chart in that area) then all three position lines should be

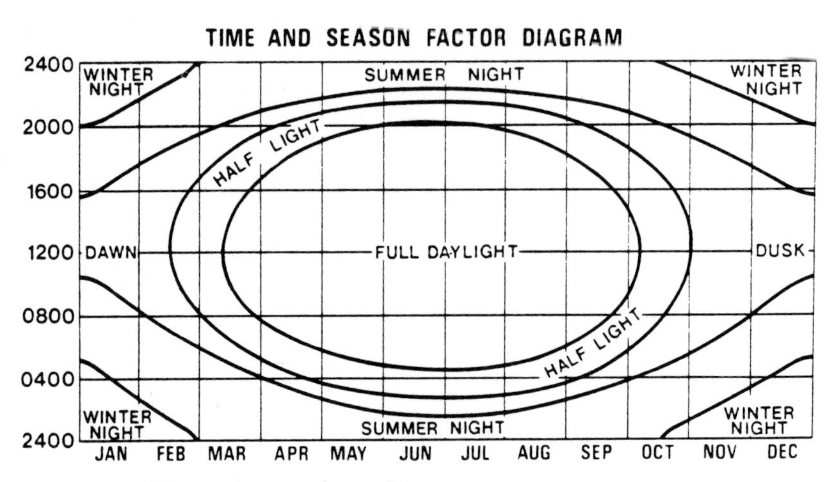

Figure 6.5 Time and season factor diagram

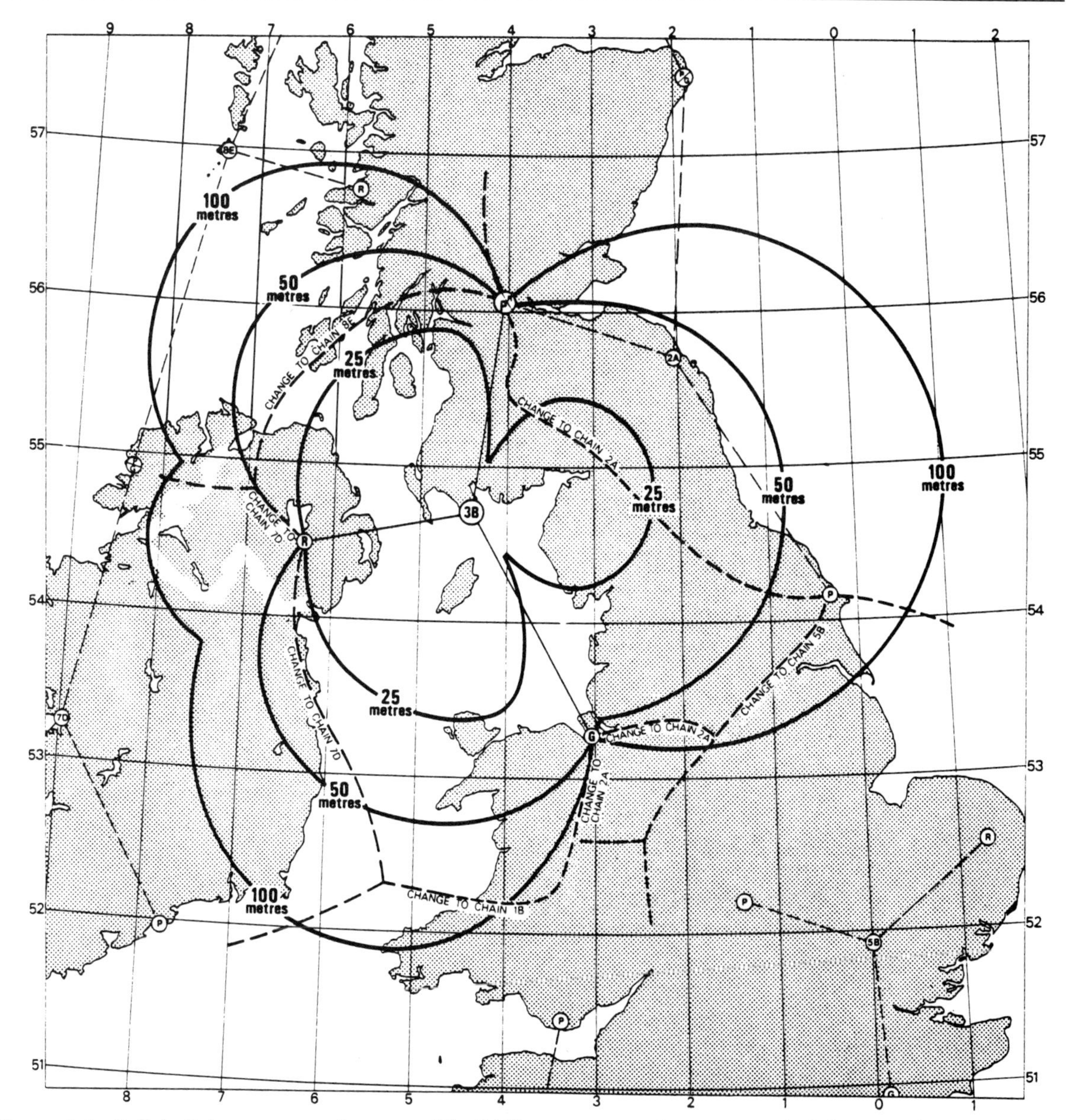

Figure 6.6 Full daylight coverage and accuracy. The full line contours enclose areas in which fix repeatability errors will not exceed the distances shown on 68 per cent of occasions during full daylight conditions of time

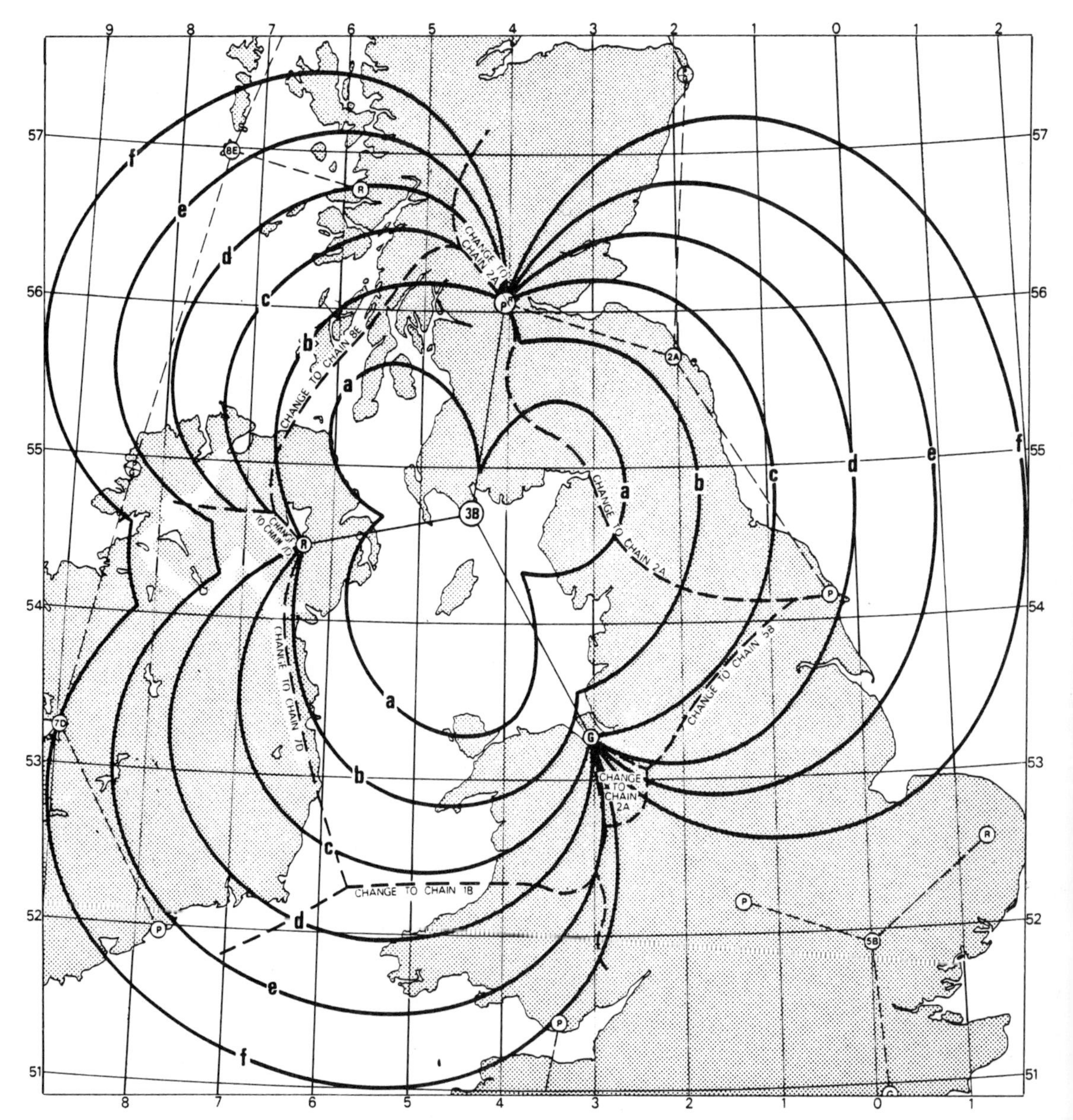

Figure 6.7 Predicted coverage and accuracy diagram (68 per cent probability level) for times other than full daylight

drawn. In the Decca data sheets there is a diagram (Figure 6.8) indicating preferred colours; these are only intended for use when a two pattern fix technique is employed and to indicate the best single line of position. This diagram should not be used as an excuse for only using two lines of position when three lines of position are available.

Example
At 1620 hours on 20 February in DR position lat. 54° 30′ N, long. 1° 15′ E and using chain 3B, the Decometer readings were as follows: red B22.85; green H31.62 and purple G76.54. Determine, using the marine data sheet extracts (Figures 6.2 to 6.8):

1 The readings to be plotted on the chart
2 The variable error for 68 and 95 per cent probability (see Table 6.1)
3 The preferred colours.

1

	Red	*Green*	*Purple*
Decometer readings	B22.85	H31.62	G76.54
Fixed error	+ 0.08	− 0.15	+ 0.15
Corrected reading	B22.93	H31.47	G76.69

2 For 68 per cent probability: the vessel will be within a circle of radius 1 nm centred on the plotted position two out of three times. For 95 per cent probability: the vessel will be within a circle of radius 2 nm centred on the plotted position nineteen out of twenty times.
3 In the event of a cocked hat the preferred colours, i.e. the position lines giving the most reliable fix, are green and purple.

When assessing the reliance which can be placed on a position obtained by Decca, the following should be taken into account:

1 The normal navigation criterion, i.e. angles of cut of the position lines (PLs).
2 The number of PLs obtained, i.e. three or, if three are not available, then the two or one preferred by the data sheets.
3 The degree of contamination of the groundwave by the skywave. This might be apparent by observing instability of the Decometer fraction pointers.
4 The knowledge of fixed errors. Areas in the fixed error diagrams without values are not necessarily free from errors; they merely mean that insufficient information has been obtained.
5 Variable errors do not guarantee that the vessel will be within the circle of uncertainty but, by frequently plotting the position in poor reception areas or where skywave is prevalent, some confidence can be achieved.
6 Where a cocked hat is obtained, reference should be made to the diagram indicating preferred colours.

In the final analysis, seamanlike criteria should be used to assess the vessel's 'accepted' position, i.e. biased towards the most reliable position line(s), nearness to danger, etc.

Good navigation practice demands that positions are cross-checked by reference to other position fixing systems.

Regular plotting of position should indicate if lane slip has occurred. Since the incoming signals only drive the Decometer fraction pointers, it is possible under certain conditions for the lane pointers to 'slip' a whole lane. Regular checking of Decometers against LI should indicate whether this has occurred, as will regular plotting of the vessel's position. With the Decca mark 52 and 53 the normal pattern and LI coordinates are compared, and if the difference exceeds 0.5 of a lane an LI alarm will be indicated.

6.5 Navigation warnings

Should any break or disturbance of normal pattern transmissions occur, these events are broadcast to mariners as Decca warnings by WT, RT or NAVTEX (see Section 19.1.3.3) but only from those coastal radio stations serving the area affected by the chain in which the irregularity occurs. Full information on this service is given:

1 In the Decca Navigator Marine Data Sheets
2 In ALRS vol. 2 (NP 282) but more completely in ALRS vol. 3 (NP 283) (see Sections 6.6.1 and 6.6.2).
3 In Admiralty Notices to Mariners, Annual Summary.

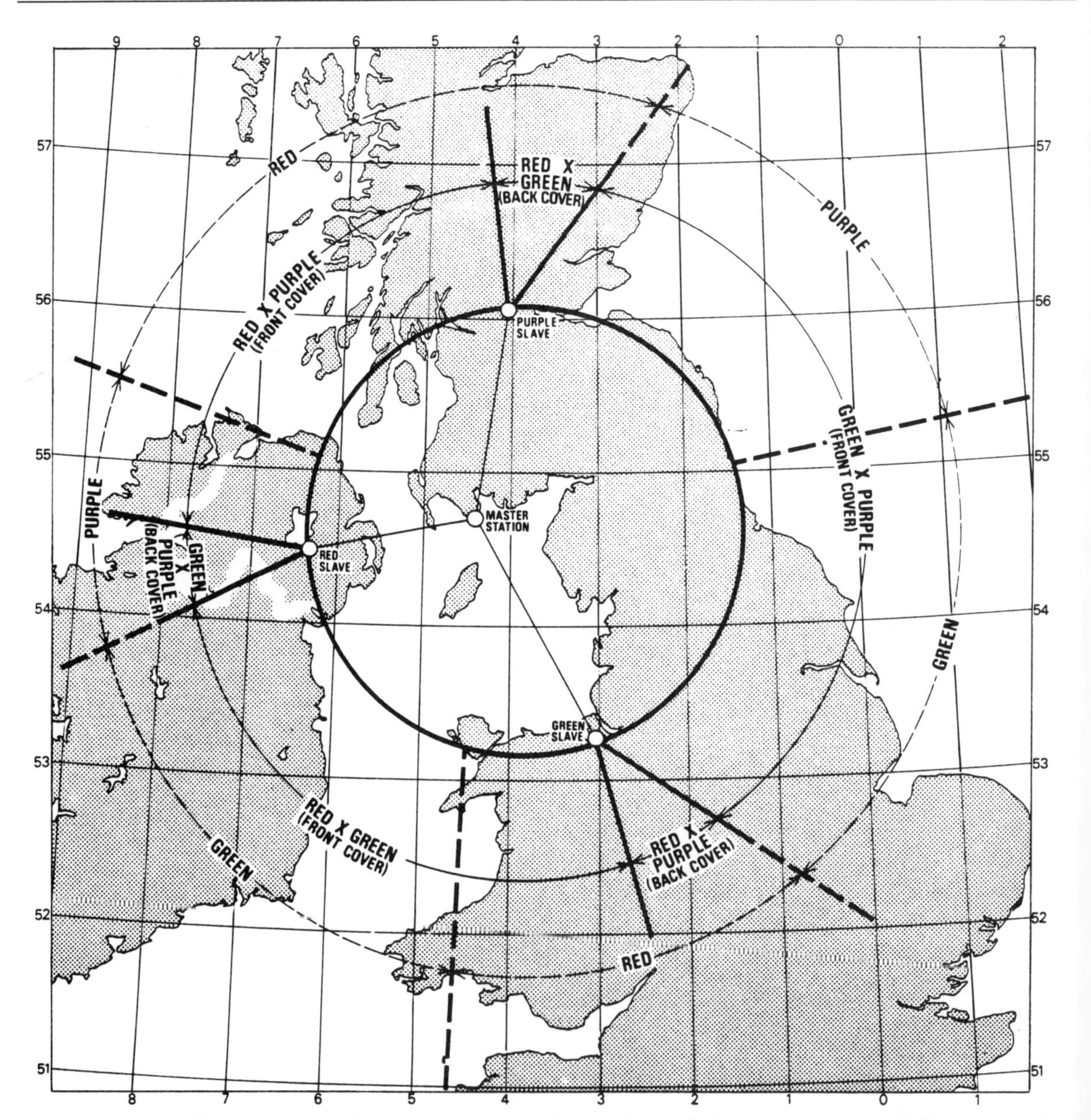

Figure 6.8 Decca lattice colours to be used for obtaining the best fix and the best position line in different parts of the operational area. Inside the inner circle drawn around the chain, the choice of colour will be obvious from inspection of the chart in use; elsewhere, the coverage is divided into sections (shown by full line radial spokes) between which different pairs of colours are used for fixing. Similarly, broken line spokes delineate areas in which different single colours give the best position line

6.6 Extracts from official publications

Within Sections 6.6.1–6.6.3, the paragraph numbering of the original publications has been retained.

6.6.1 Admiralty List of Radio Signals, vol. 3: Radio Navigational Warnings, National Practices

This section contains details of the procedures adopted by various countries for the dissemination of navigational warnings. This information is, in general, of too lengthy a nature to be included with individual station entries.

United Kingdom

Radio navigational warnings, issued by the Ministry of Defence (Navy) or other competent authorities, provide the mariner with immediate information concerning dangers to navigation. Information of a less essential nature or concerning waters within harbour limits, even though it forms the subject of a Notice to Mariners, may not be broadcast.

NAVAREA warnings Long range warnings are broadcast by Portishead (GKA) for NAVAREA I (see Figure 11.1, page 166). Messages are numbered sequentially. The text of the week's warnings together with a list of those in force is included in section III of the weekly Notices to Mariners. A printed sheet is issued weekly, in advance of the Notices to Mariners, containing the NAVAREA I warnings issued during the week and also selected important NAVAREA II, III and IV warnings.

Warnings by NAVTEX Details of the NAVTEX system are given on the previous page of this volume (in ALRS, vol. 3). United Kingdom stations Niton, Portpatrick and Cullercoats broadcast warnings by NAVTEX (for the area designated in Section 6.6.2).

Coastal warnings Coast radio stations broadcast coastal navigational warnings for specific sea regions which are shown on diagram N5 (in ALRS, vol. 3).

Warnings are broadcast at scheduled times by appropriate coast radio stations, and are repeated at scheduled times for as long as they remain in force. Vital or important warnings may be announced at any time on the distress frequencies 500 kHz, 2182 kHz and VHF ch. 16. Warnings are originated by the Ministry of Defence (Navy) and are numbered sequentially in the WZ series. Cancelling messages are not transmitted. A warning will be repeated by a coast radio station at the usual coast radio station charge for enquiry and reply.

Local warnings Some port authorities broadcast warnings relating to their areas of jurisdiction through nearby coast radio stations. Details are given in Annual Notice to Mariners no. 9.

In addition, HM Coastguard broadcasts local warnings relating to hazards which may affect craft in inshore waters outside port and harbour authority limits. These local warnings are broadcast on VHF ch. 67 after an announcement on VHF ch. 16. Warnings are not broadcast at specific broadcast schedules and do not follow a numerical sequence. They will be repeated at the discretion of the originating HM Coastguard station.

Further details of radio navigational warnings are given in the appropriate station entries in this volume, and in Annual Notice to Mariners nos 9 and 13.

Ship reports Vessels encountering dangers to navigation or severe weather conditions should notify other vessels in the vicinity and the nearest coast radio station.

6.6.2 Admiralty List of Radio Signals, vol. 3: NAVTEX

NAVTEX is a navigational telex service broadcasting safety messages on 518 kHz. Transmissions can be received by a ship's radiotelex installation but to gain full benefit from the service dedicated equipment is recommended.

The dedicated equipment comprises a small unit containing a receiver, fixed-tuned to the broadcast frequency, and a printer using 'cash-roll' paper. The equipment is switched on continuously and may be programmed to receive automatically only selected stations and/or categories of message. A microprocessor control ensures that a routine message already received will not be reprinted on subsequent transmissions, and also that messages will not be printed unless the received signal is strong enough to guarantee a reasonable copy.

All messages are prefixed by a four-character group. The first character denotes the identity of the transmitting station. The second character indicates the category of message, whilst the third and fourth characters are serial numbers, from 01 to 99 and then starting again at 01. A serial number 00 denotes urgent traffic and will always be printed, regardless of the programming of the receiving equipment.

The following subject indicator letters are used to show the category of message:

A Navigational warning
B Gale warning
C Ice report
D Initial distress information
E Weather forecast
F Pilot service message
G Decca message
H Loran-C message
I Omega message
J Satnav message
L Oil and gas rig information
Z 'No messages on hand'

NAVAREA I East Atlantic (north of 48°27′N), North Sea, Baltic. Co-ordinator: United Kingdom. Station: Portishead (GKA).

A coordinated NAVTEX service is operational in NAVAREA I from stations with identity letters as follows:

Bodø [B]
Cullercoats [G]
Härnösand [H]
Stockholm (Gislövshammar) [J]
Rogaland [L]
Portpatrick [O]
Netherlands Coast Guard [P]
Reykjavik [R] (pre-operational)
Niton [S]
Oostende [T]
Tallin [U]
Vardø [V]

There are further NAVTEX stations located outside NAVAREA I. Details of these are given in the station entries.

6.6.3 UK DTp Merchant Shipping Notice M.1158: The Use of Radar and Electronic Aids to Navigation

4 The use of the Decca Navigation system

General

4.1 Investigation of casualties by the Department has shown that in some cases the Decca Navigator equipment on board has not been used in a proper manner or that corrections and allowances for errors have not been applied.

4.2 When using the Decca Navigator system it is important that the mariner should appreciate the errors inherent in radio position fixing systems and not accept the readings as being of absolute accuracy without first consulting the Decca data sheets which are available for every installation. It should be borne in mind that normal precautions should not be neglected when using the Decca Navigator system.

4.3 Errors of Decometer readings are generally quite small and may not be of practical significance in the best areas of chain coverage. Nevertheless it must be appreciated that errors are present and can assume significance. To avoid inaccuracies which might lead to dangerous situations the data sheets should always be consulted. There are two types of error to which the system is subject:

(a) *Fixed errors* Fixed errors can be corrected by the application of the pattern corrections indicated on the data sheets. In areas where there is no information about fixed errors the charted Decca lattices should be used with caution, especially near the coast and in restricted waters.
(b) *Variable errors* Variable errors result in inaccuracies for which allowances should be made. As the name implies, the magnitude of the error at a given location is not constant. Diagrams which

give an indication of the accuracy of a fix are included in the data sheets and should always be consulted. The errors given in the tables based on a 68 per cent probability level are not likely to be exceeded on more than one occasion in three. In earlier editions of the data sheets the errors in the most and least accurate directions in the ellipse of error were tabulated for a number of points; contours showing average errors throughout the chain coverage are now provided instead. *The Admiralty List of Radio Signals, Vol. 5*, also provides diagrams indicating the accuracy coverage of Decca chains. The errors given in the diagrams are normally based on the 95 per cent probability level, so that errors quoted should not be exceeded on more than one occasion in twenty.

Lane slip

4.4 Particularly at night there is a possibility of slipping lanes due to interference such as excessive Decca skywave signals, external radio interference or electrical storms. The chance of this happening will be small at short range but will increase as distance from the centre of the chain increases. Other possible causes of lane slip are fouling of the Decca antenna and interruption of the electrical supply for a significant period. The best way of revealing an inaccurate fix due to lane slip is by plotting Decca fixes at frequent intervals and by comparing them with positions found by other navigational methods.

Using Decca lanes as tracks

4.5 In poor visibility there is danger in using Decca lattice lines as tracks along which to steer. Other ships may be doing this in the opposite direction along the same line, with a consequent risk of collision.

4.6 If a Decometer is being used as a compass on a fixed Decca reading, frequent and regular Decca readings of another colour should be taken and the position plotted on the chart. Otherwise there is the danger that lane slip will not be noticed; vessels steering by a single Decca reading are known to have been steered ashore.

7 Loran C

7.1 System introduction

Like the Decca Navigator system, the Loran C is a medium range electronic position fixing system based on hyperbolic principles. In general, the base line and coverage area of a chain is slightly greater than Decca. Most of the chains are operated by the US Government which has indicated its intention to discontinue Loran C transmitting stations, established for military use, that do not serve the North American continent. Chains are now being developed by Saudi Arabia, North West Europe and Japan.

A chain comprises a master and from two to four secondary stations (sometimes referred to as slaves). The difference in arrival time of pulsed transmissions is a measure of the vessel's position within the hyperbolic pattern (Figure 7.1).

Detailed information regarding the transmissions of the various chains is promulgated in the Admiralty List of Radio Signals (ALRS), vol. 2 (NP 282), as are coverage diagrams.

There is a wide variety of on-board Loran C receivers, but in general the information is displayed in one of two ways:

1 As latitude and longitude, in which case it can be plotted in the normal way on a standard navigational chart; or
2 As a time difference or TD (usually two are displayed simultaneously). In this case a special lattice chart is needed. Some of the lattice charts are normal navigational charts with a lattice overlay printed on them. These are kept corrected via Notices to Mariners and so can be used for navigation in the normal way. Other charts are just plotting charts where, after the position lines have been plotted, the position obtained has to be transferred to the navigational chart.

All Loran C stations transmit at the same frequency (100 kHz) and so, in order to distinguish the signals of one chain from those of an adjacent chain, the pulse repetition rate of stations in the one chain is different from that in the adjacent chain. The interval between the pulse transmissions for a particular chain or group of stations is known as the group repetition interval (GRI) and is used to identify or designate the chain. Within the chain, the transmission sequence is always master followed by secondaries in the order W, X, Y, Z. Each secondary's transmission is delayed by carefully calculated amounts such that, no matter where a vessel is within the coverage area, the signals will always arrive in the order in which they are transmitted. In order to distinguish the master transmission from that of the secondaries, the master transmits a group of nine pulses, while each secondary transmission comprises an eight pulse group. The numbering of the lattice lines is lowest on the secondary baseline extension and increases toward the master baseline extension.

7.2 Setting-up procedure

With there being so many manufacturers in the market it is not possible to give a procedure which will satisfy all equipments, but certain operations apply to the majority of sets.

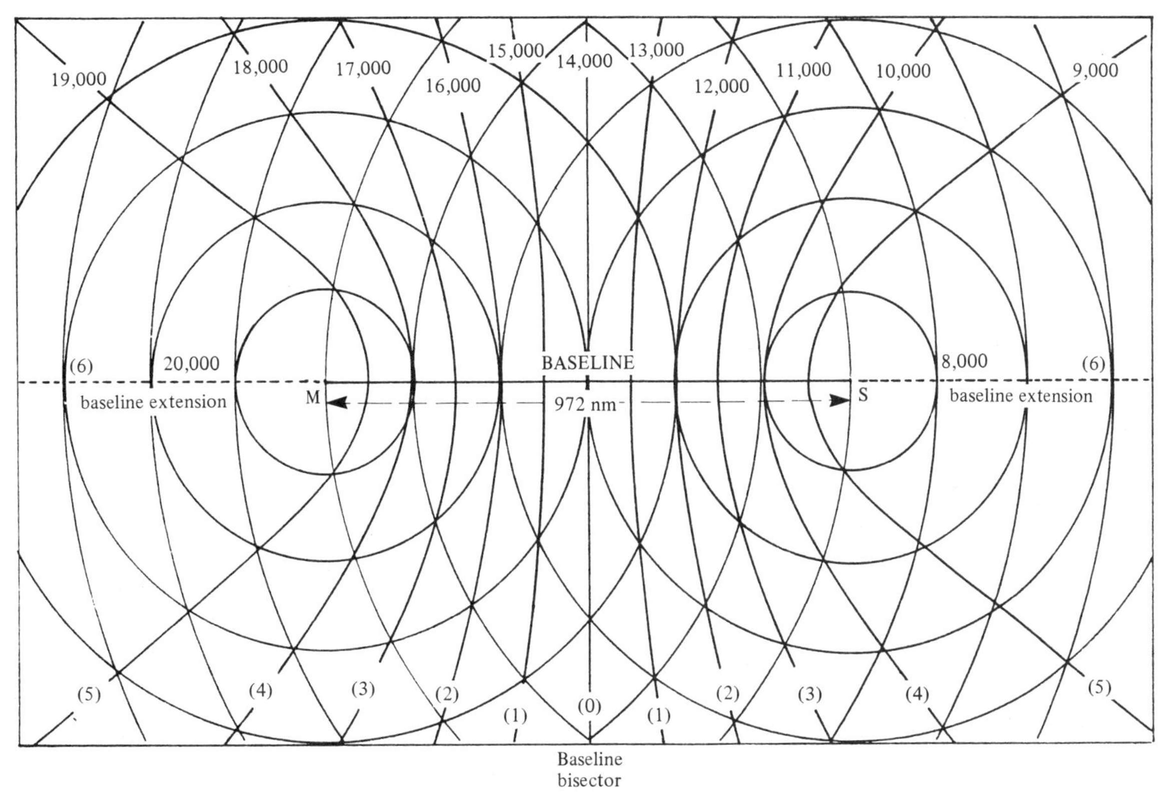

Figure 7.1 Loran C station pair

7.2.1 *Notch filters*

These are provided on most sets to allow the operator to suppress signals at frequencies near to the 100 kHz Loran C transmissions. Interfering signals may come from equipment already on board the vessel, while others may be inherent in the area through which the vessel is passing. In both cases it is best if the source of the signals can be removed. For signals generated ashore this is not the task of the ship, but such interference should be reported. Setting the notch filters involves determining the frequencies at which the interference is occurring and setting the filters (or suppressors) to those frequencies. The procedure usually involves some form of tuning dial (say, from 70 kHz to 130 kHz) and a meter to indicate signal strength. While tuning across the band of frequencies, the signal strength is noted and the filters (usually two are provided) set to eliminate the more serious of the interfering signals which are found. It is important that the filters are not set too close to 100 kHz as they could then eliminate or impair the incoming Loran signals. Interfering signals should be checked periodically and the notch filters adjusted as appropriate.

7.2.2 *Setting the GRI*

Since *all* Loran transmissions are on 100 kHz, the receivers are pretuned to that frequency. The chain selector is set with the group repetition interval (GRI), e.g. 7970; all stations in that chain transmit at intervals of 79 700 μs. Thus the receiver can look among the welter of signals coming in at 100 kHz for signals which recur at that rate, and lock on to and extract them for further processing and finer measurement of the time difference between the master and each of the various secondaries (Figure 7.2).

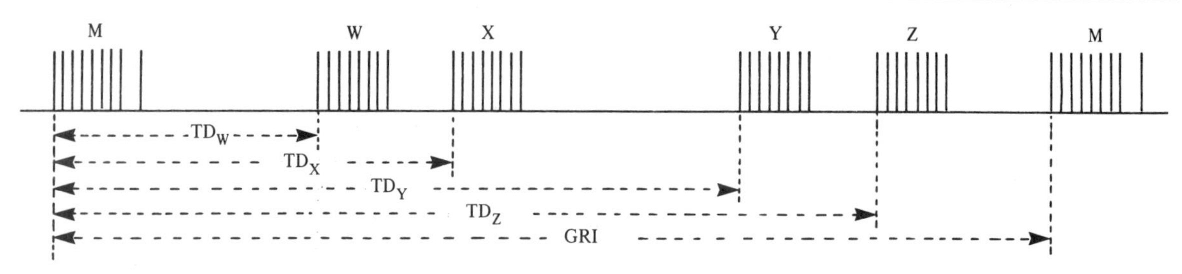

Figure 7.2 Time difference measurement

Information regarding which chain(s) is appropriate for the area can be obtained from one of the publications mentioned in Section 7.1.

7.2.3 Obtaining the TDs

When the operator has switched on, set the notch filters and selected the GRI, the receiver should automatically lock to and commence tracking the Loran signals. The display of the information derived from the incoming signals can take many forms – from simultaneous lat./long. display to a single time difference with others being requested sequentially. Perhaps the most common method at present is to display the two TDs selected by the receiver on the basis of angle of cut of the position lines, signal strength, etc. In any case, it is always possible to override the automatic facilities and select other TDs, change to an alternative chain or 'extend' the range by being prepared to accept less precise matching techniques.

7.3 System faults

In addition to the receiver's internal testing procedure, the overall transmission network is self-testing. When transmission timing falls outside tolerance, or other system faults are detected, a 'blink' procedure is ordered. This was so named because, on the early visual (CRT) displays used for matching signals and measuring time differences, the signals were made to blink or flash in order to warn the operator of a fault in the system. By careful observation of which signals were flashing, exactly where the fault lay could be determined, and unaffected parts of the system might still be used. For example, if the fault lay with only one of the secondaries, then the master and other secondaries could still be used whereas, if the fault was with the master, then no TDs could be obtained. In recent times the original blink procedures have been modified, so suffice it to say that modern receivers recognize blink conditions and warn the operator accordingly.

7.4 Error sources and accuracy

Perhaps the most serious source of error is contamination of the groundwave by skywave. In all cases it is desired to match the phase of the groundwave of the master with the phase of the groundwave of the secondary (Figure 7.3), but towards the edge of the coverage area and at night, anywhere in the coverage area, the skywave can cause problems with the automatic matching and tracking of the two signals. In some areas, skywave can even take over completely. Since the skywave will have travelled by a slightly longer route and therefore have taken a longer time, the measured time difference will be in error. For typical accuracies see Section 10.2. Much has and is being done in modern receivers to eliminate skywave problems, but with automatic receivers it can be very difficult to know when skywave problems are present.

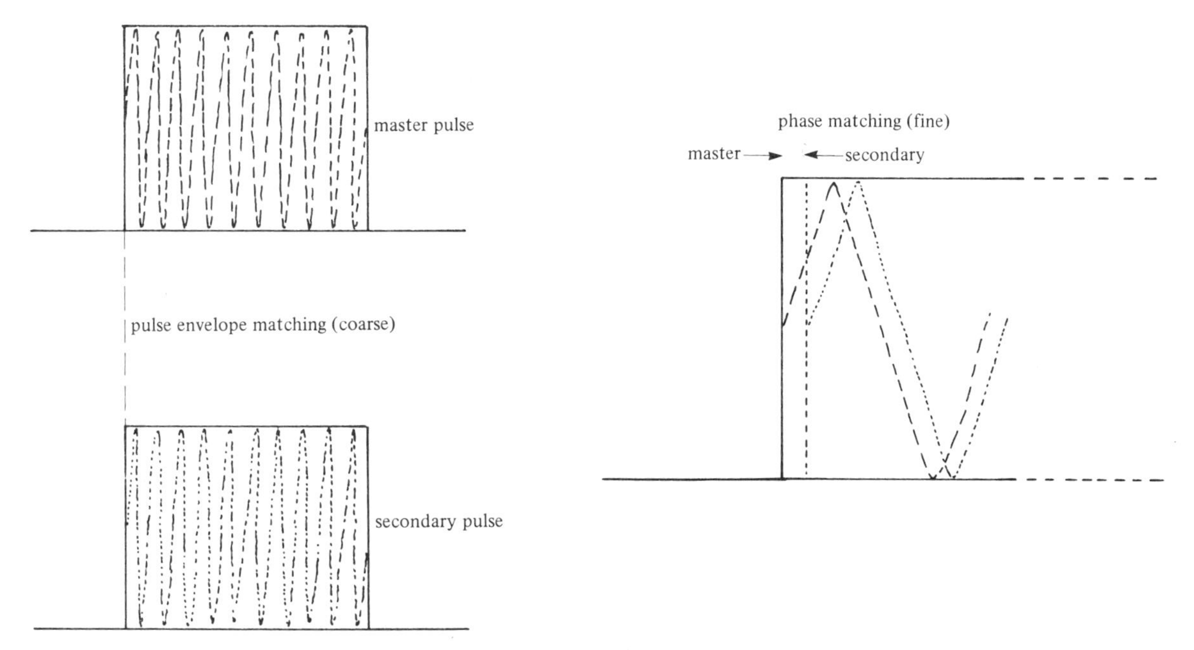

Figure 7.3 Precision time difference measurement by phase matching

7.5 Fixing position

7.5.1 *Using an ocean chart with Loran C lattice*

Under normal circumstances, the TDs taken from the receiver can be plotted on the chart using an interpolating (Decca-type) ruler. The caution on the chart should be noted when working in areas where the lattice lines are pecked, indicating that skywave signals could be present and so positions become less reliable.

Note There may be lines from more than one chain on the same chart.

7.5.2 *Using conventional navigational charts*

In order to convert TD readings to position lines for plotting on a conventional navigational chart, a series of publications (HO pub. no. 221) is needed, one volume for each master/secondary pair which is being used. The method is tedious and should be regarded as the exception rather than the rule, but may be necessary in areas for which no lattice charts have yet been produced.

7.5.3 *Using Loran C plotting charts*

These are lattice charts having a minimum of navigational detail. After the chart has been used to plot the TDs, the position is transferred to the normal navigational chart. The chart is overlaid with a grid, some 4° by 4° in lat./long. At each grid intersection is a series of corrections tabulated for day (D) and night (N) and for ground/skywave (G/S) comparison, which should be applied to the observed TDs before they are plotted.

7.5.4 *Using Loran C navigational charts for coastal areas*

The evolution of these charts has meant that charts now exist (although not necessarily currently on sale) on which:

1 Lattice lines are theoretical.

2 *All* lattice lines have been adjusted by some fixed amount.
3 The adjustment varies over the area of the chart.

The adjustment is to compensate for propagation path delays of the groundwave. These corrections, known as additional secondary phase factor (ASF) corrections, are derived from analytical and empirical models. A Loran C lattice chart should have a note on it indicating whether corrections have been applied or not. Various versions of the note exist, but there is a move to standardize the note so as to indicate whether the applied corrections have been verified by observed data or are purely theoretical. There are tables of ASF corrections which should be applied to TDs before plotting, except where the note on the chart indicates otherwise.

7.5.5 Co-ordinate conversion

Facilities in some receivers will convert the measured TDs to lat./long. and display this information. The accuracy of the displayed latitude and longitude will depend on a number of factors, primarily:

1 The ability to recognize skywave and apply the appropriate corrections
2 The knowledge and application of ASF corrections
3 The geodetic datum to which the position is referred.

The total correction will vary with time and position.

7.5.6 Waypoint navigation

Because of the added uncertainties in co-ordinate conversion, repeatable accuracy is generally higher than geodetic accuracy; that is, if the TDs at particular locations are known then it is possible to navigate from one to the next. The US Coastguard has published the listings from several Loran C waypoint surveys; these give the TDs at positions fixed simultaneously and independently, usually by visual reference to a fixed point. Loran C receivers with a waypoint navigation facility can be programmed with waypoints in TDs which will then give course and distance to steer to the next waypoint, as with normal lat./long. waypoint navigation.

Where waypoint co-ordinates are not published in TD form, they can be determined while on passage by noting the TD and independently fixing the vessel's position, preferably visually, in relation to some land feature.

It has been estimated that, using this method, accuracies of the order of ±40 metres are attainable as opposed to the normal ±0.25 nm generally quoted for the Loran C system.

8 Omega

8.1 System introduction

The Omega navigation system is a long range electronic position fixing system based on hyperbolic principles. There are no 'chains' as with Decca and Loran, but eight transmitting stations provide worldwide coverage working in pairs (any station being able to pair with any other station), thus effectively providing 28 combinations and therefore 28 separate lattices. The transmitting stations have been so placed that at any position there will be at least three lattice lines giving a satisfactory angle of cut (Figure 8.1).

The system is operated and maintained by the US Government, which has indicated its intention to support the system until at least the year 2000.

Features of the system are:

1 Transmissions are at VLF, e.g. 10.2 kHz, and so in effect the waves will follow the earth's surface.
2 The transmission format is quite complex but effectively each station transmits continuous waves at the same frequency but in its own time slot, over a period of 10 seconds. Like Decca, the lattice lines are based on phase measurements.
3 Transmission synchronization is achieved in absolute terms, by use of an atomic clock at each station.
4 Long baselines with effectively straight hyperbolas.

On switch on, the receiver is automatically locked to the Omega format and displays two or three digital readouts derived from preselected transmitters. Propagation corrections from tables must be applied to the readings before plotting on lattice charts, although newer receivers are equipped for co-ordinate conversion.

The system has been regarded as under development while practical experience of operating the system has been gained. When the US Government is satisifed that the declared accuracy requirements (2–4 nm with 95 per cent confidence) can be met, they declare the system operational. At the present time (1992) the North Atlantic, South Atlantic and North Pacific are the only areas which have been declared operational, so the system should be used with caution in other areas.

Charts intended for use with Omega may be either of the plotting type, where the plotted position has to be transferred to a conventional navigation chart, or a conventional chart overprinted with an Omega lattice. Omega stations are designated A to H, and lattice line numbering is arbitrary with the baseline bisector being numbered 900. The width of a lane on the baseline is approximately 8 nm (at a frequency of 10.2 kHz).

8.2 Setting-up procedure

Again, since there are so many different receivers on the market, only general guidance can be provided on setting up.

On switch on, there is a need for the receiver to synchronize with the Omega transmission format. This is usually done automatically, but it may be necessary to indicate to the receiver which station is the nearest and it will then lock to that signal (under normal circumstances the receiver will always work with the 10.2 kHz signal, the tuning having been preset to this frequency).

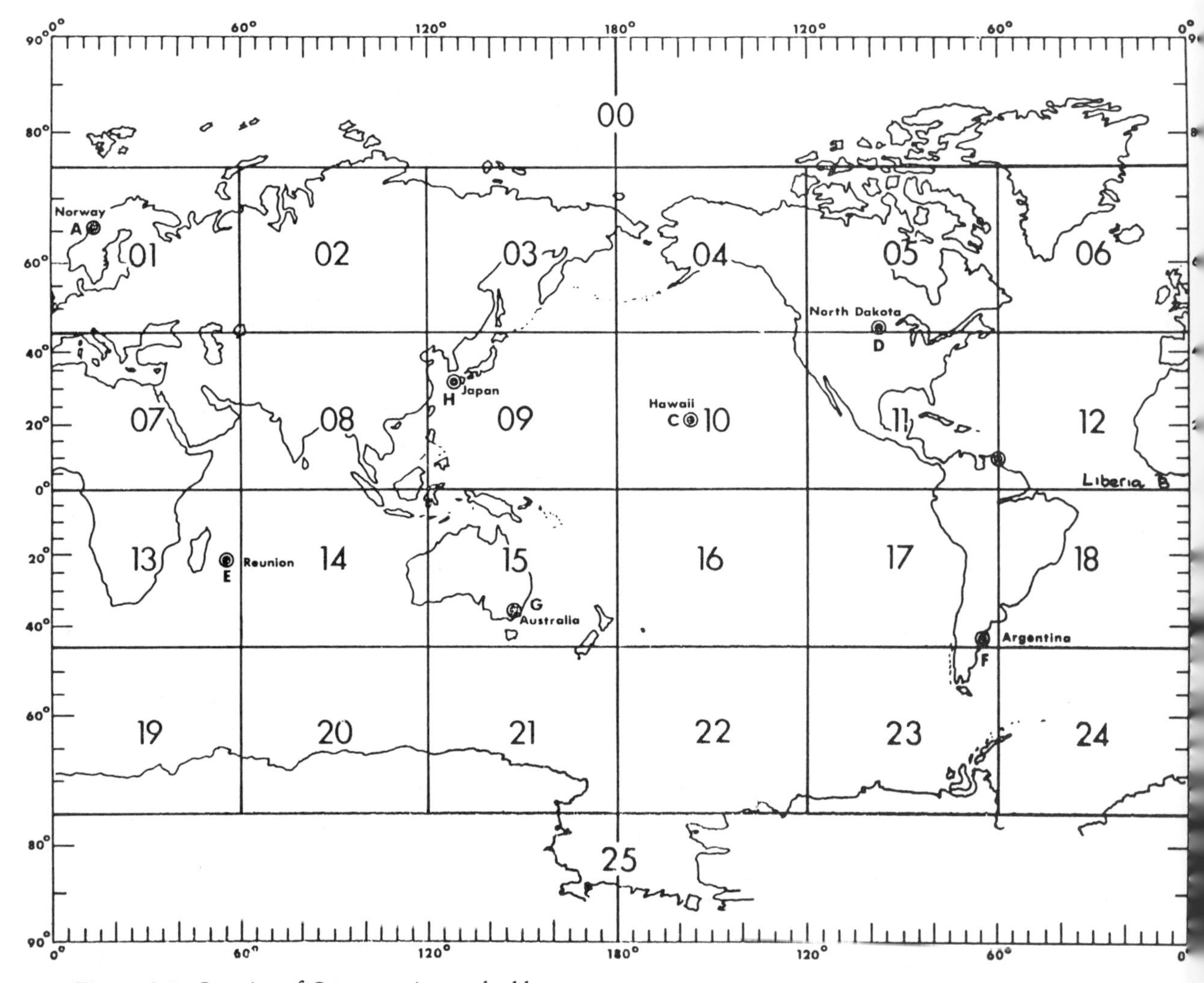

Figure 8.1 Location of Omega stations and table areas

8.2.1 Selection of station pairs

This is probably best done by observing on the chart the lattice pattern in the vicinity of the vessel. The following points should be considered.

1 Stations should be more than 500 miles away.
2 There should be a good angle of cut on the lattice lines.
3 The path should be preferably all over water.
4 The path should be all daylight or all night.

Having set in the station pairs (usually three pairs) which it is intended to use, the expected readings (at the vessel's estimated position on the Omega chart) may have to be fed in. After this the receiver should automatically track the changes in reading.

The readings may be displayed continually on some form of screen, with updating taking place every few seconds as the vessel steams across the lanes.

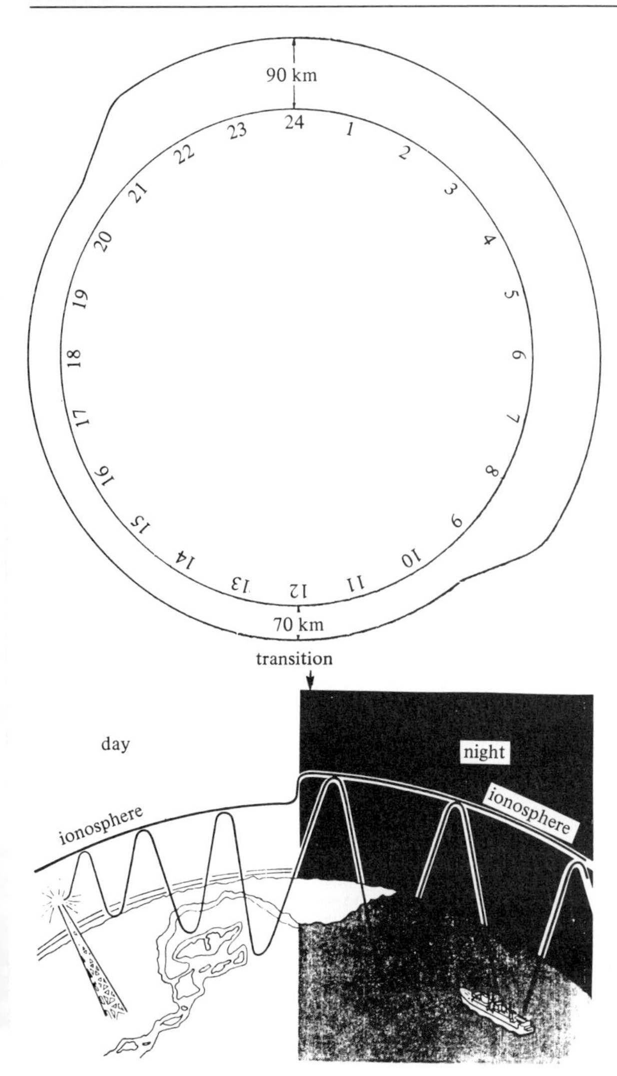

Figure 8.2 Ionospheric height over a 24 hour period

8.3 Correction of readings

The duct between earth and ionosphere, which constrains the Omega waves to follow the earth's curvature, varies in height between day and night (Figure 8.2). This change in height affects the phase of the signal arriving at the receiver, and hence the position line. The effect of this change has been calculated and is promulgated in table 2, Omega Propagation Correction Tables for 10.2 kHz (HO pub. no. 224). There is one set of tables for each station in each of the 26 areas depicted in Figure 8.1.

The corrections are tabulated for each hour GMT for each two weekly interval thoughout the year, with a table for each geographical area of about 4° of latitude and longitude. The use of the tables is most easily demonstrated with an example.

Example

On 19 September at 0100 GMT in DR position latitude 52° 00′ N, 6° 00′ W, an Omega reading of the station pair A–B indicated 737.52. Determine the reading to be plotted on the chart.

Using location 52° N, 6° W, station A, date 16–30 September, 01 GMT = –31. For 52° N, 6° W, station B, date 16–30 September, 01 GMT = –82.

Correction = (correction for A) – (correction for B) (Table 8.1).

Therefore reading to be plotted is 737.52 + [(–0.31) – (–0.82)] = 737.52 + 0.51 = 738.03.

From the table it can be seen that the period of validity of the corrections is of the order of an hour.

When the readings have been taken from the receiver, the corrections may be applied manually prior to plotting (as in the example) or may be fed into the receiver for automatic application. In the latter case there is usually an alarm after one hour to indicate that it is time for the corrections to be updated.

8.4 Sources of Omega status warnings

8.4.1 *Off-air schedule for annual maintenance*

Each year, each station is allocated a month into which it will schedule its annual maintenance and during which it can be expected to be off the air. These months are published but are apt to change from year to year. For example, a typical schedule might be as follows:

Station	*Month*	*Station*	*Month*
A Norway	August	E La Réunion	September
B Liberia	February	F Argentina	March
C Hawaii	June	G Australia	November
D N. Dakota	July	H Japan	October

Table 8.1 Omega correction tables (extracts)

LOCATION 52.0 N 6.0 W
STATION A NORWAY

DATE	GMT 00	01	02	03	04	05	06	07	08	09	10	11	12	13	14	15	16	17	18	19	20	21	22	23	24
1-15 JAN	-28	-29	-29	-28	-27	-27	-27	-28	-20	-15	-5	-14	-16	-15	-15	-18	-22	-24	-26	-28	-29	-30	-33	-31	-28
16-31 JAN	-33	-34	-34	-33	-33	-33	-32	-30	-15	-19	-6	-10	-11	-16	-12	-14	-24	-26	-30	-28	-35	-33	-34	-34	-33
1-14 FEB	-33	-34	-34	-33	-33	-34	-34	-31	-15	-16	-15	-16	-19	-17	-19	-20	-22	-27	-31	-32	-33	-36	-35	-35	-33
15-29 FEB	-32	-32	-31	-31	-31	-30	-30	-16	-12	-13	-10	-12	-15	-14	-15	-15	-15	-21	-27	-29	-32	-35	-34	-32	-32
1-15 MAR	-29	-29	-29	-27	-30	-30	-22	-14	-8	-10	-6	-10	-12	-12	-10	-12	-12	-12	-18	-26	-27	-33	-29	-29	-29
16-31 MAR	-34	-31	-29	-28	-29	-26	-12	-11	-10	-9	-7	-10	-10	-11	-11	-12	-12	-14	-17	-28	-32	-35	-37	-34	-34
1-15 APR	-34	-33	-34	-33	-29	-18	-13	-12	-11	-9	-12	-13	-15	-15	-14	-14	-14	-16	-19	-23	-28	-30	-35	-35	-34
16-30 APR	-34	-33	-33	-27	-26	-15	-13	-14	-13	-10	-13	-14	-14	-14	-14	-14	-15	-15	-17	-18	-25	-29	-36	-33	-34
1-15 MAY	-29	-30	-27	-25	-16	-12	-12	-15	-14	-13	-14	-14	-15	-14	-14	-14	-15	-15	-16	-18	-24	-27	-31	-32	-29
16-31 MAY	-25	-24	-26	-16	-13	-14	-14	-12	-16	-13	-15	-16	-16	-15	-13	-13	-13	-17	-17	-19	-22	-22	-26	-27	-25
1-15 JUN	-24	-26	-24	-19	-15	-13	-15	-16	-14	-13	-15	-15	-11	-15	-12	-12	-12	-17	-14	-13	-16	-26	-22	-25	-24
16-30 JUN	-27	-22	-24	-16	-14	-12	-15	-14	-15	-13	-16	-16	-16	-16	-16	-16	-17	-18	-17	-18	-20	-26	-26	-29	-27
1-15 JUL	-28	-25	-23	-18	-13	-12	-14	-14	-15	-12	-14	-15	-15	-16	-15	-16	-16	-17	-16	-18	-19	-25	-25	-29	-28
16-31 JUL	-26	-25	-23	-19	-12	-11	-12	-13	-13	-11	-13	-13	-13	-13	-13	-14	-14	-15	-15	-16	-18	-22	-25	-27	-26
1-15 AUG	-31	-30	-30	-20	-13	-12	-13	-14	-14	-12	-14	-14	-14	-14	-14	-14	-15	-15	-16	-17	-24	-25	-29	-32	-31
16-31 AUG	-33	-33	-32	-27	-22	-14	-10	-13	-12	-12	-13	-14	-13	-14	-14	-13	-14	-14	-15	-17	-26	-28	-35	-33	-33
1-15 SEP	-32	-32	-32	-31	-25	-14	-6	-10	-10	-7	-10	-10	-10	-10	-10	-11	-11	-12	-15	-24	-29	-33	-36	-33	-32
16-30 SEP	-30	-31	-30	-29	-25	-19	-5	-6	-7	-4	-6	-9	-8	-8	-8	-9	-9	-10	-17	-26	-28	-32	-32	-31	-30
1-15 OCT	-32	-31	-31	-30	-31	-30	-15	-9	-11	-8	-10	-14	-12	-14	-14	-14	-13	-21	-32	-34	-36	-37	-35	-32	-32
16-31 OCT	-31	-32	-32	-30	-28	-30	-24	-9	-10	-9	-11	-12	-14	-12	-15	-19	-20	-25	-34	-34	-30	-34	-30	-31	-31
1-15 NOV	-33	-32	-33	-34	-33	-33	-33	-19	-12	-13	-14	-16	-16	-17	-16	-20	-22	-28	-33	-34	-36	-36	-36	-35	-33
16-30 NOV	-30	-31	-31	-32	-30	-30	-30	-28	-10	-13	-13	-16	-17	-18	-14	-19	-24	-27	-29	-31	-35	-30	-32	-31	-30
1-15 DEC	-32	-33	-32	-33	-32	-31	-31	-31	-19	-19	-12	-20	-21	-21	-20	-22	-25	-30	-34	-34	-35	-35	-32	-32	-32
16-31 DEC	-29	-30	-30	-30	-30	-30	-29	-30	-22	-16	-8	-17	-18	-18	-18	-19	-23	-28	-32	-31	-32	-32	-31	-30	-29

LOCATION 52.0 N 6.0 W
STATION B LIBERIA

DATE	GMT 00	01	02	03	04	05	06	07	08	09	10	11	12	13	14	15	16	17	18	19	20	21	22	23	24
1-15 JAN	-81	-82	-82	-82	-82	-82	-82	-55	-13	-18	-18	-16	-14	-13	-14	-16	-21	-33	-55	-73	-78	-80	-81	-82	-81
16-31 JAN	-81	-82	-82	-82	-82	-82	-82	-56	-11	-18	-18	-15	-13	-13	-13	-15	-19	-29	-50	-71	-78	-80	-81	-82	-81
1-14 FEB	-81	-82	-82	-82	-82	-82	-82	-44	-10	-19	-17	-14	-12	-12	-12	-14	-17	-24	-45	-69	-77	-80	-81	-82	-81
15-29 FEB	-81	-82	-82	-82	-82	-82	-82	-22	-12	-20	-16	-13	-11	-10	-11	-13	-15	-21	-39	-67	-76	-79	-81	-82	-81
1-15 MAR	-81	-82	-82	-82	-82	-82	-82	-10	-14	-19	-14	-12	-10	-10	-10	-12	-14	-19	-34	-65	-75	-79	-81	-81	-81
16-31 MAR	-81	-82	-82	-82	-82	-82	-71	-9	-17	-17	-13	-11	-9	-9	-9	-11	-14	-17	-31	-61	-73	-78	-80	-81	-81
1-15 APR	-81	-82	-82	-82	-82	-81	-44	-11	-20	-15	-12	-10	-8	-8	-9	-10	-13	-16	-28	-56	-72	-78	-80	-81	-81
16-30 APR	-81	-82	-82	-82	-82	-74	-30	-13	-19	-15	-11	-9	-7	-7	-8	-10	-13	-16	-25	-51	-71	-77	-80	-81	-81
1-15 MAY	-81	-82	-82	-82	-80	-65	-22	-16	-18	-14	-11	-8	-7	-7	-8	-10	-12	-16	-22	-47	-69	-76	-79	-81	-81
16-31 MAY	-81	-81	-82	-82	-77	-59	-21	-16	-18	-14	-10	-8	-7	-7	-8	-9	-12	-15	-20	-43	-66	-75	-79	-81	-81
1-15 JUN	-81	-81	-82	-81	-74	-56	-23	-17	-17	-13	-10	-8	-7	-7	-7	-9	-12	-15	-19	-40	-64	-74	-78	-80	-81
16-30 JUN	-81	-81	-82	-81	-75	-56	-23	-17	-18	-14	-10	-8	-7	-7	-7	-9	-11	-15	-19	-38	-62	-73	-78	-80	-81
1-15 JUL	-81	-81	-82	-82	-77	-58	-27	-17	-18	-14	-11	-8	-7	-7	-7	-9	-11	-15	-18	-38	-62	-73	-78	-80	-81
16-31 JUL	-81	-81	-82	-82	-79	-63	-28	-16	-18	-14	-11	-8	-7	-7	-7	-9	-11	-15	-19	-39	-64	-74	-78	-80	-81
1-15 AUG	-81	-82	-82	-82	-81	-68	-31	-15	-19	-14	-11	-9	-7	-7	-8	-9	-12	-15	-19	-43	-67	-75	-79	-81	-81
16-31 AUG	-81	-82	-82	-82	-82	-75	-35	-13	-19	-15	-11	-9	-8	-7	-8	-10	-12	-16	-23	-49	-70	-77	-80	-81	-81
1-15 SEP	-81	-82	-82	-82	-82	-81	-40	-11	-20	-15	-12	-9	-8	-8	-9	-11	-13	-17	-30	-59	-73	-78	-80	-81	-81
16-30 SEP	-81	-82	-82	-82	-82	-82	-49	-11	-20	-16	-12	-10	-9	-9	-10	-12	-14	-18	-37	-68	-75	-79	-81	-82	-81
1-15 OCT	-81	-82	-82	-82	-82	-82	-66	-8	-18	-17	-13	-11	-10	-10	-11	-13	-16	-22	-46	-71	-77	-80	-81	-82	-81
16-31 OCT	-81	-82	-82	-82	-82	-82	-80	-8	-17	-18	-14	-12	-11	-11	-12	-14	-17	-27	-54	-72	-78	-80	-81	-82	-81
1-15 NOV	-81	-82	-82	-82	-82	-82	-81	-13	-15	-19	-16	-13	-12	-12	-13	-15	-20	-34	-59	-74	-79	-81	-81	-82	-81
16-30 NOV	-81	-82	-82	-82	-82	-82	-82	-26	-13	-19	-17	-14	-13	-13	-14	-16	-23	-37	-61	-75	-79	-81	-81	-82	-81
1-15 DEC	-81	-82	-82	-82	-82	-82	-82	-39	-13	-19	-18	-15	-14	-13	-14	-17	-24	-38	-61	-75	-79	-81	-81	-82	-81
16-31 DEC	-81	-82	-82	-82	-82	-82	-82	-48	-14	-19	-18	-16	-14	-14	-14	-17	-23	-36	-59	-74	-79	-81	-81	-82	-81

Reproduced by courtesy of the Defense Mapping Agency Hydrographic/Topographic Center

An additional reminder with more precise times and dates is given under navigation warnings in Notices to Mariners.

8.4.2 Omega system status and known propagation anomalies

Summaries of the Omega system status and known propagation anomalies are broadcast by Fort Collins, Colorado (WWV); Kekaha (Kauai), Hawaii WWVH); and Rogaland (LGQ). For frequencies and times, see ALRS, vol. 2. A recorded announcement can also be obtained by telephoning USA (703) 866 3801.

Unpredictable irregular propagation variations occur; these are sudden phase anomalies (SPAs) or sudden ionospheric disturbances (SIDs), and polar cap absorption (PCA) (see Section 8.5.2). They adversely affect the system accuracy. Information about these effects is promulgated in Hydrolant and Hydropac messages and are also repeated where necessary in UK long range radio navigational warnings.

8.5 Error sources and accuracy

Omega is primarily classified as an ocean navigational aid with accuracy in the range 2–4 nm.

8.5.1 Predictable errors

As a result of extensive validation and testing, the corrections contained in the Propagation Correction Tables (HO pub. no. 224) endeavour to take into account all the sources of predictable error, such as day/night variation and water/land propagation path. This has meant revising the tables from time to time. When taking over the library of these tables, it is important to ensure that they are the latest editions.

8.5.2 Unpredictable errors

Sudden phase anomalies, also referred to as sudden ionospheric disturbances, result from solar flare generated X-rays. These upset the stability of the ionosphere, which is the upper boundary of the Omega duct. The duration is generally not greater than one hour, but can result in position lines being shifted several miles.

Polar cap absorption results from proton bombardment of the magnetic polar regions. This is an infrequent occurrence, but it generally lasts for several days and can shift position lines by as much as 6 to 8 miles while it lasts. It only occurs where signals follow polar transmission paths. It can be of sufficient duration to justify promulgation in Notices to Mariners.

8.5.3 Intermodal interference

On transmission, more than one signal mode is radiated, but beyond some 600 miles from the station only one signal mode is dominant. Within this range from a station, interference may cause phase shifting or signal fading, especially during the change from night to day and vice versa, and lane slipping may result. Mariners are advised not to use stations when they are within some 450 miles of the transmitter (see Section 8.2.1) and, as a reminder, lattice lines within 450 nm of a transmitter on an Omega chart are dashed.

8.6 Fixing position

8.6.1 Plotting on Omega charts

Omega charts are of two types:

1. Navigational charts overprinted with an Omega lattice, and which are kept corrected via Notices to Mariners; and
2. Plotting charts, having a minimum of navigational detail, the plotted position being transferred to a conventional navigational chart.

Having obtained the reading for each position line and corrected it from the tables, it is then plotted, using an interpolating ruler, on the chart.

8.6.2 *Co-ordinate conversion*

Receivers displaying latitude/longitude are becoming much more common, but it is important to establish:

1 What corrections (if any) have been applied when making the conversion; and
2 The geodetic datum to which the latitude and longitude refer.

8.6.3 *Differential Omega*

By monitoring Omega readings ashore at a site where the correct Omega co-ordinates are accurately known, the current errors can be precisely determined (rather than predicted as in the tables) and broadcast, so that vessels in the area can use the information to correct their raw Omega readings. In this way, it is possible to achieve accuracies of the order of 0.25 nm when within 50 nm of the monitoring station, and 1 nm by day and 1.5 nm by night at some 300 nm from the monitoring station.

Monitoring stations providing this service are listed in ALRS, vol. 2. It should be noted that special receiving equipment will be required to receive these transmissions, but it will normally be an extension to the standard Omega receiver.

8.6.4 *Omega/Transit integration*

Receivers are now available which use Omega signals for continuous position indication between Transit satellite fixes. This means that the satellite system can cross-check for lane slip on the Omega, and the Omega will provide better position indication than DR, between satellite fixes (see 9.4.2).

8.7 Extract from official publication

Within Section 8.7.1, the paragraph numbering of the original publication has been retained.

8.7.1 *IMO Resolution A479(XII): Performance Standards for Shipborne Receivers for Use with Differential Omega*

1 Introduction

1.1 Receivers for differential Omega intended for navigational purposes on ships with maximum speeds not exceeding 35 knots should comply with the following minimum performance standards.

1.2 In addition to the requirements given in this Recommendation, the system should comply with the general requirements for shipborne navigational equipment (IMCO Assembly resolution A.281(VIII)).

1.3 Differential Omega requires both Omega signals and differential correction signals for correct operation. Receivers used for the reception of the differential correction signals should preferably be combined with the receivers used for reception of the Omega signals. Where separate receivers are used, care should be taken to ensure that the installation meets the overall system performance standards.

2 Performance standards for the reception of Omega signals

2.1 Signal reception

2.1.1 The system should provide for reception of Omega transmissions on the frequency of 10.2 kHz. It may additionally provide for the reception of one or more of the other Omega frequencies.

2.1.2 The antenna should be capable of receiving Omega signals from any direction of the horizontal plane at all times.

2.2 Positional information extraction

2.2.1 Means should be provided for synchronizing the system to the Omega transmission format. Automatic and/or manual means may be used, but in any case it should be possible to monitor the synchronization state continuously.

2.2.2 The system should be capable of processing information from at least four Omega stations simultaneously.

2.3 System performance

When a ship is stationary, the instrumental error introduced by the receiver to the measurement of uncorrected phase difference (line of position, LOP) on any selected pair of Omega signals should not exceed 0.02 lane width (2 centilanes). When sailing on a constant heading at speeds up to 35 knots, instrumental error should not exceed 0.04 lane width (4 centilanes).

2.4 Display of positional information

2.4.1 Equipment which gives positional information in terms of LOPs should be capable of displaying at least three operator selected LOPs either simultaneously or sequentially with the following facilities:

.1 A display of at least two whole lane digits and providing a readout to 0.01 lane width for each preselected pair of stations;
.2 Means for setting up initially the whole lane digit counts;
.3 Identification of the selected Omega stations;
.4 Where LOP information is displayed sequentially, provision should be made for holding any one pair of stations on display for as long as required without interruption to the continuous updating of LOP counts. Separate visual indication that the display is in the hold condition should be provided; and
.5 Where provision is made for manually entering corrections in order to display corrected LOP counts, the applied correction with its polarity sign should be separately displayed at the same time as the corrected LOP.

2.4.2 An alternative method of displaying the positional information may be used, provided that such method conforms in principle to the recommendations of paragraph 2.4.1. In the case where a latitude and longitude display is used, presentation should be as a minimum in the form of degrees, minutes and tenths of minutes. The display should also clearly indicate north, south, east and west. The readout values of latitude and longitude should be based on the World Geodetic System 1972 (WGS-72).

2.4.3 Means may be provided to transform the computed position based on WGS-72 into data compatible with the datum of the navigational chart in use. Where this facility exists, positive indication should be provided to indicate that the facility is currently in use and means should be provided to indicate the transformation correction.

2.4.4 When a system is designed for operation on a single Omega frequency only it should be provided with means of identifying lane slip sufficient to assist the reestablishment of the correct lane information.

2.5 Displays and indicators

2.5.1 The brilliance of all illumination, except for any warning light, should be adjustable; a common control may be used. The range of adjustment should be such that the display of positional information is clearly readable in bright diffused daylight, and that at night the brightness is the minimum necessary to operate the equipment.

2.5.2 Where the figures of a digital display are built up of individual parts (e.g. segments) then a facility should be provided which makes it possible to check all the segments of each figure. During such checking the operation of the equipment, except for the display, should not be interrupted.

2.6 Power supply

2.6.1 It should be possible to supply the receiver from the usual power supplies available on board ships: alternating current 100-115-220-230V ±15%, 50 or 60 Hz; direct current 24-32V ±15%.

2.6.2 The receiver should be fitted with a built-in emergency supply which should be capable of being automatically substituted with no break to the normal supply described in 2.6.1 above. This emergency supply should be capable of supplying the equipment during at least 10 minutes.

2.7 Warning devices

2.7.1 If the receiver is of the type which requires the operator to select the Omega stations whose signals will be employed to generate position information, a warning device should be provided to indicate the absence of a signal from a selected station.

2.7.2 If the receiver is of the type which automatically selects the most suitable Omega signals from those received, a warning device should be provided to indicate the lack of sufficient usable signals for normal equipment operation.

2.7.3 Provision may be made to indicate which

Omega signals are being received at a strength sufficient to be employed in position fixing.

2.7.4 The equipment should be fitted with a warning device for indicating main power supply failure which remains active until reset by the operator.

2.8 Controls

2.8.1 All controls should be of such size as to permit normal adjustments to be made easily. The controls should be clearly identified.

2.8.2 Where the inadvertent operation of a control could lead to failure of the equipment or false position-fixing information, the control should be protected from accidental operation.

2.9 Human errors

The number of manual calculations needed to transform the uncorrected Omega signals into a charted position should be kept to a minimum. Reliable automatic correction of Omega data is preferable. For navigation purposes, a reliable automatic transformation of Omega information into geographical co-ordinates is preferable. In this case due regard should be taken of possible additional errors which may be introduced by this process.

2.10 Auxiliary equipment

Single frequency (10.2 kHz) receivers should, and other receivers may, have an output to peripheral equipment, e.g. LOP or co-ordinate recorder, or path plotter. For this output, position data should be in digital form according to the format defined in CCITT Opinion V24.

3 Additional performance standards for the reception of differential Omega

3.1 Reception of signals

3.1.1 The system should provide for reception of differential Omega corrections for the basic frequency of 10.2 kHz. It may additionally provide for the reception of corrections for one or more of the other Omega frequencies.

3.1.2 The receiving equipment for differential Omega corrections should be able to receive corrections transmitted in accordance with the performance standards for differential Omega correction transmitting systems (resolution A425(XI)) and should indicate the Omega transmissions for which differential corrections are available.

3.1.3 Correction receivers should operate satisfactorily when the electric field received from the transmitting station is 10 μV/m or greater, day and night, in the conditions for atmospheric noise as defined by CCIR for the band 285–415 kHz. Correction receivers should have a selectivity, or protection devices, allowing acceptable reception of correction information when interfering signals are present. Operation should also be possible when the interfering signal is a non-modulated carrier frequency, at a level 20 dB above the wanted signal, on any frequency outside a band of ±200 Hz centred on the nominal frequency of the correction transmitting station.

3.1.4 The antenna for the reception of differential Omega corrections may be combined with the antenna described in paragraph 2.1.2. The antenna for the reception of differential Omega corrections (whether the same as the one described in paragraph 2.1.2 or not) should provide satisfactory reception of correction signals in the conditions described above and from any direction in the horizontal plane.

3.2 Extraction of position data

3.2.1 Means should be available for the synchronization of the system with the differential Omega correction transmission format. It is possible to use automatic or manual means but, in any case, it should be possible to monitor the state of synchronization.

3.2.2 The system should be capable of processing information relating to at least four Omega stations simultaneously.

3.3 System operation

3.3.1 Instrumental errors introduced by the correction receiving equipment should not be greater than those accepted for Omega receivers, according to paragraph 2.3 above.

3.4 Position information display

3.4.1 The system Omega and differential Omega may be in two forms:

.1 Separate Omega and differential Omega receivers:
  .1.1 The user may only add the differential Omega corrections to the raw data from his Omega receiver before reporting his position on the chart.
  .1.2 The user may enter differential Omega corrections into the Omega receiver under the conditions described in paragraph 2.4.1.5.
.2 Combined Omega and differential Omega receivers:
  .2.1 The combined receiver may separately display Omega and differential Omega data. The user may combine them as described in paragraph 3.4.1.1.
  .2.2 The combined receiver may, under the control of the user, automatically add differential Omega corrections to raw Omega data.

3.4.2 Where the differential Omega receiver gives correction information for LOPs, it should be able to display the corrections for at least three LOPs selected by the user, either simultaneously or sequentially in the following manner:

.1 Display from 0 to 99 centilanes of correction, providing reading for 1 centilane for each station pair selected.
.2 If found necessary, display combined with the display described in paragraph 3.4.2.1; of the integer part of the correction.
.3 Identification of the selected Omega stations.
.4 Where LOP information is displayed sequentially, provision should be made for holding any one pair of stations on display for as long as required without interruption to the continuous updating of LOP counts. Separate visual indication that the display is in the hold condition should be provided.
.5 Where provision is made for manually entering corrections in order to display corrected LOP counts, the applied correction with its polarity sign should be separately displayed at the same time as the corrected LOP. In addition the user should be clearly advised whether corrections are applied or not.
.6 Where means are provided for automatically entering the differential Omega corrections, the user should be clearly advised whether corrections are applied or not.
.7 Means should also be provided to make sure that differential Omega corrections can only be applied to raw Omega data.

3.4.3 Alternative methods of displaying the positional and correction information may be used as mentioned in paragraphs 2.4.2 and 2.4.3, provided that such methods conform in principle to the recommendations of paragraphs 2.4.1 and 3.4.2.

3.4.4 Where automatic receiving systems are used:

.1 The selection of Omega stations in such a system should be automatic. The system should be capable of evaluating the quality of Omega signals directly received as well as that of the corrections for each Omega station. It should establish the position information through the use of all available information from the various stations while taking account of the quality of each one. The operator should, however, have the possibility to control the choice of stations manually.
.2 Position data should be automatically obtained when a position estimated from dead reckoning or another means has been introduced. The acceptable uncertainty on the estimated initial position is essentially related to the number of Omega frequencies that the system may directly receive on board. This acceptable uncertainty should be clearly known by the operators.
.3 Even if it uses Omega corrections only on the frequency 10.2 kHz, an automatic receiver should preferably be capable of directly receiving Omega signals on the frequencies 10.2 kHz and 13.6 kHz. It could also, although it is not essential, work with the frequencies 11.33 kHz and 11.05 kHz.
.4 An automatic system of differential Omega should preferably be capable of correcting the dispersion which results, at a distance from the correction transmitting station of more than 200 nautical miles, from variations of the propagation velocity of Omega waves between day and night.
.5 An automatic system should be so designed that differential Omega corrections can only be applied to raw Omega data.

.6 It is desirable for the system to give an indication of quality of the positional data displayed.

3.5 Displays and indicators
Indication and display devices should conform with the recommendations of paragraph 2.5.

3.6 Power supply
Power supply devices should conform with the recommendations of paragraph 2.6.

3.7 Warning devices
3.7.1 The Omega and differential Omega systems should be fitted with the warning devices mentioned in paragraph 2.7.

3.7.2 Warning should be given:

.1 When the correction transmitting station transmits no correction for any of the selected stations;
.2 When correction information for any of the selected stations is not correctly received on board;
.3 When correction information has not been updated during the last period of 6 minutes for any of the selected stations.

3.7.3 A warning may be given when the 8 Hz modulation is not present.

3.7.4 For those receivers mentioned in paragraph 3.4.4, the recommendation of paragraph 3.7.2 is replaced by an alarm if the quality of position data is unacceptable.

3.8 Controls
Controls should conform with the recommendations of paragraph 2.8.

3.9 Human errors
3.9.1 The number of manual calculations needed to transform the uncorrected Omega signals into a charted position should be kept to a minimum.

3.9.2 Differential Omega correction should be directly applied to raw Omega data, excluding the usual corrections applicable to Omega use.

3.9.3 Automatic correction of raw Omega data by the corrections received from differential Omega stations is preferable. As for Omega alone, due consideration should be given to possible additional errors resulting from the transformation into geographical co-ordinates.

3.10 Auxiliary equipment
3.10.1 Omega and differential Omega systems may be fitted with an output for connection with peripheral equipment such as LOP or co-ordinate recorders, or path plotters.

3.10.2 Such a facility is desirable with receivers working only with frequency 10.2 kHz and with automatic equipment. On this output position data should be in the form of a digital message according to the format defined in CCITT Opinion V24.

9 Satellite navigation systems

9.1 The Transit satellite system

The Transit satellite navigation system has been available for commercial use since 1967. The system comprises some five operational satellites in near circular polar orbits at a height of approximately 600 nm and having an orbital period of about 105 minutes. The orbits form an almost stationary birdcage in space, within which the earth rotates.

The satellites continually transmit data relating to their orbits. The shipboard receiver continually searches the satellite transmission band(s) and, as soon as a satellite comes above the horizon, locks on to the signal and measures the Doppler shift as the satellite approaches and recedes (the duration above the horizon can be up to some 18 minutes). The receiver decodes the orbital data transmitted by the satellite and, using that and the Doppler count over a short period (2 minutes or less), determines the line of position on which the ship must lie. This is repeated for a number of successive periods while the satellite is above the horizon. Each position line is reduced to a common time and the most probable position displayed or printed out in latitude and longitude.

9.2 Setting-up procedure

This initialization procedure varies from maker to maker, but initially the receiver must be programmed with the best estimate of the vessel's latitude, longitude, date and time (GMT to within 15 minutes), and antenna height. The order of setting up is usually in response to a prompt, and input format is usually indicated in some way, e.g. XXX.XX.XXW, on the display.

Other information may be required, such as which area of the world, course and speed input method, whether a printer is to be connected, and how often the output is required.

Once initialized, the receiver will complete its tasks automatically, e.g. search for and lock on to the satellites as they rise, obtain the Doppler count and decode the satellite's orbital data, and finally compute and display the position along with a reliability indicator.

In the period between satellite fixes, the computer will use the course and speed inputs to update and display the vessel's position every few seconds. It is important to appreciate that the position being displayed (except at the time of a fix) at any instant is an estimated position based on the accuracy of the course and speed sensors and on whether any allowance has been made for tidal drift or current. In order to check how long the vessel has been running on DR, it is normally possible to recall the time of and information relating to the last fix.

9.3 Error sources and accuracy

9.3.1 Errors in velocity during the period of a fix

Since the satellites are travelling virtually north/south, the Doppler count will be affected by the north/south component of the vessel's velocity; thus this must be removed before the Doppler count is used to determine the position line. Any error in the north/south component of velocity will therefore displace the position line. It can be seen that an east/west error will have little or no effect on the count.

Because a series of position lines is obtained during the period that the satellite is above the horizon, it is necessary to do what is in effect a running fix calculation. This involves using the course and speed to run up each position line to a common time. Again, any error in velocity will result in an error in position.

9.3.2 *An error in antenna height*

The antenna height is needed to calculate the height above the geoid, which is in turn used to calculate how the position surface intersects with the earth's surface and hence the position line. Care should be taken to note whether the antenna height to be fed in is the height above sea level or above the geoid. In the latter case, a special diagram is needed.

9.3.3 *Other errors*

These include errors in orbital predictions, propagation path anomalies resulting from ionospheric and tropospheric refraction, and possibly oscillator drift. All can result in some degree of error in the final position. In the case of satellite oscillator drift, though, the errors are outside the operator's control and, except in cases where the highest precision is required, should not give cause for concern.

9.3.4 *Accuracy*

When using a satellite in the usable range (i.e. between about 10° and 70° altitude) and in the absence of velocity errors, a single channel receiver will give positions to an order of ±100 metres (see reference datum in Section 9.4).

9.4 Fixing position

With all receivers, the position is displayed in latitude and longitude and so can be plotted directly on an ordinary navigational chart. However, in recent years chart users will have observed that on some charts there is a note regarding positions obtained from satellite navigation systems. Before plotting on the particular chart, the position obtained from the satellite needs correcting. Usually the amount is quite small, but not always so. The reason for this is that the geodetic datum to which the latitude and longitude refer is called WGS-72 (World Geodetic System 1972), which in effect assumes a slightly different shaped earth to that used for the chart. This is becoming a problem with all systems involving latitude/longitude conversion (see also Section 7.5.5, point 3; and Section 8.6.2).

9.4.1 *Interval between fixes*

Perhaps the main disadvantage with the Transit satellite navigation system is the interval between fixes, during which time the vessel must proceed on dead reckoning. If the satellites had gone into and maintained the orbits as theoretically planned, a position could have been fixed each 90 minutes or so, but there is no guidance system on board the satellite and so the orbits have tended to change with time and in particular to precess (this should not be the case with the NAVSTAR system). As a result of changes in orbit, there can be long periods (in the order of three to four hours) during which no position is obtained. The time interval is not standard and is (slightly) less in higher latitudes as the satellites converge at the poles.

There have been a number of occasions when a satellite's transmissions have had to be suspended (sometimes for a number of months) because of:

1 Its proximity to another satellite
2 Shadowing of solar panels and resultant power loss.

It is interesting to note that some of the present satellites are over fifteen years old, which is far in excess of the lifespan originally envisaged.

A new model (NOVA) satellite is being used to replace some of the satellites which have ceased to function and which have had to be turned off, but there are quite a number of the original type remaining which are also being used. The shipboard receiver cannot distinguish between the types of satellite from which the signals originate (other than by their identity number). The position fixing potential is the same for *all* satellites, be they old or new.

9.4.2 *Omega/Transit integration*

Because of the possibly long intervals between Transit fixes, DR navigation techniques are used in the intervening period (see Section 8.6.4). Receivers are now available which use Omega signals for continuous position indication between fixes by satellite.

9.5 Global positioning system – NAVSTAR

The global positioning system (GPS) is satellite based and is being implemented by the US Department of Defense. Although the system is not yet fully operational, it already provides many hours of coverage daily. It will eventually enable land, sea and air users to determine a three-dimensional position, velocity and time, continuously, anywhere in the world and in all weathers.

The system when complete will comprise 24 satellites (21 operational and 3 spares) orbiting at some 10 900 nm above the earth in a period of 12 hours. Each satellite will transmit data from which its position can be determined at any instant. Each satellite has small rockets which enable its orbit to be controlled and precisely maintained.

The user's receiver will measure (in effect, simultaneously) the distance from three (or more) satellites whose positions are known and from this information the observer's position can be calculated.

9.6 GPS receivers

The marketplace has seen a sudden spate of manufacturers producing receivers but what has been particularly noticeable is how small some of the receivers are. They appear very similar to the larger hand-held calculators. Also, as more receivers have come on to the market, the price has dropped and is now below £1000.

9.7 Accuracy

The readout of position is in latitude and longitude to a datum of WGS-84. (It should be noted that all position fixing systems will work towards giving positions based on this datum but there is a long way to go before widespread conformity.)

Positional accuracy for civil marine users will be of the order of 100 metres or better for 95% of the time. The planned introduction of differential GPS should improve the accuracy to some 30 m. In the case of differential GPS, shore stations (in the US marine radio beacons will be used) determine the errors locally and transmit corrections which are received by the augmented GPS receiver and incorporated in the position calculation.

9.8 System status

Since planned satellite deployment is expected to continue until mid-1994 and the system will continue to be developed, up to date information can be obtained by addressing questions to:

Commanding Officer
GPS Information Center (GPSIC)
US Coast Guard ONSCEN
7323 Telegraph Road
Alexandria, VA22310 – 3988
Tel: (703) 866 3806

There also are Operational Advisory Broadcasts.

10 Navigation system comparison

10.1 Frequencies for electronic navigation systems

Electronic navigation systems have to share a cluttered frequency spectrum with numerous other marine sources and, of course, with a host of general communications. Figure 10.1 summarizes the use of the electromagnetic radiation spectrum in both marine and general applications.

10.2 Position fixing systems: comparison

Table 10.1 provides a summary and comparison of the position fixing systems described in earlier chapters.

10.3 Extracts from official publications

Within Sections 10.3.1 and 10.3.2, the paragraph numbering of the original publications has been retained.

10.3.1 SOLAS (74) as amendment to 1991. Chapter V, Regulation 12: Shipboard navigational equipment

(a) For the purpose of this regulation, 'constructed' in respect of a ship means a stage of construction where:

- (i) The keel is laid; or
- (ii) Construction identifiable with a specific vessel begins; or
- (iii) Assembly of that ship has commenced comprising at least 50 tonnes or 1 per cent of the estimated mass of all structural material, whichever is less.

(b) (i) Ships of 150 tons gross tonnage and upwards shall be fitted with:

- (1) A standard magnetic compass, except as provided in subparagraph (iv);
- (2) A steering magnetic compass, unless heading information provided by the standard compass required under (1) is made available and is clearly readable by the helmsman at the main steering position;
- (3) Adequate means of communication between the standard compass position and the normal navigation control position to the satisfaction of the administration; and
- (4) Means for taking bearings as nearly as practicable over an arc of the horizon of 360°.

(ii) Each magnetic compass referred to in subparagraph (i) shall be properly adjusted and its table or curve of residual deviations shall be available at all times.

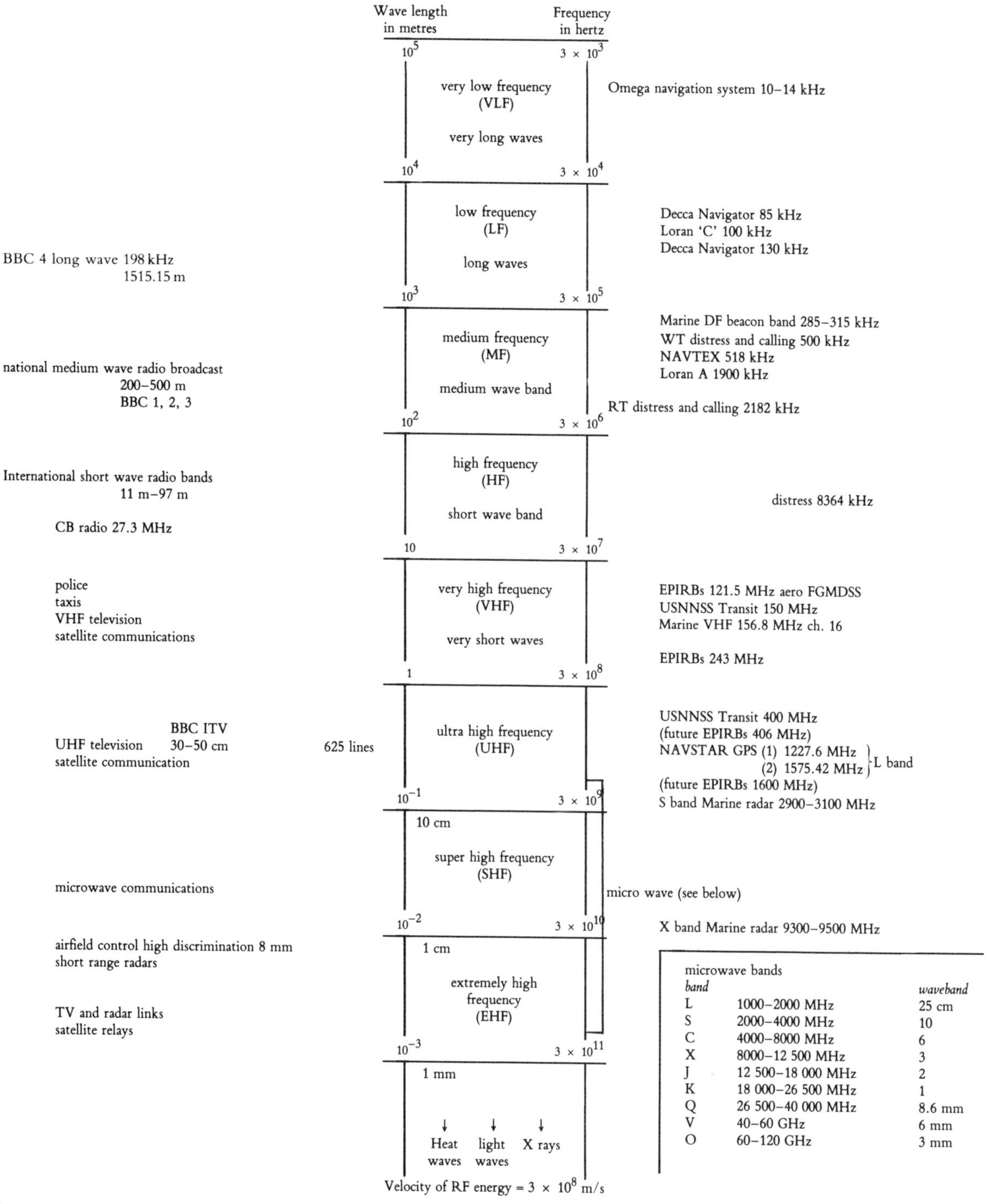

microwave bands

band		waveband
L	1000–2000 MHz	25 cm
S	2000–4000 MHz	10
C	4000–8000 MHz	6
X	8000–12 500 MHz	3
J	12 500–18 000 MHz	2
K	18 000–26 500 MHz	1
Q	26 500–40 000 MHz	8.6 mm
V	40–60 GHz	6 mm
O	60–120 GHz	3 mm

Figure 10.1 Frequencies used for electronic navigation systems

Table 10.1 Position fixing systems compared

System	*WT DF*	*Decca*	*Loran C*	*Omega*	*Transit*	*NAVSTAR*	*Inertial systems*	*Radar/ARPA*
Principle	Directional antenna	Phase comparison	Pulse/time difference	Phase comparison (Time difference)	Range difference by Doppler shift during satellite pass	Direct ranging from at least 3 satellites in known positions	Improved deduced reckoning	Ranging by echo principle
Coverage (range)	50 nm (some to 200 nm)	240 nm	1200 nm	Worldwide	Worldwide	Worldwide	Worldwide	Line of sight 10 miles + depending on heights
Position lines	Bearings (great circles)	Hyperbolic	Hyperbolic	Hyperbolic (range circles)	Neo-hyperbolic	Range surfaces	N/A	Range circles bearing lines
Transmission frequency	250–550 kHz 1600–3000 kHz	70–130 kHz	100 kHz	10–14 kHz	150 and 400 MHz	1227.6 MHz 1575.42 MHz	N/A	2900–3100 MHz 9300–9500 MHz
Emission type	Amplitude modulated continuous waves (some in grouped sequence)	Interrupted continuous waves	Pulse modulated continuous waves	Time shared continuous waves	Phase modulated continuous waves	Coded continuous waves	N/A	Pulsed
Data format	CRT visual and audible signal or meter and goniometer	Analogue 'clock' display giving Decca lane ref. (lat./long.)	Automatic time difference display (some lat./long.)	Omega station pair and lane number (some lat./long.)	Lat./long.	(Lat./long.)	Lat./long.	Range bearing
Position fix	Apply corrections, lay off on normal chart	Apply corrections, plot on special lattice chart	Plot on special plotting or lattice chart	Plot on special plotting or lattice chart	Plot on normal chart	Plot on normal chart	Plot on normal chart	Lay off on normal chart
Time for fix	5–10 min	1 min	1 min	1 min	30 s	30 s	30 s	1 min
Typical accuracy	5 nm at 50 nm	< 0.25 nm	0.25 nm	1–3 nm	0.1 to 0.25 nm	25 m – 100 m	Cumulative – needs updating	.18 nm on the 12 nm range scale
24 hour continuous availability	Most	Yes	Yes	Yes	Fix intervals 90 min–5 hrs	Yes	Yes	Yes
All weather availability	Skywave problems at twilight	Skywave problems at limits of coverage	Skywave problems at limit of coverage	Modal interference close to station, polar cap absorption, sudden ionospheric disturbance	Yes	Yes	Yes	Chart comparison. Problems with: rain clutter, sea clutter, radar to radar interference
Typical cost	£4500	£1500 (rental)	> £5000 for highest accuracy	£6000	£2000 (or more) for full usage	£1000+	> £80 000 for marine use	£7000 to £50 000

(iii) A spare magnetic compass, interchangeable with the standard compass, shall be carried, unless the steering compass mentioned in subparagraph (i)(2) or a gyro compass is fitted.

(iv) The administration, if it considers it unreasonable or unnecessary to require a standard magnetic compass, may exempt individual ships or classes of ships from these requirements if the nature of the voyage, the ship's proximity to land or the type of ship does not warrant a standard compass, provided that a suitable steering compass is in all cases carried.

(c) Ships of less than 150 tons gross tonnage shall, as far as the administration considers it reasonable and practicable, be fitted with a steering compass and have means for taking bearings.

(d) Ships of 500 tons gross tonnage and upwards constructed on or after 1 September 1984 shall be fitted with a gyro compass complying with the following requirements:

(i) The master gyro compass or a gyro repeater shall be clearly readable by the helmsman at the main steering position.

(ii) On a ship of 1600 tons gross tonnage and upwards a gyro repeater or gyro repeaters shall be provided and shall be suitably placed for taking bearings as nearly as practicable over an arc of the horizon of 360°.

(e) Ships of 1600 tons gross tonnage and upwards, constructed before 1 September 1984, when engaged on international voyages shall be fitted with a gyro compass complying with the requirements of paragraph (d).

(f) Ships with emergency steering positions shall at least be provided with a telephone or other means of communication for relaying heading information to such positions. In addition, ships of 500 tons gross tonnage and upwards constructed on or after 1 February 1992 shall be provided with arrangements for supplying visual compass readings to the emergency steering position.

(g) Ships of 500 tons gross tonnage and upwards constructed on or after 1 September 1984 and ships of 1600 tons gross tonnage and upwards constructed before 1 September 1984 shall be fitted with a radar installation. From 1 February 1995, the radar installation shall be capable of operating in the 9 GHz frequency band. In addition, after 1 February 1995, passenger ships irrespective of size and cargo ships of 300 tons gross tonnage and upwards when engaged on international voyages, shall be fitted with a radar installation capable of operating in the 9 GHz frequency band. Passenger ships of less than 500 tons gross tonnage and cargo ships of 300 tons gross tonnage and upwards but less than 500 tons gross tonnage may be exempted from compliance with the requirements of paragraph (r) at the discretion of the Administration, provided that the equipment is fully compatible with the radar transponder for search and rescue.

(h) Ships of 10 000 tons gross tonnage and upwards shall be fitted with two radar installations, each capable of being operated independently of the other. From 1 February 1995, at least one of the radar installations shall be capable of operating in the 9 GHz frequency band.

(i) Facilities for plotting radar readings shall be provided on the navigating bridge of ships required by paragraph (g) or (h) to be fitted with a radar installation. In ships of 1600 tons gross tonnage and upwards constructed on or after 1 September 1984, the plotting facilities shall be at least as effective as a reflection plotter.

(j) (i) An automatic radar plotting aid shall be fitted on:

(1) Ships of 10 000 tons gross tonnage and upwards, constructed on or after 1 September 1984;

(2) Tankers constructed before 1 September 1984 as follows:

(aa) If of 40 000 tons gross tonnage and upwards, by 1 January 1985;

(bb) If of 10 000 tons gross tonnage and upwards, but less than 40 000 tons gross tonnage, by 1 January 1986;

(3) Ships constructed before 1 September 1984, that are not tankers, as follows:
(aa) If of 40 000 tons gross tonnage and upwards, by 1 September 1986;
(bb) If of 20 000 tons gross tonnage and upwards, but less than 40 000 tons gross tonnage, by 1 September 1987;
(cc) If of 15 000 tons gross tonnage and upwards, but less than 20 000 tons gross tonnage, by 1 September 1988.

(ii) Automatic radar plotting aids fitted prior to 1 September 1984 which do not fully conform to the performance standards adopted by the organization may, at the discretion of the administration, be retained until 1 January 1991.

(iii) The administration may exempt ships from the requirements of this paragraph, in cases where it considers it unreasonable or unnecessary for such equipment to be carried, or when the ships will be taken permanently out of service within two years of the appropriate implementation date.

(k) When engaged on international voyages, ships of 1600 tons gross tonnage and upwards constructed before 25 May 1980, and ships of 500 tons gross tonnage and upwards constructed on or after 25 May 1980, shall be fitted with an echo sounding device.

(l) When engaged on international voyages, ships of 500 tons gross tonnage and upwards constructed on or after 1 September 1984 shall be fitted with a device to indicate speed and distance. Ships required by paragraph (j) to be fitted with an automatic radar plotting aid shall be fitted with a device to indicate speed and distance through the water.

(m) Ships of 1600 tons gross tonnage and upwards constructed before 1 September 1984 and all ships of 500 tons gross tonnage and upwards constructed on or after 1 September 1984 shall be fitted with indicators showing the rudder angle, the rate of revolution of each propeller and, in addition, if fitted with variable pitch propellers or lateral thrust propellers, the pitch and operational mode of such propellers. All these indicators shall be readable from the conning position.

(n) Ships of 100 000 tons gross tonnage and upwards constructed on or after 1 September 1984 shall be fitted with a rate-of-turn indicator.

(o) Except as provided in regulations 1/7(b)(ii), 1/8 and 1/9, while all reasonable steps shall be taken to maintain the apparatus referred to in paragraphs (d) to (n) in an efficient working order, malfunctions of the equipment shall not be considered as making a ship unseaworthy or as a reason for delaying the ship in ports where repair facilities are not readily available.

(p) When engaged on international voyages, ships of 1600 tons gross tonnage and upwards shall be fitted with a radio direction-finding apparatus. The Administration may exempt a ship from this requirement if it considers it unreasonable or unnecessary for such apparatus to be carried or if the ship is provided with other radionavigation equipment suitable for use throughout its intended voyages.

(q) Until 1 February 1999, ships of 1600 tons gross tonnage and upwards constructed on or after 25 May 1980 and before 1 February 1995, when engaged on international voyages, shall be fitted with radio equipment for homing on the radiotelephone distress frequency.

(r) All equipment fitted in compliance with this Regulation shall be of a type approved by the administration. Equipment installed on board ships on or after 1 September 1984 shall conform to appropriate performance standards not inferior to those adopted by the organization. Equipment fitted prior to the adoption of related performance standards may be exempted from full compliance with those standards at the discretion of the administration, having due regard to the recommended criteria which the organization might adopt in connection with the standards concerned.

(s) A rigidly connected composite unit of a pushing vessel and associated pushed vessel, when designed as a dedicated and integrated tug and barge combination, shall be regarded as a single ship for the purpose of this regulation.

(t) If the application of the requirements of this regulation necessitates structural alterations to a ship constructed before 1 September 1984, the administration may allow extension of the time limit for fitting the required equipment not later than 1 September 1989, taking into account the first scheduled dry docking of such a ship required by the present regulations.

(u) Except as provided elsewhere in this regulation, the administration may grant to individual ships exemptions of a partial or conditional nature, when any such ship is engaged on a voyage where the maximum distance of the ship from the shore, the length and nature of the voyage, the absence of general navigation hazards, and other conditions affecting safety are such as to render the full application of this regulation unreasonable or unnecessary. When deciding whether or not to grant exemptions to an individual ship, the administration shall have regard to the effect that an exemption may have upon the safety of all other ships.

Table 10.2 provides a summary of the UK government's legislation, consequent upon the IMO requirements for the fitting of navigational equipment. For the full text, reference should be made to Statutory Instrument 1984, 1203 or IMO Regulation 12, Chapter V of the International Convention for the Safety of Life at Sea (SOLAS) 1974 as amended by the expanded maritime safety committee.

10.3.2 UK Statutory Instrument 1984, no. 1203

Provision of navigational equipment installations

3(1) Every ship shall be fitted with a magnetic compass installation and comply with part I of these regulations.

(2) Every ship of 500 tons or over but less than 1600 tons constructed on or after 1 September 1984 shall:

(a) Be fitted with a gyro compass installation and comply with part II of these regulations;
(b) Be fitted with a radar installation and comply with part III of these regulations;
(c) Be fitted with indicators showing the rudder angle, the rate of revolution and direction of thrust of each propeller and, if fitted with variable pitch propellers or lateral thrust propellers, the pitch and operational mode of such propellers. All these indicators shall be readable from the normal navigation control position.

(3) Every ship of 500 tons or over but less than 1600 tons when engaged on an international voyage shall:

(a) If constructed on or after 25 May 1980, be fitted with an echo sounder installation and comply with part IV of these regulations;
(b) If constructed on or after 1 September 1984, be fitted with a speed and distance measuring installation and comply with part V of these regulations.

(4) Every ship of 1600 tons or over, whenever constructed, shall:

(a) Be fitted with a gyro compass installation and comply with part II of these regulations: provided that this requirement shall apply to ships constructed before 1 September 1984, only when engaged on international voyages;
(b) Be fitted with a radar installation (or, if the ship is of 10 000 tons or over, two radar installations), and comply with part III of these regulations;
(c) Be fitted with indicators showing the rudder angle, the rate of revolution and direction of thrust of each propeller and, if fitted with variable pitch propellers or lateral thrust propellers, the pitch and operational mode of such propellers. All these indicators shall be readable from the normal navigation control position.

(5) Every ship of 1600 tons or over when engaged on an international voyage shall:

(a) Whenever constructed, be fitted with an echo

Table 10.2 UK requirements to carry navigational equipment

G.R.T Tonnage	Constrn Date	Magnetic Comp. Installation				Gyro Comp. Installation		Indicators					Radar Installation					ARPA		D.F. Installation			Log		R/T Homg	Echo SDR
		Std Comp	Strg or Spare	Emerg Strg	Comm	Master	Reptr	Rudder Angle	Revs	Trans Thrust	Var Pitch	Rate of Turn	Number	Display Size	Plotg Facility	Qual obs	Log Kept		Qual obs	Rx	Comm	Com Ant	Spd	Dist		
> 150T		●	●	‡	●																					
150 → 1600		●	●	‡	●																					
500 to 1600T	Before 1.9.84	●	●	‡	●																					Ic
	After 1.9.84	●	●	‡	●	●		●	●	‡	‡		1	180mm	●	●	●						I	I		I
1600T Plus	Before 1.9.84	●	●	‡	●	I	I	●	●	‡	‡		1	NS ø	●	●	●			I	I	Ic			Ic	I
	After 1.9.84	●	●	‡	●	●	●	●	●	‡	‡		1	250mm	R	●	●			I	I	Ic	I	I	Ic	I
10,000T Plus	Before 1.9.84	●	●	‡	●	I	I	●	●	‡	‡		2	NS ø	●	●	●			I	I	Ic			Ic	I
	Before 1.9.84 (TK)	●	●	‡	●	I	I	●	●	‡	‡		2	NS ø	●	●	●	1.1.86 ●	●	I	I	Ic			Ic	I
	ALL After 1.9.84	●	●	‡	●	●	●	●	●	‡	‡		2	250mm 340mm	R	●	●	●	●	I	I	Ic	I	I	Ic	I
15,000 to 20,000	Before 1.9.84 (Not TK)	●	●	‡	●	I	I	●	●	‡	‡		2	NS ø	●	●	●	1.9.88 ●	●	I	I	Ic			Ic	I
	Before 1.9.84 (TK)	●	●	‡	●	I	I	●	●	‡	‡		2	NS ø	●	●	●	1.1.86 ●	●	I	I	Ic	I	I	Ic	I
20,000 to 40,000	Before 1.9.84 (Not TK)	●	●	‡	●	I	I	●	●	‡	‡		2	NS ø	●	●	●	1.9.87 ●	●	I	I	Ic			Ic	I
	Before 1.9.8 (TK)	●	●	‡	●	I	I	●	●	‡	‡		2	NS ø	●	●	●	1.1.86 ●	●	I	I	Ic	I	I	Ic	I
40,000 Tons Plus	Before 1.9.84 (Not TK)	●	●	‡	●	I	I	●	●	‡	‡		2	NS ø	●	●	●	1.9.86 ●	●	I	I	Ic			Ic	I
	Before 1.9.84 (TK)	●	●	‡	●	I	I	●	●	‡	‡		2	NS ø	●	●	●	1.1.85 ●	●	I	I	Ic	I	I	Ic	I
100,000 Plus	After 1.9.84	●	●	‡	●	●	●	●	●	‡	‡	●	2	250mm 340mm	R	●	●	●	●	I	I	Ic	I	I	Ic	I

Remarks

● Fitting required

‡ If provided

I On international voyages

c Only applies to ships constructed after 25.5.80

TK Tankers

NS Not specified

R At least reflection plotter equivalent

Passenger Ships

In general, for v/l's of their tonnage and date of construction with the exception of Class V (in smooth waters) and not International) for which – no specification

ø If installed AFTER 1.9.84 340mm/250mm

Display diam.

sounder installation and comply with part IV of these regulations;

(b) If constructed on or after 1 September 1984, be fitted with a speed and distance measuring installation and comply with part V of these regulations;

(c) Whenever constructed, be fitted with a direction finder installation and comply with part VI of these regulations;

(d) If constructed on or after 25 May 1980, be fitted with an installation for homing on the radiotelephone distress frequency (2182 kHz) and comply with part VII of these regulations.

(6)

(a) Every ship of 10 000 tons or over constructed on or after 1 September 1984 shall be fitted with an automatic radar plotting aid and comply with part VIII of these regulations.

(b) Ships constructed before 1 September 1984 shall be fitted with an automatic radar plotting aid and comply with part VIII of these regulations, as follows:

(i) Tankers of 40 000 tons or over, by 1 January 1985;

(ii) Tankers of 10 000 tons or over but less than 40 000 tons, by 1 January 1986;

(iii) Ships, other than tankers, of 40 000 tons or over, by 1 September 1986;

(iv) Ships, other than tankers, of 20 000 tons or over but less than 40 000 tons, by 1 September 1987;

(v) Ships, other than tankers, of 15 000 tons or over but less than 20 000 tons, by 1 September 1988.

(7) Every ship of 100 000 tons or over constructed on or after 1 September 1984 shall be fitted with a rate of turn indicator and comply with part IX of these regulations.

11 Publications

11.1 Written data accessibility levels

There is a mass of published navigational information aimed at the professional ship navigator. Mentioned in this chapter are those official and semi-official publications likely to be carried on board the average foreign-going merchant ship. Some are compulsory, others optional. It does not include those publications designed for use on specialized ships or ships on specialized trades, or the abbreviated versions produced to satisfy the requirements of local and pleasure craft, though they may be mentioned in passing.

At any one time the majority of this written information is superfluous to the navigator's immediate needs. Unless a conscious effort is made to integrate all these publications into the normal navigation operation, many important sources can be habitually overlooked. The Sailing Directions are a case in point, the ignoring of which, though carried, has contributed to two notable incidents (see 'Stranding of MV Sealuck', *Seaways*, August 1985, p. 15; and 'Stranding of VLCC Aguila Axteca', *Seaways*, August 1985, p. 17).

The persons responsible for navigation have a duty to:

1 Make themselves aware of what written navigational information sources are available;
2 Obtain and study these publications in order that the relevant parts can be accessed at short notice;
3 Develop a routine that ensures that all corrections are obtained and made promptly and a system of documentation that indicates clearly to all users the date of the latest correction. (A publication that cannot be relied upon will either be totally rejected or used with varying degrees of risk by unsuspecting navigators.)

Information from these publications is used at various stages of the navigation process, some more frequently than others, and it is useful therefore to categorize the information into accessibility levels and locate them on the bridge accordingly.

Written data accessibility levels

Level 1 instantly necessary – commit to memory
Level 2 instantly accessible – on display
Level 3 readily accessible – on the bridge ready for immediate use
Level 4 available on the bridge – may take time to access.

Reference will be made to these accessibility levels throughout this chapter, using the abbreviation AL.

11.2 Chart catalogue and folios

11.2.1 *Admirality Chart Catalogue (NP 131)*

The catalogue is an integral part of a vessel's navigational chart outfit which is required by Statutory Instrument (see Section 11.26.2). It is published by the Hydrographer of the Navy with new editions yearly, and is obtainable through chart agents.

The catalogue lists and displays diagramatically all the navigational charts currently produced by the British Admiralty, including lattice charts and plotting charts for use with radio navigation aids. Also listed are other navigational publications such as Sailing Directions, Light Lists, Distance Tables, etc.

Consultation of this catalogue is essential if the best chart in terms of scale and coverage for a particular situation is to be identified.

Similar publications are available from the hydrographic offices of other nations.

AL 3

11.2.2 Folios and indexes

Navigation charts are grouped into folios, each covering a particular geographical area. Standard British Admiralty folios are in common use, although some shipping companies and chart agencies have their own standard folios.

The folios carried on board will, in general, reflect the intended trading areas. Economic reasons and the difficulty of keeping them all fully up to date makes the carriage of worldwide folios unlikely if not pointless.

If working with standard folios it remains the responsibility of the navigator to ensure that each folio includes all the large scale charts of the areas to be navigated (as the amount of detail shown varies with scale), and that the lattice charts relevant to the radio navigation aids are carried. It is important that a corrected index is kept with the folio to identify these additions.

AL 2 The current working chart
AL 3 The passage charts in sequential order and including other charts of the same area but of other scales
AL 4 All other charts.

11.2.3 Chart management and correction log

Maintenance of the chart outfit has a high priority amongst the functions of the navigating officer. It is essential to the safety of the ship that a high standard of efficiency is achieved in terms of both provision and correction of these charts.

Over the past 10–15 years, there has been a vastly increased amount of underwater activity around offshore oilfields and the like, much of it resulting in amendments to charts. Consequently, the task of keeping the charts up to date has become more time consuming. This has had two unfortunate side effects. The more serious is the failure by some navigators to obtain the latest editions and to apply the latest available corrections before using the chart. Secondly, the navigator may allow the demands of chart correcting to supersede his look-out function whilst on watch – potentially a most dangerous action. A considerable number of navigating officers believe that chart correcting on watch is necessary and that this is condoned by some masters and companies. This surely is a misapprehension, and warnings about keeping a proper look-out must be emphasized. However, it should also be recognized by companies and masters just how time consuming chart corrections can be, and the chart management function must be allocated its proper place amongst all the other jobs to be carried out by the deck officers.

When setting up a chart management system, the following points must be borne in mind:

1 There is no real benefit to be gained from carrying more chart folios than is necessary for the intended trading area, unless they are kept fully corrected. If a vessel is to change its trading area it is generally more efficient to obtain a complete updated folio before diverting.
2 Arrangements will need to be made for the supply of extra charts and new editions at short notice.
3 Arrangements will need to be made for the expeditious, certain and sequential supply of corrections (e.g. Notices to Mariners) both for the British Admiralty charts and any foreign charts carried.
4 A system of documentation is needed that shows quickly and clearly that all relevant corrections have been received and applied and also that all new editions have been ordered or obtained. A chart correction log is published by the Hydrographic Office which contains a correction sheet for each standard chart folio in numerical order, with spaces for logging the notices affecting them. (NP 133A.)
5 Arrangements will need to be made in the on-board organization for the time necessary to complete the chart correcting. Relevant charts must be up to date before the passage plan can be made.

AL 3 The chart correction log.

11.2.4 *Foreign charts*

Many foreign states now conduct surveys of their own coastal areas and consequently publish charts and plans for these areas. Although the survey information is passed on to the British Hydrographic Office for inclusion in British charts, time delays can be long.

For most areas covered by foreign charts, the British Hydrographic Office policy is to keep charts to the smallest scale which will adequately show the dangers and navigational aids. Chart correcting information, promulgated by Notices to Mariners, is normally restricted to those items considered essential for safe navigation. As a consequence some of these small scale charts may not meet the requirements of certain states (Canada, for example) for vessels visiting their waters in terms of scales and being up to date.

In general, where foreign charts are available it is advisable to make use of them for the greater detail and better scales that they provide. International standardization of chart symbols enables the foreign charts to be used with little difficulty, although some discrepancies still exist between British and some foreign charts as a result of different charting data. The main problem with using foreign charts and publications is generally the difficulty of obtaining new editions and the corrections that apply to them. Some correcting systems may be printed only in the national language.

The management of these foreign charts should be the same as for the rest of the vessel's chart outfit.

AL 2 The current working chart
AL 3 Passage charts
AL 4 All other charts.

11.3 Sailing Directions and Ocean Passages of the World

11.3.1 *Sailing Directions (NP 1 to 72)*

The texts of the pilot books (as they are more colloquially known) amplify the charted information, drawing the mariner's attention to the most important features of coastal areas. The most frequently used parts deal with port approaches, where the mariner is concerned with regulations, communications, signals, anchorages, tidal peculiarities and recommended approach paths. Also included are tables of expected radar detection ranges of prominent landmarks and views from seaward either as line drawings or increasingly, in new editions, as photographs of harbour approaches.

The pilot books should always be consulted in conjunction with the chart, in particular during the passage planning stage where a safe route is being sought. In general this source of information tends to be underused, probably owing to the sheer volume of information which needs to be sifted before the relevant parts can be found. Steps are currently under way to revise the format of these books to make them easier to use. New editions are published at intervals of ten to fifteen years.

AL 2 Data extracted for use in the passage plan
AL 3 The pilot book covering the passage in question
AL 4 All other pilot books.

11.3.2 *Ocean Passages of the World (NP 136)*

Ocean Passages of the World deals with all those sea areas not covered by the Sailing Directions. It details recommended ocean routes based on:

1 The shortest distance
2 Climatic conditions, e.g. prevailing winds, storm frequency, ice, fog, and currents likely to be encountered.

Its information is most useful when trying to optimize an ocean crossing by weather routeing, and would be used in conjunction with pilot and routeing charts (Section 11.8.2). It should also be consulted during normal passage planning for load line restrictions and recommended landfall areas.

AL 3 During the ocean passage planning stage
AL 4 At all other times.

11.3.3 Supplements and updating

Corrections to the Sailing Directions are issued as Notices to Mariners. The usual procedure is to keep a file of corrections and enter them in pencil in the page margin. The corrections are included in a supplement which is updated approximately every 18 months. Supplements should be kept with the parent volume at all times. Care must be taken when using Sailing Directions to consult the latest supplement. A list of corrections in force, i.e. issued since the last supplement, is published in section IV of the weekly Notices to Mariners for the last week of each month, and in addition those in force at the end of the year are reprinted in the Annual Summary of Notices to Mariners.

Ocean Passages of the World is updated by new editions. This occurs very infrequently as the information contained therein is liable to only a very gradual change. The last new edition (1987) reflected the increased power and speed of vessels presently plying the oceans and also the new trade routes that have been developed over the past 20 years.

11.4 Notices to Mariners

11.4.1 Weekly Notices to Mariners

The weekly edition of Notices to Mariners forms the principal method of keeping Admiralty charts and publications up to date. Without the conscientious use of these notices the charts quickly become dangerously out of date as the positions and characteristics of navigation marks are changed and new shoal depths are found. Although the notices are issued weekly, mariners should bear in mind that the data may take several weeks to reach the Admiralty, according to its source. Hence, even if a chart is corrected to the latest notice, there can never be a 100 per cent guarantee of completeness without the addition of the radio navigation warnings (see Sections 11.4.3 and 19.1.9).

The Notices to Mariners are divided into six sections:

- I The index, which allows a rapid identification of those publications affected by that edition
- II Notices to mariners, including advance notice of new charts and publications, corrections to charts, and temporary and preliminary notices
- III Radio navigational warnings: a summary of the important navigation warnings that were promulgated by radio during the previous week (see Section 11.11.1)
- IV Corrections to Sailing Directions (see Section 11.3.3)
- V Corrections to Admiralty Lists of Lights and Fog Signals, designed to be cut out and pasted in the appropriate page of the list (see Section 11.5)
- VI Corrections to Admiralty Lists of Radio Signals, applied as in V.

Efforts are constantly being made to reduce the time that it takes to put this information into the hands of the mariner. Developments in communications technology are being studied. NAVTEX and satellite links (see Section 19.1.3) are two possible future means of keeping the all important publications up to date.

AL 1–4 The booklet is designed to be dismembered and each section applied to the appropriate publication or file.

11.4.2 Annual Summary of Notices to Mariners

This summary is published on 1 January each year and contains three sections:

1 Notices to Mariners of an important and semi-permanent nature, the majority of which are republished each year with little or no change. They will include, amongst others, addenda and corrigenda for the Admiralty Tide Tables, advice on distress and rescue, submarines, AMVER, sources of Admiralty publications, long range navigational warnings and traffic separation schemes. This list of traffic separation schemes includes all schemes whether IMO adopted or not.

2 The summary of temporary and preliminary notices in force. Note that these will include the current operational status of the radio navigational aids.

3 A summary of corrections to the Sailing Directions still in force.

AL 2 The temporary and preliminary notices and long range warnings related to the passage will be extracted and included in the passage plan (see Part II)
AL 3 All other annual notices
AL 4 The remainder of the publication.

11.4.3 NAVTEX

Since 1982 certain European coast radio stations have been broadcasting weather and navigational information by teleprinter. It is a co-ordinated service and is expected eventually to operate worldwide. A dedicated NAVTEX receiver and printer is pretuned to 518 kHz and left operational continuously.

The types of messages presently include:

A Navigational warnings which affect the particular coverage area, issued by the Hydrographic Office
B Gale warnings issued by the Meteorological Office
C Ice reports (where applicable)
D Distress alerting
E Weather bulletins
F Pilot service messages
G–J Navigational aids.

The normal schedule for transmissions for a particular area is four hourly, but message types B and D are always sent out immediately on receipt by the coast station. The receiver may be programmed to receive only the relevant parts of the transmissions.

The fitting of NAVTEX is compulsory for most vessels after 1993 under the GMDSS requirements (see 'Sections 17.2.2 and 19.1.3.2).

AL 2 The printout; for immediate attention of the OOW.

11.5 Admiralty Lists of Lights and Fog Signals (NP 74–84, vols. A–L)

These volumes, the carriage of which is compulsory on British ships, provide details of all navigation lights except buoy lights under 8 metres in height. They include lights not necessarily shown on small scale charts. In the leading pages to each volume are graphs that enable dipping and visibility ranges to be estimated.

The data in these volumes tends to change frequently, and consequently a complete section (section V) of the weekly Notices to Mariners is devoted to these changes and each volume has a new edition approximately every 15 months.

It is worth remembering that because of the time needed to prepare and distribute blocks or new editions of charts, a change to a navigation light may be applied to the Light Lists several weeks before the corresponding change to the chart appears (also some alterations of a temporary nature are promulgated only in the List of Lights). Consequently, the Light Lists should be the most up to date and must invariably be consulted when using lights for navigation.

AL 2 The relevant volume when on a coastal passage at night
AL 3 The relevant volume at other times
AL 4 Other volumes.

11.6 Admiralty Lists of Radio Signals (NP 281–286)

There is a compulsory carriage requirement on British ships for these volumes.

11.6.1 Volumes for navigators

The navigator is primarily concerned with two of the five volumes:

Volume 2 gives details of radiobeacons including aero radiobeacons in coastal regions, radio direction finding stations, coast radio stations providing a QTG service. (i.e. DF bearings on request), calibration stations, Racons and Ramarks. Also included are electronic position fixing systems, satellite navigation systems, standard times and radio time signals, and diagrams showing radio beacon locations and electronic position fixing system coverage.

Volume 6 (parts 1 and 2) provides particulars of radio stations working in the port operations and information services, services to assist vessels requiring pilots, and services concerned with traffic management. Diagrams to be used in conjunction with these services are included.

Amendments to the data are fairly numerous and appear in section VI of the weekly Notices to Mariners. Supplements of corrections are issued regularly and must be consulted when using all Lists of Radio Signals. New editions occur every one to two years.

AL 2 All data extracted for use in the passage plan, such as radiobeacon characteristics and VHF working channels

AL 3 All remaining data and volumes.

11.6.2 *Volumes for radio operators*

Volumes 1, 3 and 4 of the Admiralty List of Radio Signals are more the concern of the communications specialist than the navigator. However, as more modern and (from an operator's point of view) less complex communications systems are developed, the navigator is likely to become directly involved with the information contained in these volumes (see Section 19.1.4):

Volume 1 contains particulars of coast radio stations, medical advice by radio, quarantine and pollution reports, maritime satellite service, regulations regarding the use of radio, distress, search and rescue procedure and reporting systems.

Volume 3 gives details of radio weather services and navigational warnings, including meteorological codes.

Volume 4 contains a list of meteorological observation stations.

Each volume includes associated diagrams.

The updating system for volumes 1, 3 and 4 is identical with that of the volumes mentioned in 11.6.1.

AL 2 and **AL 3** for the communications specialist.

11.7 The Mariner's Handbook (NP 100)

This book, which must be carried on British ships, is a wide ranging source of information for the mariner.

Chapters deal with:

1. Navigational publications and their correction
2. The use of charts and other navigational aids including buoyage
3. Operational information and regulations including pilot ladders and pollution
4. The sea, its tides, currents and anomalies
5. General maritime meteorology with advice on operating in adverse weather conditions such as tropical revolving storms
6. Ice – descriptions and photographs
7. Operations in polar regions. Precautions when navigating through ice
8. Observing and reporting hydrographic information
9. The IALA Buoyage System
10. Collision Regulations and Glossary.

New editions are published infrequently with major alterations and additions appearing as a supplement.

AL 4 Available as a reference at all times, although the bridge watchkeeper should be familiar with the majority of the contents of this book.

11.8 Passage planning charts, routeing charts and DTp publications

11.8.1 *Passage planning charts*

Passage planning charts are a new concept, designed to assist the navigator in choosing a route through difficult navigational areas, for example the English Channel and Dover Straits. They are small scale charts showing the whole area, and include many additional references such as recommended routes, crossing points, radio stations and broadcast times, reporting procedures and other advice to mariners.

AL 2 During the passage planning stage
AL 3 During the actual passage.

11.8.2 *Routeing charts*

These charts are the ocean equivalent of the passage planning charts previously mentioned, but have been published for very many years. They show at a glance important information relevant to the planning of an ocean crossing and are compiled for each month of the year. They include details of the principal cross-ocean routes, current directions and rates, load line zones, and other climatic and meteorological information such as wind directions and frequency, and storm, fog and ice areas.

New editions are published occasionally but very little of the data is subject to change. They should be used in conjunction with Ocean Passages of the World (Section 11.3.2).

Similar charts are published by the United States.

AL 2 During the passage planning stage
AL 3 During the actual passage.

11.8.3 *Guide to the Planning and Conduct of Sea Passages (DTp)*

This guide restates and elaborates on the passage planning recommendations contained in Notice M.854, Navigation Safety, and its annex (see Section 12.4.2). It includes worked examples of parallel indexing and the planning of a coastal passage. Annex III gives a useful checklist of items for passage appraisal (see Section 14.7.1). This booklet forms a handy chartroom reference for coastal passage planning.

AL 2 Annex III, the passage planning checklist
AL 3 The remainder.

11.9 Tide tables

11.9.1 *Admiralty Tide Tables. Volume 1: European Waters (NP 201)*

Part I of this annual publication gives daily predictions for high and low water for a number of standard ports for which ample data is available. Part II gives the differences from one of the standard ports for many other secondary ports. The data for some secondary ports is incomplete and should be regarded as approximate. The predictions are, however, accurate enough for normal navigational purposes provided the usual safety margins are allowed.

Before using these tables the navigator must note the corrections that appear pasted inside the front cover and also the addenda and corrigenda in the Annual Notice to Mariners no. 1. These are the corrections and additions found necessary in the time interval between the volume being sent for publication and coming into force – a period of several months.

Included in the volume are comprehensive examples of the method of use and warnings of the various factors, such as tidal surges, that could result in the actual height being different from that predicted.

AL 2 Extracted predictions for use with the passage plan
AL 3 All the remaining data.

11.9.2 *Admiralty Tide Tables. Volumes 2 and 3: Atlantic, Pacific and Indian Oceans (NP 202 and 203)*

The majority of the predictions in these volumes are based on data provided by appropriate foreign authorities and laid out in a similar manner to volume 1. However, because of the often very large distances between a secondary port and the selected standard port, the method of predicting the secondary port tides is different and accuracy is generally lower. Consequently, when foreign tide tables for a secondary port outside Europe are available, they should be consulted.

Volumes 2 and 3 also include tables for estimating tidal streams at selected points.

Corrections to these volumes are provided by the same method as for volume 1.

AL 2 Extracted predictions for use with the passage plan
AL 3 All remaining data.

11.9.3 Local tide tables

Harbour authorities of many ports publish their own daily predictions and much of this data is in fact used in the compilation of the Admiralty Tide Tables (ATT). If these local tide tables are available, the navigator might find them more convenient to use than predicting from the ATT. In addition, with foreign secondary ports, these local tables may be more accurate.

AL 2 When relevant
AL 4 Otherwise.

11.10 Tidal stream atlases

11.10.1 Admiralty Tidal Stream atlas (NP 209, 217–21, 233, 249–53, 256–7, 264–5, 337)

These atlases give the rates and directions of tidal streams around the British Isles and Hong Kong. Data is given for both spring and neap tides and covers a period of six hours either side of high water at some standard port within the area. They are a very convenient source of tidal information for vessels in coastal areas, but not in port approaches. No details of heights are included.

AL 2 Whilst relevant
AL 4 Otherwise.

11.10.2 Current atlas of the world

Illustrative information concerning the currents of the world's oceans can be found in several publications. The data is generally given in a similar form to the Admiralty Tidal Stream Atlas, although as changes with time are only very gradual, they are illustrated on a monthly or seasonal basis.

Monthly routeing charts also include current information, and Ocean Passages of the World comments in detail on all significant ocean currents.

AL 2 During the passage planning stage
AL 3 At other times.

11.10.3 Tidal charts

These charts are published by the Hydrographic Office and show curves of equal tidal range and equal times of tides for certain areas around the UK, Malacca Strait and Arabian Gulf. They allow the navigator to estimate the rise of tide offshore and are based on a given standard port, showing the corrections to apply for the coastal area. They can be particularly useful to deep draught vessels operating in relatively shallow offshore areas. A certain amount of interpolation is needed when using these charts and, depending on the amount of data that was available when drawing up the chart, the predictions arrived at may not be very accurate.

AL 2 During the passage planning stage for deep draught vessels
AL 4 For other vessels and at other times.

11.11 Navigation warnings

11.11.1 Radio navigation warnings

The Radio Navigation Warnings is a co-ordinated worldwide system. It is designed to promulgate quickly important navigational information to mariners. Such warnings relating to recent dangerous wrecks, drifting mines, ice reports, alterations to navigational aids, etc. are all urgent enough to warrant immediate issue. Many of these warnings are of a temporary nature, but others are more permanent and may be superseded by a Notice to Mariners.

There are three principal types of radio navigation warning:

1 Those broadcast for a Navarea (sixteen Navareas cover the world) (see Figure 11.1)
2 Coastal warnings broadcast by a local radio station in the particular coastal area to which the warning applies
3 Local warnings, usually broadcast by port radio stations and applying only to the pilotage area of that port.

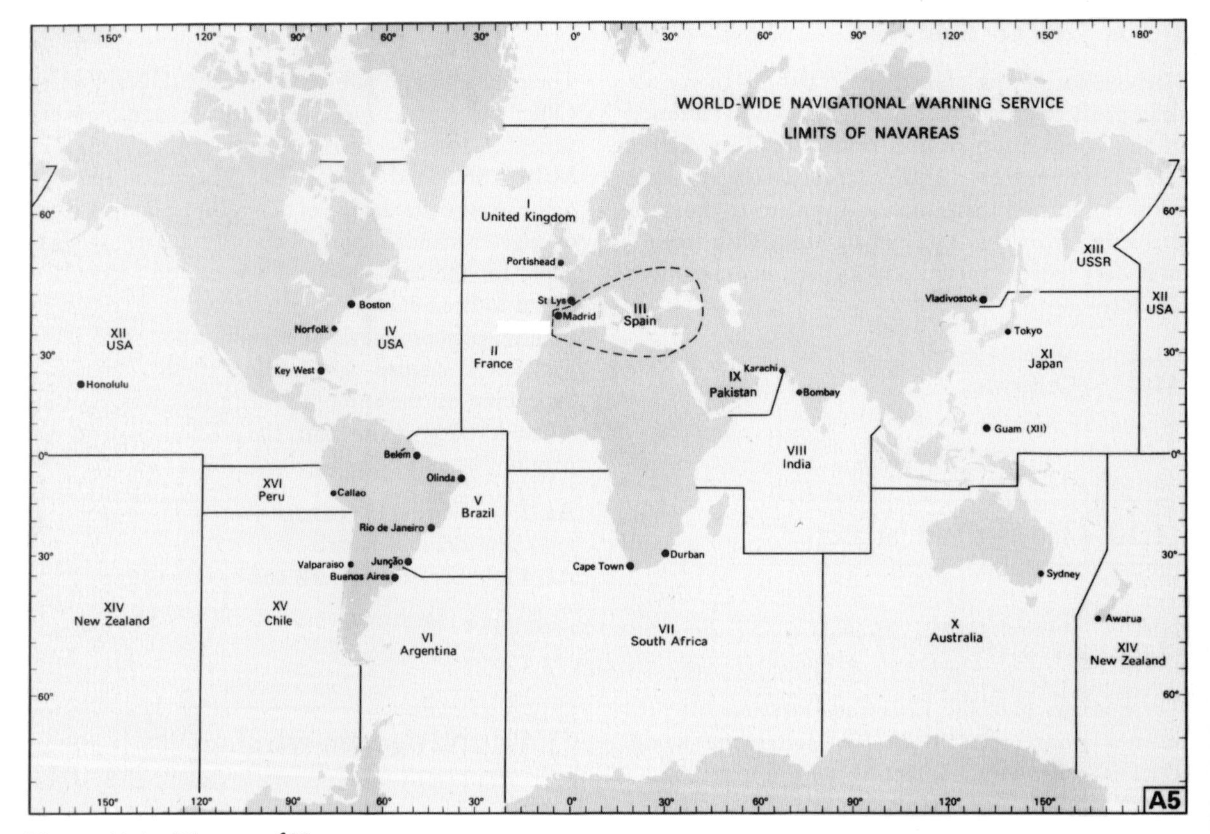

Figure 11.1 Diagram of Navareas

The United States also continues to broadcast navigation warnings by radio for the Atlantic (Hydrolant) and Pacific (Hydropac) Oceans (see Figure 11.2).

Because of the serious content of these warnings, navigators must make every effort to receive them, and details of the service and broadcast times and frequencies are published in Annual Notices to Mariners nos. 9 and 13 and in ALRS, vol. 3 (see also Sections 11.4.3, 19.1.3.3 and 19.1.9).

On receipt they must be immediately brought to the notice of the master or OOW, and when in port the current warnings must be obtained before sailing.

A file should be kept of all warnings and the details should be entered on the navigation chart in pencil until such time as it becomes permanent (by Notice to Mariners) or is cancelled.

AL 2 When relevant to the passage plan
AL 3 On file.

11.11.2 Published navigation warnings

All radio navigation warnings are subsequently published and are made available from various offices around the world. All the warnings are prefixed with a sequential number so that the navigator will know when he has them all. Missing warnings should be obtained before departure from port.

The Hydrographic Office reprints all Navarea 1 warnings and other selected important warnings for other Navareas, Hydrolant and Hydropac series in section III of the weekly edition of Notices to Mariners. Those warnings of a permanent nature will eventually be issued as a Notice to Mariners and be fully incorporated into the appropriate chart or other publication.

AL 2 When relevant to the passage plan
AL 3 On file.

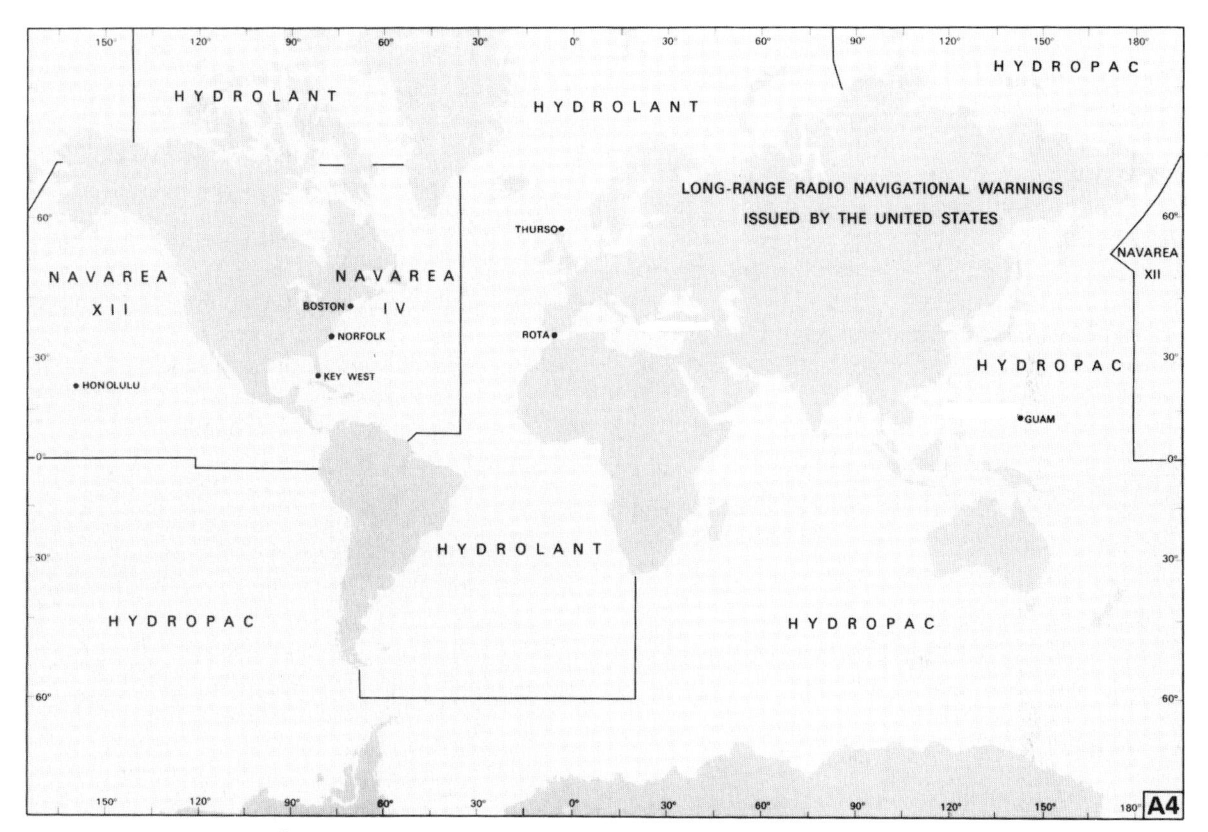

Figure 11.2 Diagram of long range navigation warning areas

11.12 Weather reports

11.12.1 Radio broadcasts

The current and forecast weather conditions are of fundamental interest to the mariner, from both a safety and an operational point of view. Radio broadcasts form the principal method of promulgation of this data, and full details of transmissions are contained in ALRS, vol. 3. Important information concerning safety such as gales, fog and ice are also broadcast as radio navigation warnings (Section 11.11.1; see also Section 19.1.9).

As much of this information can affect passage planning, it is important to receive all weather messages for the area concerned and to receive them as early as possible, so that they may be acted upon in good time. Whilst in port, local forecasts are available by telephone from the local meteorological station.

It is worth remembering that a fair proportion of the data that is used to compile these reports and forecasts derives from the current weather observations sent in by merchant ships on passage. Although in recent years the weather satellites have provided much valuable data for ocean areas, without merchant ship reports the accuracy of weather forecasts would be very much worse.

AL 2 On file.

11.12.2 Facsimile charts

Weather maps, wave height prognoses and other information of interest to mariners can be transmitted by radio from certain coast stations for reception by a special dedicated receiver (sometimes referred to as

Mufax, which is the name of the receiver commonly fitted on British ships). This machine can print out the maps, albeit rather slowly, line by line (see also Section 19.1.3.5).

The weather maps and prognoses are an essential part of ocean weather routeing, and details of transmissions are given in ALRS, vol. 3. Developments in satellite communications are gradually improving the quality of this data source.

Details of facsimile transmissions of navigation warnings and ice charts are given in ALRS, vol. 3. In coastal areas the NAVTEX system (Section 11.4.3) promulgates this information but not in diagrammatic form.

AL 2 At specified times when relevant to the passage.

11.13 Merchant Shipping (M) Notices

11.13.1 M Notices in general

Merchant Shipping (M) Notices are issued as and when required by the Department of Transport. They advise on the interpretation of various shipping rules and regulations and also make recommendations concerning the safe operation of ships, and should therefore be seen as statements of good practice. They are the usual method of introducing IMO resolutions to British ships, although the more important resolutions do result in Statutory Instruments (SI) being issued.

M Notices cover all aspects of ship operations and several are related directly to navigation and bridge operations. The latter group should be filed on the bridge and hence available to the watchkeeping officers at all times. Because of the increasingly large number of these notices, care must be exercised in keeping this file up to date and watchkeepers need to make a conscious effort to read and assimilate their contents. A summary of all notices currently in force is issued approximately once per year.

AL 3 All navigational notices
AL 4 All other notices.

11.13.2 Notable navigational M Notices

By late 1991 there were over 30 notices relating to navigational subjects. In the context of this book, the following all offer useful guidance (asterisks indicate that extracts from these notices are included in this manual):

M.845* Dangers in the Use of VHF Radio in Collision Avoidance: refers to language and identification problems and the risk of agreeing actions contrary to the Collision Regulations (see Sections 19.1.8 and 19.3.6)

M.854* Navigation Safety: gives detailed guidance on the planning and conduct of passages (see Section 12.4.2)

M.930 Interaction between Ships: describes the various interaction dangers

M.982 Use of EPIRBs on Frequencies 121.5 MHz and 243 MHz: lays down the code of practice for their use and points out their limitations

M.988* Report to Harbour Authorities by Tankers Entering or Leaving United Kingdom Ports (see Sections 23.5 and 23.9.5)

M.1016 Principles and Operational Guidance for Deck Officers in Charge of a Watch in Port; and

M.1102* Operational Guidance for Officers in Charge of a Navigational Watch: 1016 and 1102, give recommendations on the items that should appear in either companies' or masters' standing orders (see Section 18.3.1)

M.1026* Proper Use of the VHF Channels at Sea: sets out the correct procedures to be used with VHF and gives several example exchanges (see Sections 19.1.5 and 19.3.3)

M.1119* Radiotelephone Distress Procedure (see Sections 19.1.7 and 19.3.5)

M.1158* The Use of Radar and Electronic Aids to Navigation: sets out what is considered to be the current good practice in the use of navigational aids (see Sections 5.11.5 and 6.6.3)

M.1192* Merchant Ship Position Reporting: draws attention to the AMVER system (see Sections 23.5 and 23.9.4)

M.1219 Magnetic compasses – operation and maintenance: gives recommendations to be followed to ensure accurate and reliable operation of magnetic compasses.

M.1235* Use of the Radiotelegraph Automatic Keying Device by Unskilled Persons in an Emergency (see Sections 19.1.7 and 19.3.4)

M.1252* Standard Marine Navigational Vocabulary (and amendment): has been compiled to assist in the greater safety of navigation by standardizing the language used in communication for navigation at sea and port approaches (see Sections 19.1.5 and 19.3.2)

M.1448* Observance of Traffic Separation Schemes: gives advice on the interpretation of rule 10 (see Sections 23.6 and 23.9.6)

M.1449* Navigation in the Dover Strait (see Sections 23.6 and 23.9.3)

M.1471* Use of Automatic Pilot: draws attention to the requirements of SI 1981, no. 571 regarding the testing of steering gear (see Section 1.6.7)

AL 3 On file.

11.14 Navigation systems manuals

Every navigation system fitted to a ship must have with it the appropriate operating manual as supplied by the manufacturer. These manuals lay down the correct method of operation so as to achieve optimum performance and to avoid damage or misuse.

Included in the manuals for position fixing systems will be reminders of the accuracy attainable and notification of any inherent errors to which the system is liable. Some systems such as Decca, Loran and Omega have comprehensive correction tables and diagrams to enable the optimum accuracy to be achieved. Naturally, the operators of these systems will require the operating manual to hand whenever the system is in use. The modern ARPA radar tends to be very sophisticated, and as a consequence, if the operator expects to get the best performance from it, the manual will need to be studied extensively (see Sections 5.8, 5.9).

Most navigators would benefit from revising the contents of the systems manuals occasionally simply to prevent bad habits and short cuts creeping into their technique and to avoid overlooking the maintenance requirements that occasionally apply.

Several of these manuals are subject to amendments as the system is modified and developed. These amendments to the manual are obtainable from the manufacturer.

AL 2 Correction tables when applicable
AL 3 At other times.

11.15 Mersar Manual (IMO)

The Merchant Ship Search and Rescue (Mersar) Manual is published by the IMO as guidance for those in distress at sea and requiring assistance or those rendering assistance. In particular it is designed to aid the master of any vessel called on to conduct search and rescue (SAR) operations.

It is convenient to use and, although not compulsory, it has a sure place on the bridge of every vessel. Much of the contents should be familiar to all bridge watchkeepers (see Chapter 17).

AL 2 During SAR operations
AL 3 At other times.

11.16 Guide to Helicopter/Ship Operations (International Chamber of Shipping (ICS))

Over the last decade, helicopters have become increasingly common as a means of transport, with many of their operations involving the marine environment and contact with shipping. Operations in the past have been made more difficult and dangerous by a general lack of understanding by both helicopter pilots and ships' crews of each other's problems, and a lack of standardized communications procedure and equipment.

This booklet is produced as guidance for the ship's crew to rectify these omissions by setting out standard procedures for working and communicating with helicopters.

Sections include:

1 Advice to ship managers on the need to employ helicopters of the correct type for a particular job
2 Operating details of helicopters such as areas of responsibility, regulations and limiting weather conditions
3 Facilities required on the ship for landing and winching operations
4 Operating procedures including communications
5 Emergency procedures involving helicopters such as Medevac and emergency landings
6 Advice for helicopter passengers.

AL 2 Derived checklists
AL 3 All other data.

11.17 Bridge Procedures Guide (ICS)

Most shipping companies and ship operators have, for many years, included in their standing orders various instructions to masters and officers concerning routine bridge procedures. This guide is an attempt to standardize these various procedures by taking a consensus of those at present in use and including all those items that are recognized as good practice.

These standard procedures are now available as good advice to those masters who have the responsibility of drawing up their own procedures. In addition, if the procedures are adopted by the majority of ships as recommended, officers transferring between companies will be immediately familiar with them.

Naturally it would be impossible for these recommended procedures to be all embracing and to apply fully to every type of ship. However, they do form a useful basis from which a particular ship's procedures can be developed, e.g. certain items not applicable would be deleted and other items would be included because of the particular ship's characteristics.

AL 2 Checklist for the current situation
AL 3 All other data.

11.17.1 Guidance to officers

The first part of the ICS Bridge Procedures Guide gives masters and officers a brief reminder of those matters of relevance in routine bridge operations including bridge organization, duties of the officer of the watch (OOW) and good practice in the operation and maintenance of navigation equipment.

AL 3

11.17.2 Routine and emergency checklists

Part B of the ICS Bridge Procedures Guide comprises bridge checklists for routine operations. These include, among others:

Preparation for sea
Embarkation/disembarkation of pilot
Changing over the watch
Preparation for arrival in port
Restricted visibility
Heavy weather
Master/pilot information exchange.

Part C gives emergency checklists including:

Main engine failure
Steering failure
Imminent collision/collision
Stranding
Fire
Man overboard

For example, the quoted man overboard procedure is:

1 Lifebuoy with light, flare or smoke signal released
2 Avoiding action taken
3 Position of lifebuoy as search datum noted
4 Master informed
5 Engine room informed
6 Look-outs posted to keep person in sight
7 Ship manoeuvred to recover person as recommended on wheelhouse poster
8 Three long blasts sounded and repeated as necessary

9 Rescue boat's crew assembled
10 Position of vessel relative to person overboard plotted
11 Vessel's position available in radio room, updated as necessary.
12 Man overboard warning broadcast

AL 2 When applicable
AL 3 Otherwise.

11.18 Handbook for Radio Operators (Lloyds of London Press)

This is intended as a guide for persons operating radio equipment in the marine radio frequency bands. Most of the details included are subject to international regulations, and strict observance of the provisions and procedures covered is essential for the efficient exchange of communications with other ships and shore stations (see also Section 19.1.5).

Where a specialist radio operator is carried, the majority of communications will be under his control. However, with modern equipment there is an increasing tendency for bridge personnel to conduct their own radiotelephone communications on the MF and HF bands and on Satcoms in addition to the well established VHF band. To do this, the person concerned must hold a GMDSS General Operators Certificate and be familiar with the relevant sections of the handbook (see also Section 11.6). Particular attention should be paid to chapter 7 of the handbook which deals with distress, urgency and safety communication by radiotelephone (see Section 19.1.7).

The book is updated by occasional new editions, the latest of which, incorporating the GMDSS requirements, is due for publication in 1993.

AL 3

11.19 Ships' Routeing (IMO)

This publication is intended primarily for administrations responsible for planning and supporting traffic routeing systems for use by international shipping. It gives details of every IMO approved routeing scheme, including diagrams, although mariners should bear in mind that additions and changes to these schemes will probably appear as navigation warnings, Notices to Mariners and corrections to charts long before amendments are issued for this volume.

It is also important to appreciate that the conduct of vessels within or near these schemes is prescribed by rule 10 of the International Collision Regulations (see Sections 23.6 and 22.3.1). Some routeing schemes not adopted by IMO but instituted by local authorities may be included on the charts, and details of these appear in the Annual Summary of Notices to Mariners.

AL 4

11.20 International Code of Signals (DTp)

The purpose of the International Code of Signals is to provide methods of communication in safety related situations, especially where language difficulties arise. It is suitable for transmission by all means of communications including radiotelephony, and is based on the principle that every single, two or three letter signal will have a complete and unambiguous meaning.

When being used on the radiotelephone, all letters and numbers must be spelt out using the phonetic pronunciations, and all messages should be preceded by the word INTERCO to warn the receiving station that the following message will be using International Code (see also Sections 19.3.2 and 19.3.3).

All bridge watchkeepers should have committed to memory the single letter 'urgent' codes and have a good working knowledge of the two letter codes and their layout, by topic, in the book. The three letter codes are concerned with medical matters, and begin with letter M.

Carriage of the Code is compulsory, and it is advisable to keep it within reach of the radiotelephone.

Lists of amendments are published at infrequent intervals.

AL 3

11.21 The Nautical Almanac (HMSO) and navigation tables

Carriage of the annual edition of the Nautical Almanac (NP 134) is compulsory and essential for celestial position fixing. Although compiled for one particular year it can be used in emergency for the subsequent year for sun and star sights with slightly lower accuracy. An explanation of how to use the almanac under these circumstances is included. Other frequent uses for the almanac are compass error checks and sunrise and sunset calculations.

Navigation tables are also compulsory and have various publishers. Their use with celestial position fixing and the 'sailings' calculations is gradually being superseded by sight reduction tables and hand held calculators. However, there remain many other sections essential to accurate and safe navigation.

AL 3

11.22 Port information books

11.22.1 By port authorities

Information for shipping concerning port entry is included in the Sailing Directions. In addition to the data related to navigation in the port approach, there is detailed guidance on local regulations and a description of port facilities. ALRS, vol. 6 also gives details of traffic management and communications.

Most major ports publish similar guides to ships' masters, but these generally concentrate on facilities and regulations. The information tends to be very detailed and, provided the latest edition is available, it is likely to be the most up-to-date source on the subject of regulations and facilities (see also Section 23.8).

AL 2 Extracted data for inclusion in the passage plan
AL 3 All other data.

11.22.2 Guides to port entry

Several publishers have produced complete collections of information regarding world port facilities. Very little of this information is of direct navigational interest, and hence these books are perhaps more likely to be on the desk of the operations manager than on the ship. However, some of the guides do include useful plans and mooring diagrams. Updating may be either by supplement or by yearly new edition.

Three well known guides are:

World Port Directory (Fairplay Publications)
Ports of the World (Lloyds)
Guide to Port Entry (Shipping Guides Limited)

AL 4

11.22.3 By own ship

Considerable benefit may be gained by keeping brief notes of any peculiarities relating to port approach and entry procedures for your particular ship. Much of this information comes from the pilot and by observing actual operations, e.g. the point at which tugs are made fast, turning areas, height of pilot ladder, conditions of berths, competence of line handlers, etc. This information can all be useful on a subsequent visit, especially if the officers have been changed since the previous visit.

AL 2 Data included in the passage plan
AL 3 All other data.

11.23 Own ship manoeuvring and stopping data

11.23.1 Trials data

After completion of fitting out, and before entering

service, all vessels undergo proving trials to make sure that the performance meets the specification. Included in these trials are measurements of the vessels turning circles and stopping characteristics. These measurements will be used by the ship handler to assess the searoom needed to execute emergency manoeuvres.

This information is essential to whoever has control of the vessel. Consequently, the master and bridge watchkeepers should memorize it. For the benefit of the pilot the data should be presented in diagrammatic form and displayed at the control position. Some countries have regulations requiring the display of manoeuvring data. IMO Resolution A601(15) recommends a format for this display (see Section 11.26.1).

AL 1 For the master and watchkeeping officers
AL 2 For the pilot.

11.23.2 Compiled from own observations

The manoeuvring data gathered during the trials generally appertains to only one set of conditions, e.g. calm weather, no current, deep water, clean hull and light draught; and relates primarily to emergency manoeuvres. In practice, variables such as draught, trim, weather conditions, under-keel clearance and how the engines and helm are used in combination can have a marked effect on these characteristics.

During service it is advisable to gather supplementary information at every opportunity to add to the trials data. In particular, data is needed concerning the characteristics at other states of loading and trim and for angles of helm other than maximum. Significant changes in characteristics caused by wind should be recorded (see also Section 15.3.3). Critical revolutions values, and speed at which steerage way is lost, should also be ascertained and recorded.

AL 2

11.24 International Collision Regulations

No bridge watchkeeper should need reminding of the existence or importance of these regulations (see also Chapter 22). They are subject to occasional revision and amendment, and the current (1985) version appears as UK Statutory Instrument 1983, no. 709, which should be to hand on the bridge.

The vast majority of the included data should be committed to memory, although not necessarily word for word, to allow the watchkeeper to make the immediate correct responses in an emergency.

Instant recognition of all navigation light and shape configurations is essential to the safe conduct of the vessel. Some of the more uncommon arrangements might usefully be put on to a display diagram to assist in memorizing.

AL 1 The steering and sailing rules
AL 3 Technical details of lights and sound signals as contained in the annex to the regulations.

11.25 Standing orders

11.25.1 Company standing orders

Although not compulsory, company standing orders are the generally accepted method of standardizing the operation of a company's fleet of ships. They lay down the standard practices to be observed and detail the areas of responsibility of each officer. This is particularly important where officers may not be company trained.

If the standing orders are to remain as a positive contribution to the efficient running of the ship they must be updated periodically to introduce modern practices and remove outmoded ones.

AL 1 Areas of responsibility of the persons concerned
AL 3 All other data.

11.25.2 Master's standing orders

These set the general standards expected by the master of his officers and in particular how the bridge is to be operated. They should not conflict with company standing orders or any statutory requirements.

The master's standing orders can be most effective in producing an efficient bridge team, provided they are carefully compiled. Too often the purpose of these standing orders is misunderstood by junior watchkeepers, and there is merit in discussing the contents with the officers concerned (see also Section 19.2.1).

It is important not to write one thing and then give the impression of meaning something else. For example, the master who leaves orders to call him if in doubt and then grumbles at the first false alarm is sowing the seeds of eventual disaster.

AL 1

11.25.3 Night orders

Night orders are supplementary to the standing orders and give specific instructions to the officer of the watch for the period while the master is in bed (see also Section 19.2.1).

There is no need for a separate document where a passage plan is in use; the master's orders for the night should appear in the bridge note book, thus avoiding a multiplicity of executive documents.

AL 2

11.26 Extracts from official publications

Within Sections 11.26.1 and 11.26.2, the paragraph numbering of the original publications has been retained.

11.26.1 IMO Resolution A601(15): Recommendation on the Provision and the Display of Manoeuvring Information on Board Ships

1 Introduction

1.1 In pursuance of the Recommendation on Data Concerning Manoeuvring Capabilities and Stopping Distances of Ships, adopted by resolution A160(ESIV), and paragraph 10 of regulation II/1 of the International Convention on Standards of Training, Certification and Watchkeeping for Seafarers, 1978, Administrations are recommended to require that the manoeuvring information given herewith is on board and available to navigators.

1.2 The manoeuvring information should be presented as follows:

.1 Pilot card
.2 Wheelhouse poster
.3 Manoeuvring booklet.

2 Application

2.1 The Administration should recommend that manoeuvring information, in the form of the models contained in the appendices, should be provided as follows:

.1 for all new ships to which the requirements of the 1974 SOLAS Convention, as amended, apply, the pilot card should be provided;
.2 for all new ships of 100 metres in length and over, and all new chemical tankers and gas carriers regardless of size, the pilot card, wheelhouse poster and manoeuvring booklet should be provided.

2.2 The Administration should encourage the provision of manoeuvring information on existing ships, and ships that may pose a hazard due to unusual dimensions or characterics.

2.3 The manoeuvring information should be amended after modification or conversion of the ship which may alter its manoeuvring characteristics or extreme dimensions.

3 Manoeuvring information

3.1 *Pilot card (appendix 1).* The pilot card, to be filled in by the master, is intended to provide information to the pilot on boarding the ship. This information should describe the current condition of the ship, with regard to its loading, propulsion and manoeuvring equipment, and other relevant equipment. The contents of the pilot card are available for use without the necessity of conducting special manoeuvring trials.

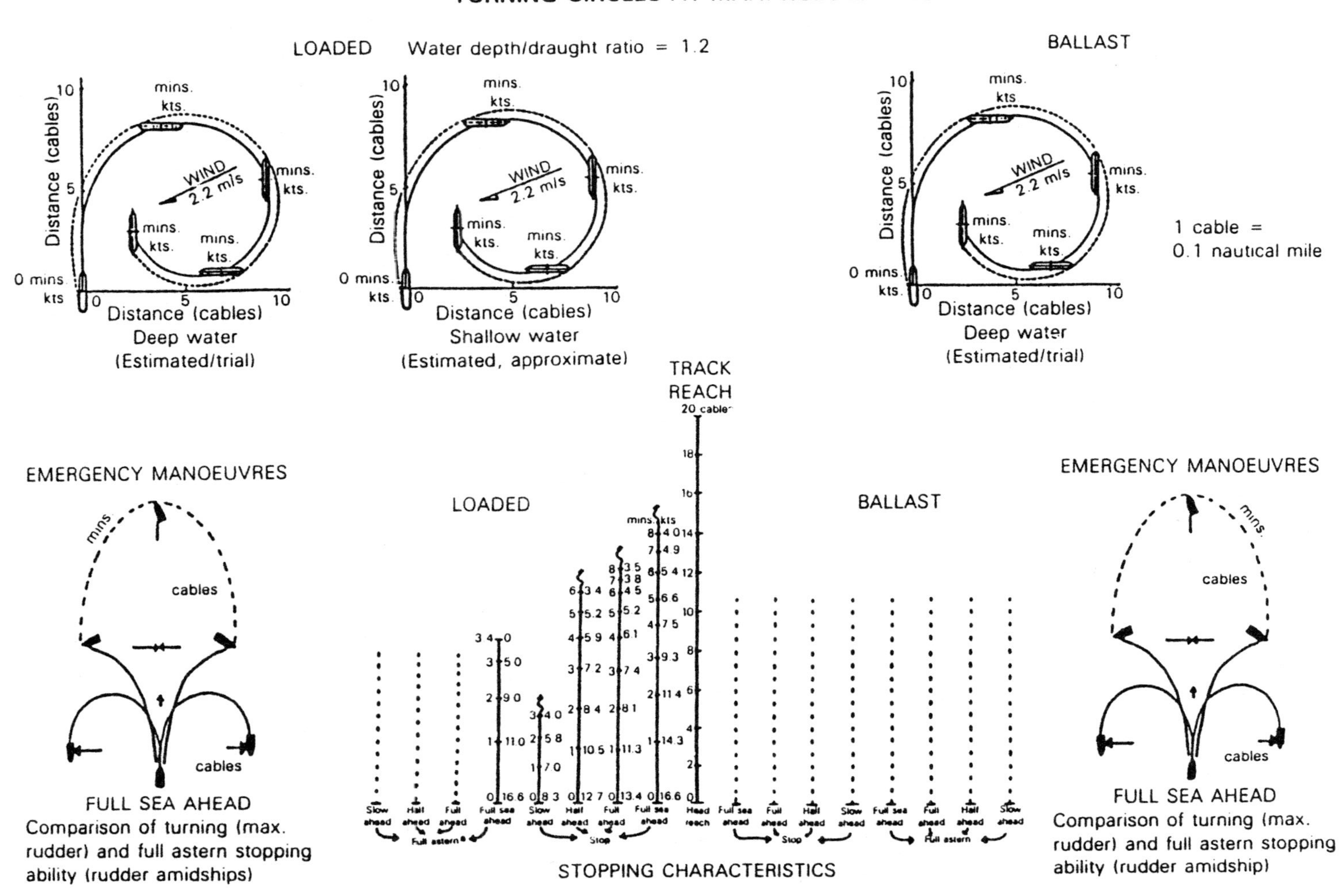

Part of the wheelhouse poster

3.2 *Wheelhouse poster (appendix 2).* The wheelhouse poster should be permanently displayed in the wheelhouse. It should contain general particulars and detailed information describing the manoeuvring characteristics of the ship, and be of such a size to ensure ease of use. The manoeuvring performance of the ship may differ from that shown on the poster due to environmental, hull and loading conditions.

3.3 *Manoeuvring booklet (appendix 3).* The manoeuvring booklet should be available on board and should contain comprehensive details of the ship's manoeuvring characteristics and other relevant data. The manoeuvring booklet should include the information shown on the wheelhouse poster together with other available manoeuvring information. Most of the manoeuvring information in the booklet can be estimated but some should be obtained from trials. The information in the booklet may be supplemented in the course of the ship's life.

11.26.2 UK Statutory Instrument 1975, no. 700: The Merchant Shipping (Carriage of Nautical Publications) Rules 1975

The Secretary of State in exercise of powers conferred by section 86 of the Merchant Shipping Act 1970 (c. 36) and now vested in him (see SI 1970, no. 1537: 1970 III, p. 5293) and of all other powers enabling him in that behalf, hereby makes the following rules:

1(1) These rules may be cited as the Merchant Shipping (Carriage of Nautical Publications) Rules 1975 and shall come into operation on 1 June 1975.
(2) The Interpretation Act 1889 (c. 63) shall apply for the interpretation of these rules as it applies for the interpretation of an Act of Parliament.

2 These rules apply to all ships registered in the United Kingdom other than

(a) Those which are less than 12 metres in length; and
(b) Fishing vessels.

3 The Secretary of State hereby specifies the charts described in rule 4 below and the directions and information mentioned in rule 5 below, as the charts, directions and information which appear to him necessary or expedient for the safe operation of ships to which these rules apply.

4 The said charts are those

(a) Which are of such a scale and which contain sufficient detail as clearly to show
  (i) All navigational marks which may be used by a ship when navigating the waters which are comprised in the chart;
  (ii) All known dangers affecting the waters; and
  (iii) Information concerning any traffic separation schemes, two-way routes, recommended tracks, inshore traffic zones and deep water routes applicable to those waters and areas therein which are to be avoided.
(b) Which are either published by the Hydrographer of the Navy or, if not so published, are of a similar scale to those so published and contain equivalent detail.
(c) Which, in all cases, are of the latest obtainable edition; and
  (i) In the case of charts published by the Hydrographer of the Navy have been corrected and kept up to date from the latest relevant obtainable Notices to Mariners and radio navigational warnings; and
  (ii) In the case of charts not so published, have been otherwise adequately corrected and kept up to date.

In paragraph (b) of this rule the reference to the Hydrographer of the Navy includes a reference to any authority in any country other than the United Kingdom duly exercising functions similar to those of the Hydrographer.

5 The said directions and information are such as is contained in the publications mentioned in column 1 of the schedule to these rules, being publications which, in all cases, are of the latest obtainable edition and which incorporate the latest relevant supplements and, in the case of any such publication which is published otherwise than by the publisher specified opposite thereto in column 2 of the said schedule, is of equivalent standard and content.

6 Every ship to which these rules apply which goes to sea or attempts to go to sea shall carry at least

(a) One copy of a chart which complies with the requirements specified in rule **4** above, being a chart which is appropriate for each part of the intended voyage; and

(b) Except in the case of a ship which does not intend to go to sea beyond a distance of 5 nautical miles from any coastline, one copy of each of the publications mentioned in column 1 of the schedule to these rules, with English language text, as is appropriate for that voyage.

Schedule Publications, directions and information

Column 1 Publication	*Column 2* Publisher
(*a*) International Code of Signals	Her Majesty's Stationery Office
(*b*) Merchant Shipping Notices	Department of Trade
(*c*) Mariners Handbook	Hydrographer of the Navy
(*d*) Notices to Mariners	Hydrographer of the Navy
(*e*) Nautical Almanac	—
(*f*) Navigational tables	—
(*g*) Lists of Radio Signals	Hydrographer of the Navy
(*h*) Lists of Lights	Hydrographer of the Navy
(*i*) Sailing Directions	Hydrographer of the Navy
(*j*) Tide tables	—
(*k*) Tidal stream atlases	—
(*l*) Operating and maintenance instructions for navigational aids carried by the ship	—

Explanatory note

(This note is not part of the rules)

These rules require ships registered in the United Kingdom (other than those less than 12 metres in length and fishing vessels) which go or attempt to go to sea to carry charts containing sufficient detail to show navigational marks, known dangers and other specified information which is appropriate for each part of the intended voyage. Such ships intending to go to sea beyond 5 nautical miles from any coastline are also required to carry directions and information as set out in specified publications, as is appropriate for such a voyage.

Part II
Passage Planning

12 Introduction to passage planning

The importance of preparing oneself and the vessel before setting off on a sea passage is irrefutable, especially on a coastal passage where the sheer number and proximity of hazards can quickly spell disaster for the unprepared. Reliance on memory and past experience does produce a very high proportion of successful passages, but human nature dictates that eventually something important will be overlooked. A professional mariner cannot be satisfied with less than 100 per cent success.

Formalizing the mass of procedures and techniques used during a sea passage is not easy, which is probably why regulations are necessary to encourage mariners to make the effort. The following part of this book examines all aspects of this formalizing process and suggests ways of solving the problem. Obviously these suggestions are not mandatory, but should all be capable of adaptation to suit particular requirements.

The first and biggest hurdle with passage planning is to convince oneself that it is both necessary for and capable of producing the required result – an increased proficiency for mariners and progress towards that 100 per cent goal.

12.1 Requirements to plan

Statutory Instrument 1982, no. 1699, Certification and Watchkeeping Regulations, schedule 1 (see Section 12.4.1) requires that 'the intended voyage shall be planned in advance, taking into consideration all pertinent information.'

Notice M. 854, Navigation Safety (see Section 12.4.2) recommends steps that:

1 'Ensure that all the ship's navigation is planned in adequate detail;
2 Ensure that there is a systematic bridge organization that provides for:
   (a) Comprehensive briefing of personnel
   (b) Close monitoring of position
   (c) Cross checking.'

These requirements apply to all ships regardless of size, and the plan should extend from berth to berth, i.e. it should include pilotage, open sea and coastal phases.

12.2 Advantages of planning

In all probability every shipmaster and experienced navigating officer will say, with conviction, that he already pays full attention to every item mentioned in the appraisal chapter of this book (Chapter 14). However, the full advantages of passage planning will not be realized unless the details and procedures of the plan are formalized.

The main advantages accruing from planning are:

1 Problem areas are identified and allowed for.
2 The officer of the watch is less likely to overlook important data.

3 Data is appraised more effectively using checklists.
4 Less time is needed for position fixing and a better look-out may be maintained.
5 Additional personnel are advised in good time.
6 Contingencies involving departure from the plan are covered.
7 Inadvertent deviations from the plan are quickly apparent.
8 Rest periods can be planned more effectively.

Furthermore, the existence of a well tried system for planning a passage allows a major diversion to be dealt with quickly and efficiently.

Having completed the planning in the detail suggested, the master can derive considerable comfort from the knowledge that his bridge team is properly briefed.

12.3 Involvement of all watchkeeping officers

12.3.1 Bridge watchkeepers

Although the master has ultimate responsibility, all the bridge watchkeeping officers have a collective responsibility for the safe navigation of their ship. Despite the distractions offered by other on-board duties, all bridge watchkeepers are navigating officers; chief officers should need no reminding of their role as deputy to the master, or third officers of their role in backing up the second officer.

For a passage plan to be fully effective, all the watchkeepers must use it fully and have confidence in it. This will only occur when they feel personally involved. Only one officer, normally the second officer as ship's navigator, will probably be delegated the task of completing the paperwork, i.e. marking the charts and bridge notebook; however, other watchkeepers should be involved in discussions of problems to be faced, should perhaps be delegated various aspects of the appraisal, and should certainly check the completed plan.

For a first time passage the amount of work involved in planning can be rather daunting as well as time consuming, and is therefore perhaps best tackled by team effort. Subsequent repeats of the same passage generally reduce the workload to the modifications required by changed circumstances.

12.3.2 Others

The bridge watchkeeper is not able to execute a passage plan totally in isolation. Many parts of the plan involve assistance from other members of the crew, such as engine room staff and seamen. This assistance is always the result of direct orders, but they are unable to co-operate fully without some degree of briefing and an awareness of the situation being met. Not even direct orders can cope with key personnel being absent at the critical moment.

It is neither practical nor necessary for everyone involved in a passage plan to be present at its inception, but the chief engineer and chief petty officer will need early warning of navigational requirements to avoid conflict with the everyday running of their departments.

12.4 Extracts from official publications

Within Sections 12.4.1 and 12.4.2, the paragraph numbering of the original publications has been retained.

12.4.1 UK Statutory Instrument 1982, no. 1699: Certification and Watchkeeping Regulations

Schedule 1 (regulation 4): Principles of watchkeeping arrangements for navigational watch

In paragraphs 1 and 5 of this schedule in their application to small ships without a bridge, the word 'bridge' shall be construed as meaning the position from which the navigation of the ship is controlled.

1 *Watch arrangements*

(a) The composition of the watch shall at all times be adequate and appropriate to the prevailing circumstances and conditions and shall take into account the need for maintaining a proper look-out.

(b) When deciding the composition of the watch on the bridge which may include appropriate deck ratings, the following factors, *inter alia*, shall be taken into account:
 (i) At no time shall the bridge be left unattended;
 (ii) Weather conditions, visibility and whether there is daylight or darkness;
 (iii) Proximity of navigational hazards which may make it necessary for the officer in charge of the watch to carry out additional navigational duties;
 (iv) Use and operational condition of navigational aids such as radar or electronic position indicating devices and any other equipment affecting the safe navigation of the ship;
 (v) Whether the ship is fitted with automatic steering;
 (vi) Any unusual demands on the navigational watch that may arise as a result of special operational circumstances.

2 Fitness for duty

The watch system shall be such that the efficiency of watchkeeping officers and watchkeeping ratings is not impaired by fatigue. Duties shall be so organized that the first watch at the commencement of a voyage and the subsequent relieving watches are sufficiently rested and otherwise fit for duty.

3 Navigation

(a) The intended voyage shall be planned in advance, taking into consideration all pertinent information, and any course laid down shall be checked before the voyage commences.

(b) During the watch the course steered, position and speed shall be checked at sufficiently frequent intervals, using any available navigational aids necessary, to ensure that the ship follows the planned course.

(c) The officer of the watch shall have full knowledge of the location and operation of all safety and navigational equipment on board the ship and shall be aware and take account of the operating limitations of such equipment.

(d) The officer in charge of a navigational watch shall not be assigned or undertake any duties which would interfere with the safe navigation of the ship.

4 Navigational equipment

(a) The officer of the watch shall make the most effective use of all navigational equipment at his disposal.

(b) When using radar, the officer of the watch shall bear in mind the necessity to comply at all times with the provisions on the use of radar contained in the Collision Regulations and Distress Signals Order 1977.

(c) In cases of need the officer of the watch shall not hesitate to use the helm, engines and sound signalling apparatus.

5 Navigational duties and responsibilities

(a) The officer in charge of the watch shall:
 (i) Keep his watch on the bridge, which he shall in no circumstances leave until properly relieved;
 (ii) Continue to be responsible for the safe navigation of the ship, despite the presence of the master on the bridge, until the master informs him specifically that he has assumed that responsibility and this is mutually understood;
 (iii) Notify the master when in any doubt as to what action to take in the interest of safety;
 (iv) Not hand over the watch to the relieving officer if he has reason to believe that the latter is obviously not capable of carrying out his duties effectively, in which case he shall notify the master accordingly.

(b) On taking over the watch the relieving officer shall satisfy himself as to the ship's estimated or true position and confirm its intended track, course and speed and shall note any dangers to navigation expected to be encountered during his watch.

(c) A proper record shall be kept of the movements and activities during the watch relating to the navigation of the ship.

6 Look-out

In addition to maintaining a proper look-out for the purpose of fully appraising the situation and the risk of collision, stranding and other dangers to navigation, the duties of the look-out shall include the detection of ships or aircraft in distress, shipwrecked persons, wrecks and debris. In maintaining a look-out the following shall be observed:

(a) The look-out must be able to give full attention to the keeping of a proper look-out and no other duties shall be undertaken or assigned which could interfere with that task.
(b) The duties of the look-out and helmsman are separate and the helsman shall not be considered to be the look-out while steering, except in small ships where an unobstructed all round view is provided at the steering position and there is no impairment of night vision or other impediment to the keeping of a proper look-out. The officer in charge of the watch may be the sole look-out in daylight provided that on each such occasion:
   (i) The situation has been carefully assessed and it has been established without doubt that it is safe to do so;
   (ii) Full account has been taken of all relevant factors including, but not limited to:

      State of weather
      Visibility
      Traffic density
      Proximity of danger to navigation
      The attention necessary when navigating in or near traffic separation schemes;

   (iii) Assistance is immediately available to be summoned to the bridge when any change in the situation so requires.

7 Navigation with pilot embarked

Notwithstanding the duties and obligations of a pilot, his presence on board shall not relieve the master or officer in charge of the watch from their duties and obligations for the safety of the ship. The master and the pilot shall exchange information regarding navigation procedures, local conditions and the ship's characteristics. The master and officer of the watch shall co-operate closely with the pilot and maintain an accurate check of the ship's position and movement.

8 Protection of the marine environment

The master and officer in charge of the watch shall be aware of the serious effects of operational or accidental pollution of the marine environment and shall take all possible precautions to prevent such pollution, particularly within the framework of relevant international and port regulations.

12.4.2 UK DTp Merchant Shipping Notice M.854: Navigation Safety

1 Research into recent accidents occurring to ships has shown that by far the most important contributory cause of navigational accidents is human error, and in many cases information which could have prevented the accident was available to those responsible for the navigation of the ships concerned.

2 There is no evidence to show serious deficiency on the part of deck officers with respect either to basic training in navigation skills or ability to use navigational instruments and equipment; but accidents happen because one person makes the sort of mistake to which all human beings are prone in a situation where there is no navigational regime constantly in use which might enable the mistake to be detected before an accident occurs.

3 To assist masters and deck officers to appreciate the risks to which they are exposed and to provide help in reducing these risks, it is recommended that steps are taken to:

(a) Ensure that all the ship's navigation is planned in adequate detail with contingency plans where appropriate;
(b) Ensure that there is a systematic bridge organization that provides for
   (i) Comprehensive briefing of all concerned with the navigation of the ship;

(ii) Close and continuous monitoring of the ship's position, ensuring as far as possible that different means of determining position are used to check against error in any one system;
(iii) Cross-checking of individual human decisions so that errors can be detected and corrected as early as possible;
(iv) Information available from plots of other traffic to be used carefully to ensure against over-confidence, bearing in mind that other ships may alter course and speed.

(c) Ensure that optimum and systematic use is made of all information that becomes available to the navigational staff;
(d) Ensure that the intentions of a pilot are fully understood and acceptable to the ship's navigational staff.

4 The annex to this notice provides information on the planning and conduct of passages which may prove useful to mariners.

Annex: Guide to the planning and conduct of passages

Pilotage

1 The contribution which pilots make to the safety of navigation in confined waters and port approaches, of which they have up-to-date knowledge, requires no emphasis; but it should be stressed that the responsibilities of the ship's navigational team do not transfer to the pilot and the duties of the officer of the watch remain with that officer.

2 After his arrival on board, in addition to being advised by the master of the manoeuvring characteristics and basic details of the vessel for its present condition of loading, the pilot should be clearly consulted on the passage plan to be followed. The general aim of the master should be to ensure that the expertise of the pilot is fully supported by the ship's bridge team (see also paragraph 16).

3 Attention is drawn to the following extract from IMO resolution A 285 (VIII):

'Despite the duties and obligations of a pilot, his presence on board does not relieve the officer of the watch from his duties and obligations for the safety of the ship. He should co-operate closely with the pilot and maintain an accurate check on the vessel's position and movements. If he is in any doubt as to the pilot's actions or intentions, he should seek clarification from the pilot and if doubt still exists he should notify the master immediately and take whatever action is necessary before the master arrives.'

Responsibility for passage planning

4 In most deep sea ships it is customary for the master to delegate the initial responsibility for preparing the plan for a passage to the officer responsible for navigational equipment and publications, usually the second officer. For the purposes of this guide the officer concerned will be referred to as the navigating officer.

5 It will be evident that in small ships, including fishing vessels, the master or skipper may himself need to exercise the responsibility of the navigating officer for passage planning purposes.

6 The navigating officer has the task of preparing the detailed passage plan to the master's requirements prior to departure. In those cases when the port of destination is not known or is subsequently altered, it will be necessary for the navigating officer to extend or amend the original plan as appropriate.

Principles of passage planning

7 There are four distinct stages in the planning and achievement of a safe passage:

1 Appraisal
2 Planning
3 Execution
4 Monitoring

8 These stages must of necessity follow each other in the order set out above. An appraisal of information available must be made before detailed plans can be drawn up and a plan must be in existence before tactics for its execution can be decided upon. Once the plan and the manner in which it is to be executed have been decided, monitoring must be carried out to ensure that the plan is followed.

Appraisal

9 This is the process of gathering together all information relevant to the contemplated passage. It will of course be concerned with navigational information shown on charts and in publications such as sailing directions, light lists, current atlas, tidal atlas, tide tables, Notices to Mariners, publications detailing traffic separation and other routeing schemes, and radio aids to navigation. Reference should also be made to climatic data and other appropriate meteorological information which may have a bearing upon the availability for use of navigational aids in the area under consideration such as, for example, those areas subject to periods of reduced visibility.

10 A checklist should be available for the use of the navigating officer to assist him to gather all the information necessary for a full passage appraisal and the circumstances under which it is to be made. It is necessary to recognize that more up-to-date information, for example, radio navigational warnings and meteorological forecasts, may be received after the initial appraisal.

11 In addition to the obvious requirement for charts to cover the area or areas through which the ship will proceed, which should be checked to see that they are corrected up to date in respect of both permanent and temporary Notices to Mariners and existing radio navigational warnings, the information necessary to make an appraisal of the intended passage will include details of:

(a) Currents (direction and rate of set)
(b) Tides (times, heights and direction of rate of set)
(c) Draught of ship during the various stages of the intended passage
(d) Advice and recommendations given in sailing directions
(e) Navigational lights (characteristics, range, arc of visibility and anticipated raising range)
(f) Navigational marks (anticipating range at which objects will show on radar and/or will be visible to the eye)
(g) Traffic separation and routeing schemes
(h) Radio aids to navigation (availability and coverage of Decca, Omega, Loran and DF and degree of accuracy of each in that locality)
(i) Navigational warnings affecting the area
(j) Climatological data affecting the area
(k) Ship's manoeuvring data.

12 An overall assessment of the intended passage should be made by the master, in consultation with the navigating officer and other deck officers who will be involved, when all relevant information has been gathered. This appraisal will provide the master and his bridge team with a clear and precise indication of all areas of danger, and delineate the areas in which it will be possible to navigate safely, taking into account the calculated draught of the ship and planned underkeel clearance. Bearing in mind the condition of the ship, her equipment and any other circumstances, a balanced judgement of the margins of safety which must be allowed in the various sections of the intended passage can now be made, agreed and understood by all concerned.

Planning

13 Having made the fullest possible appraisal using all the available information on board relating to the intended passage, the navigating officer can now act upon the master's instructions to prepare a detailed plan of the passage. The detailed plan should embrace the whole passage, from berth to berth, and include all waters where a pilot will be on board.

14 The formulation of the plan will involve completion of the following tasks:

(a) Plot the intended passage on the appropriate charts and mark clearly, on the largest scale charts applicable, all areas of danger and the intended track taking into account the margins of allowable error. Where appropriate, due regard should be paid to the need for advance warning to be given on one chart of the existence of a navigational hazard immediately on transfer to the next. The planned track should be plotted to clear hazards at as safe a distance as circumstances allow. A longer distance should always be accepted in preference to a shorter more hazardous route. The possibility of main engine or steering gear breakdown at a critical moment must not be overlooked.

(b) Indicate clearly in 360° notation the true direction of the planned track marked on the charts.

(c) Mark on the chart those radar conspicuous objects, Ramarks or Racons, which may be used in position fixing.

(d) Mark on the charts any transit marks, clearing bearings or clearing ranges (radar) which may be used to advantage. It is sometimes possible to use two conspicuous clearing marks where a line drawn through them runs clear of natural dangers with the appropriate margin of safety; if the ship proceeds on the safe side of this transit she will be clear of the danger. If no clearing marks are available, a line or lines of bearings from a single object may be drawn at a desired safe distance from the danger; provided the ship remains in the safe segment, she will be clear of the danger.

(e) Decide upon the key elements of the navigational plan. These should include but not be limited to:
   (i) Safe speed having regard to the manoeuvring characteristics of the ship and, in ships restricted by draught, due allowance for reduction of draught due to squat and heel effect when turning;
   (ii) Speed alterations necessary to achieve desired ETAs *en route*, e.g. where there may be limitations on night passage, tidal restrictions, etc.;
   (iii) Positions where a change in machinery status is required;
   (iv) Course alteration points, with wheel-over positions; where appropriate on large scale charts, taking into account the ship's turning circle at the planned speed and the effect of any tidal stream or current on the ship's movement during the turn;
   (v) Minimum clearance required under the keel in critical areas (having allowed for height of tide);
   (vi) Points where accuracy of position fixing is critical, and the primary and secondary methods by which such positions must be obtained for maximum reliability;
   (vii) Contingency plans for alternative action to place the ship in deep water or proceed to an anchorage in the event of any emergency necessitating abandonment of the plan.

15 Depending on circumstances, the main details of the plan referred to in paragraph 14 above should be marked in appropriate and prominent places on the charts to be used during the passage. These main details of the passage plan should in any case be recorded in a bridge notebook used specially for this purpose to allow reference to details of the plan at the conning position without the need to consult the chart. Supporting information relative to the passage such as times of high and low water, or of sunrise or sunset, should also be recorded in this notebook.

16 It is unlikely that every detail of a passage will have been anticipated, particularly in pilotage waters. Much of what will have been planned may have to be changed after embarking the pilot. This in no way detracts from the real value of the plan, which is to mark out in advance where the ship must *not* go and the precautions which must be taken to achieve that end, or to give initial warning that the ship is standing into danger.

Execution

17 Having finalized the passage plan, and as soon as estimated times of arrival can be made with reasonable accuracy, the tactics to be used in the execution of the plan should be decided. The factors to be taken into account will include:

(a) The reliability and condition of the ship's navigational equipment

(b) Estimated times of arrival at critical points for tide heights and flow

(c) Meteorological conditions, particularly in areas known to be affected by frequent periods of low visibility

(d) Daytime versus night-time passing of danger points, and any effect this may have upon position fixing accuracy

(c) Traffic conditions, especially at navigational focal points.

18 It will be important for the master to consider

whether any particular circumstance, such as the forecast of restricted visibility in an area where position fixing by visual means at a critical point is an essential feature of the navigation plan, introduces an unacceptable hazard to the safe conduct of the passage; and thus whether that section of the passage should be attempted under the conditions prevailing, or likely to prevail. He should also consider at which specific points of the passage he may need to utilize additional deck or engine room personnel.

Monitoring

19 The close and continuous monitoring of the ship's progress along the preplanned track is essential for the safe conduct of the passage. If the officer of the watch is ever in any doubt as to the position of the ship or the manner in which the passage is proceeding he should immediately call the master and, if necessary, take whatever action he may think necessary for the safety of the ship.

20 The performance of navigational equipment should be checked prior to sailing, prior to entering restricted or hazardous waters, and at regular and frequent intervals at other times throughout the passage.

21 Advantage should be taken of all the navigational equipment with which the ship is fitted for position monitoring, bearing in mind the following points:

(a) Visual bearings are usually the most accurate means of position fixing.
(b) Every fix should, if possible, be based on at least three position lines.
(c) Transit marks, clearing bearings and clearing ranges (radar) can be of great assistance.
(d) When checking, use systems which are based on different data.
(e) Positions obtained by navigational aids should be checked where practicable by visual means.
(f) The value of the echo sounder as a navigational aid.
(g) Buoys should not be used for fixing but may be used for guidance when shore marks are difficult to distinguish visually; in these circumstances their positions should first be checked by other means.
(h) The functioning and correct reading of the instruments used should be checked.
(i) An informed decision in advance as to the frequency with which the position is to be fixed should be made for each section of the passage.

22 On every occasion when the ship's position is fixed and marked on the chart in use, the estimated position at a convenient interval of time in advance should be projected and plotted.

23 Radar can be used to advantage in monitoring the position of the ship by the use of parallel indexing techniques. Parallel indexing, as a simple and most effective way of continuously monitoring a ship's progress in restricted waters, can be used in any situation where a radar conspicuous navigation mark is available and it is practicable to monitor continuously the ship's position relative to such an object.

13 Types of passage plan

13.1 Ocean

The chief characteristic of the plan for an ocean passage is the long time involved, sometimes 10 to 15 days. It is necessary to look this far ahead for avoidable delays and diversions.

Because traffic concentrations and navigational hazards on an ocean crossing are relatively few, the main data being appraised is environmental, such as winds, currents, ice, fog, etc. (Table 13.1). This data, which can be very variable, is not necessarily very accurate. This makes the plan liable to frequent updating as each piece of new information is received, e.g. by radio weather report.

Away from navigational danger the need for high positional accuracy and frequency of fixing is reduced and indeed available position fixing systems are relatively few. However, as the landfall is approached, the emphasis moves away from environmental factors and back to position fixing.

Where severe weather is the major factor controlling the progress and safety of the ship, then weather routeing is employed either as an on-board exercise using weather forecasts and prognoses, or from shore-based agencies who advise the ship accordingly. In both cases, reassessments and adjustments to the plan are made daily.

13.2 Coastal

The time span of a coastal passage can be from a few hours to several days. Because of the large quantities of relevant navigational information that needs to be appraised, some of which is of a variable nature such as tides and visibility, it is probably unrealistic to plan in detail much more than two days ahead.

Table 13.1 Route selection factors

Environment	*Ship*	*Cargo*	*Navigational/operational*
Weather pressure patterns and storm paths	Type	Type	Schedule
Winds, seas, swells	Speed capability	Special requirements *re* temperature and humidity	Fuel economics
Tropical revolving storms	Draught, trim	Protection from sea	Insurance restrictions
Ocean currents	Stability		Load line zones
Localized weather	Deck load		Company regulations
Seasonal weather	Handling characteristics		IMO routeings
Tides	Navigational aids		Local regulations
Ice			Traffic density
			Position fixing accuracy
			Navigational data accuracy
			Navigation team experience
			Contingency

On the longer coastal passages, a reappraisal and updating of the plan can be made every 24 hours, or whenever new and significant data is received.

The coastal plan probably uses up more of the planner's time than any other. However, much of the data being appraised remains reasonably static, and consequently a subsequent passage of the same area can be planned with far less effort than was needed for the first transit. Nevertheless, the mariner must never neglect to reappraise the variable data when making that subsequent passage.

13.3 Landfall

The landfall is the transition from the ocean to the coastal plan. It only lasts for a couple of hours at most but is important enough to warrant special mention. The coasts of the world are littered with the wrecks of vessels that did not get the landfall quite right.

Choosing the place and time of landfall are most important. When appraising the charts and publications, a recommended landfall may be suggested, otherwise the navigator must seek:

1 A place where the ship's position on approach can be fixed accurately and unambiguously
2 The availability of a reliable long range position fixing system
3 The availability of long range lights for navigation if a night-time approach is contemplated
4 A place where there are no off-lying dangers such as shoals and strong currents or predictable fog banks. It is also best to avoid if possible distractions such as known fishing vessel concentrations and places where joining a coastal traffic stream could be awkward.

On approaching the landfall area it is possible that conditions may have changed, for example bad visibility or the failure of a position fixing system. In such cases the landfall would need to be either replanned for another place or delayed to another time, for example to daylight or to the next satellite pass.

13.4 Pilotage waters

Having a pilot on board does not in any way relieve the master or OOW from the responsibility of ensuring the safe progress of their vessel. Consequently, there is the need to preplan the pilotage stage of the voyage in just as much detail as the rest. The skill and knowledge of the pilot is obviously important and will be made use of when he arrives. However, it should not be treated as the only source of information.

Planning beforehand is limited to the appraisal and marking the chart with relevant information. Actual tracks to be followed are less important provided clear water areas have been defined. In appraising all the navigational problems right up to the berth, key staff can be alerted and equipment and manning can be upgraded to a level commensurate with the danger. The bridge and ship organizational details of the plan are most important. Similarly, the approach to the pilot boarding area is critical and should include identification of a safe waiting area in case the pilot is delayed.

The passage plan and bridge notebook can only be finalized after the pilot has boarded and given his approval.

13.5 Contingency

Every passage is accompanied by the risk of having to abandon the plan owing to changed conditions or circumstances. These changed conditions may arise with little or no warning, allowing no time for a full reappraisal or replanning.

Experience points towards certain events being more likely to occur than others, and hence alternative plans can be appraised, drawn up and held ready to meet such eventualities.

The following are examples of some semi-predictable events for which contingency plans might be used:

1 Delay due to reduced visibility
2 Pilot, tugs, berths not available
3 Changed tidal conditions
4 Breakdown of ship or equipment
5 Diversion to new destination
6 The need to seek refuge owing to heavy weather.

14 Appraisal

14.1 Appraisal checklist

Complete appraisal involves consulting numerous sources of information. Unless a systematic approach is adopted this process can be very time consuming or, more importantly, lead to some critical piece of data being overlooked.

To ensure an effective appraisal, a checklist is essential. A general guide can be found in annex III of the DTp Guide to the Planning and Conduct of Sea Passages (see section 14.7.1). In practice, a more detailed list, suited to the particular ship, may be needed.

14.2 Publications

A major source of advice for a coastal passage is contained in the Sailing Directions (pilot books). It is necessary to separate the relevant information from the irrelevant. To this end note from the chart the approximate route to be followed and the adjacent prominent geographical points, then to save time make full use of the index and section headings. The leading pages of the Sailing Directions include a diagram of the area covered by that volume and give an indication of the direction in which the text follows the coastline.

Having identified the relevant information, either make separate brief notes or make page and line references. The pilot book is available for reference throughout the passage, but for the moment we are looking for factors that will influence our choice of route.

The other publications that may be consulted are detailed in Chapter 11.

14.3 The chart

The navigation chart forms the principal source of navigational data and consequently must be the most up to date and of the largest scale available. Careful study in conjunction with the other publications will allow identification of difficult navigation conditions such as:

1 Danger areas requiring high position fixing accuracy
2 Areas of inadequate position fixing accuracy
3 Heavy traffic concentrations and traffic streams
4 Areas of expected low visibility where this could lead to position fixing problems
5 Areas of relatively high current or tidal rates.

In all events, an expeditious route, maintaining ample sea room, is being sought.

Following this appraisal, a list of working charts for the passage can be drawn up.

14.4 Navigation aids

Careful judgement is needed when deciding on the prime position fixing system to be used in a given area, giving due regard to what is available and its relative reliability and accuracy.

Never overlook the fact that visual bearings as a means of fixing positions are in most cases more accurate and more reliable than any radio navigational aid. To continue using a given position fixing system after its accuracy has become degraded and to ignore the fact that another system has become available and more accurate is lazy and unprofessional. We are all amused by the stories of watchkeepers lifting their faces from the radar visor to discover that the visibility is now perfect, but it serves to illustrate the point.

Factors controlling system accuracy may be constantly changing, and it is necessary to assess them for every part of the proposed track. A secondary position fixing check system must be identified in the same way.

Choice must be based on the ability of the system to do the best job. User convenience of a particular system should not be a controlling factor, as all navigators should be equally proficient on all available systems.

14.5 Tides

Tidal heights in relation to under-keel and overhead clearances have obvious significance to deep draught and large air draught vessels in coastal areas, and in extreme cases tidal windows may need to be calculated. Tidal windows are needed when the required depth is greater than that charted and the vessel depends on the rise of tide to carry it through the area in question. As high water occurs at different places at different times and the rate of rise and fall can vary, the calculation involved can be quite complex (see *Malacca/Singapore Straits Guide to Planned Passages for Deep Draught Vessels*, ICS, May 1981). In most cases, however, the water depth remains a simple cross-check of position. The rise of tide offshore seldom approaches the values quoted in the tidal tables for the adjacent coast, but confirmation of the actual rise for certain areas can be obtained from co-tidal charts if necessary.

Tide or current rates and directions are assessed with course keeping and progress in mind and, although these elements are variable, the more significant values are allowed for in choosing the route or are brought to the attention of the OOW for allowance during the passage. Tidal rate significance increases when it is large in relation to own ship's speed; therefore more attention must be paid in areas when own ship's speed is reduced, for example in pilotage areas or in reduced visibility. A tidal rate of one-tenth of own speed can at worst produce a divergence of 5° from the required course line.

In practice an actual measure of tidal effect from frequent position fixing is better than calculating the allowance beforehand, unless the vessel is on a long coastal leg with only mediocre position fixing accuracy.

14.6 The operational status of the vessel

14.6.1 Manoeuvrability and speed

There will be a safe speed for each leg of the route, bearing in mind the size of the vessel and its manoeuvring characteristics (see also Sections 22.1.2 and 22.2.2).

The ability to estimate times of certain events in the passage is an essential part of planning. When looking hours or days ahead it is necessary to have a fairly accurate estimation of expected speed over the ground, otherwise the majority of the effort put into assessing time variable data will go to waste and will need to be repeated.

An accurate knowledge of the space needed to stop the ship in safety is a prime requirement in determining the margins of safety. The greater the initial speed, the greater the space needed. Hence, when margins of safety are necessarily reduced in confined waters, speed must be reduced to a level at which the ship can be kept in safe water following an unforeseen deviation from track or other failure. In narrow channels where this requirement cannot be met fully, the high risk must be recognized by increasing the preparedness of the vessel and crew, for example by having anchors ready and possibly tugs in attendance.

In shallow waters where the depth under keel is less than the draught of the ship, there is bound to be some interactive effect (see Section 15.3.3) which reduces the manoeuvrability of the vessel. Since interactive effects vary with speed, special consideration must be given to determining a safe speed in shallow water.

14.6.2 Margin of safety

It is prudent never to approach any navigational danger closer than is necessary to fix the position and expedite the voyage. 'Rock hopping' may impress passengers, but it is an unjustifiable navigational risk.

In any area of restricted searoom, thought must be given to the distance that it is safe to pass any particular danger. The margins of safety are based on distances measured from the navigational danger rather than distances either side of the course line, and hence form an extension to the no-go or 'immediate danger' area. Consequently, the margin lines show limits beyond which it is not advisable to navigate under prevailing conditions.

The area bounded by the margin lines can still be used in an emergency provided the danger area is not entered; otherwise there arises an immediate risk of grounding (see Figures 15.1 and 15.2).

The distances involved can vary between 5 miles and 1 cable, depending on the type of ship being considered, and normally the course line should be laid off well outside this limit to provide ample searoom.

The minimum passing distance from a navigational danger can be estimated after considering:

1 Manoeuvrability – stopping distance, turning circle diameter and interactive effects
2 Engine readiness, reliability and speed
3 Steering reliability and status, i.e. hand or automatic
4 The ability to fix position accurately and the frequency of fixing
5 Environmental conditions – wind, visibility and tides
6 Traffic density and searoom needed for anti-collision manoeuvres
7 Anchoring readiness.

14.6.3 Parallel indexing and radar maps (see Sections 5.7.2.2 and 20.1.4.1)

In confined waters where traditional position fixing methods are too slow, or where the number of visible navigation marks are too few or remote to enable the navigator to react in time to changes in tidal condition or other events causing the vessel to leave the course line, parallel indexing (PI) or radar maps should be considered as a means of continuously checking progress.

Mariners are advised to practise these techniques when conditions are not hazardous so that maximum benefit can be achieved when they are really needed. It is worth bearing in mind that in areas where the searoom is extremely limited, wrongly laid off PI lines or map elements can very quickly lead to disaster. The ship's position must be fixed on the chart by normal methods as frequently as possible, because parallel indexing is not a position fixing system as such but a visual indication of proximity to course line and danger. Its prime value lies in its ability to give early warning to the navigator of positional trend.

To derive the data from the chart that enables the navigator to draw PI lines on the radar reflection plotter can take a considerable time. It is therefore not a technique that can be embarked upon at short notice; it must be preplanned.

During the appraisal stage of passage planning, the navigator must:

1 Identify the area for PI or map coverage
2 For PI, decide on the most suitable working range scale(s)
3 Identify a suitable parallel indexing target
4 Complete the chartwork, paying special attention to the margins of safety
5 Extract the data needed to PI in a suitable form ready to use with the minimum of delay when needed. Include margin of safety data as precisely as is practicable, remembering that traffic conditions cannot be accurately predicted and consequently there is always the possibility of being forced into using all the available water.

14.6.4 Equipment tests and manning level changes

As navigation conditions change, certain routine tests and changes in manning levels must be carried out. An early identification of suitable points at which to make these changes is necessary. For example:

1 The change from automatic to manual steering. This is required *before* entering areas of restricted searoom, reduced visibility or heavy traffic. It should not be left until other navigational matters are pressing.
2 Testing the rudder hard over each way. This must be carried out in an area where a jammed rudder will not spell disaster.
3 Advising engine room (ER) personnel of impending use of main engines. Advice on *possible* manoeuvres *before* entering difficult areas and adequate warning of *probable* manoeuvres. This will allow the ER manning level to be adjusted and engines put into a state of readiness for manoeuvres.
4 Testing of engines for astern movement. This should be carried out when clear of traffic and before entering restricted waters.
5 Testing of ancillary equipment before entering harbour. This should be made at a point where the tests will not distract from the navigation.
6 Additional personnel to the bridge. They should be alerted and summoned with sufficient time to acclimatize (e.g. for night vision) and to be briefed on the present situation.
7 Look-outs to be posted and anchors cleared.
8 Changing charts. This should be marked on the current working chart.

14.7 Extract from official publication

14.7.1 Guide to the Planning and Conduct of Sea Passages (DTp)

Annex III: Checklist of items for passage appraisal

1 Select largest scale appropriate charts for the passage.
2 Check that all charts to be used have been corrected up to date from the latest information available.
3 Check that all radio navigational warnings affecting the area have been received.
4 Check that sailing directions and relevant lists of lights have been corrected up to date.
5 Estimate the draught of the ship during the various stages of the passage.
6 Study sailing directions for advice and recommendations on route to be taken.
7 Consult current atlas to obtain direction and rate of set.
8 Consult tide tables and tidal atlas to obtain times, heights and direction and rate of set.
9 Study climatological information for weather characteristics of the area.
10 Study charted navigational aids and coastline characteritics for landfall and position monitoring purposes.
11 Check the requirements of traffic separation and routeing schemes.
12 Consider volume and flow of traffic likely to be encountered.
13 Assess the coverage of radio aids to navigation in the area and the degree of accuracy of each.
14 Study the manoeuvring characteristics of the ship to decide upon safe speed and, where appropriate, allowance for turning circle at course alteration points.
15 If a pilot is to be embarked, make a careful study of the area at the pilot boarding point for preplanning intended manoeuvres.
16 Where appropriate, study all available port information data.
17 Check any additional items which may be required by the type of ship, the particular locality, or the passage to be undertaken.

15 Making the plan

Having appraised all the data sources and reached decisions on all the relevant items on the check list, the navigator can now write up the plan as an instruction to the bridge team.

15.1 Marking the chart

15.1.1 Courses to make good

These should be laid off on the working charts and marked clearly in 360° notation. On lattice charts, the course lines can be made more conspicuous by drawing double lines closely spaced.

When using construction lines to help with the positioning of alter-course positions, e.g. clearing distance radii, leading lines, etc., care should be taken that they are not able to be construed as course lines, particularly as they almost invariably lead towards a navigational hazard.

15.1.2 Adjacent danger areas

The danger areas adjacent to the vessel's proposed track should be highlighted (Figure 15.1). Shoal waters for a particular draught may not necessarily be clearly defined on the chart. Therefore, define the area or limit and cross-hatch in pencil the no-go area. This should be done carefully, as highlighting an area that is in fact not dangerous brings the whole system into disrepute. Also, try not to obscure important printed data.

Isolated dangers should also be brought to the attention of the OOW, but generally the insertion of a margin of safety line does this adequately.

15.1.3 Margins of safety

These only apply if the vessel is liable to stray into a danger area during the course of an anti-collision manoeuvre or between successive fixes (Figure 15.2). If they apply, they should be positioned relative to the danger area at a distance that takes into account the ship speed and manoeuvrability, position fixing accuracy and frequency, etc. (see Section 14.6.2).

The safety margin lines are bound to be something of a compromise. In one extreme they could mirror the danger area spaced at a safe distance off it. In the other extreme, they could parallel the vessel's proposed track at a distance dictated by the closest approach of the danger. The first is unrealistic, being too time consuming and in most events totally unnecessary; the second is inefficient, as it does not show the true searoom situation and confines the vessel without good reason. The compromise should reflect the clear water that the vessel might reasonably be expected to make use of during anti-collision manoeuvres etc.

15.1.4 Navigation marks

The attention of OOW should be drawn by circling, or by drawing a large arrow towards, those navigation marks expected to be used either visually or by radar or for DF (Figure 15.3). Often a perfectly usable navigation mark at the far end of the chart in use will be neglected unless attention is drawn to it.

This process can, if followed, considerably shorten the time needed to identify and fix the position on unfamiliar coasts.

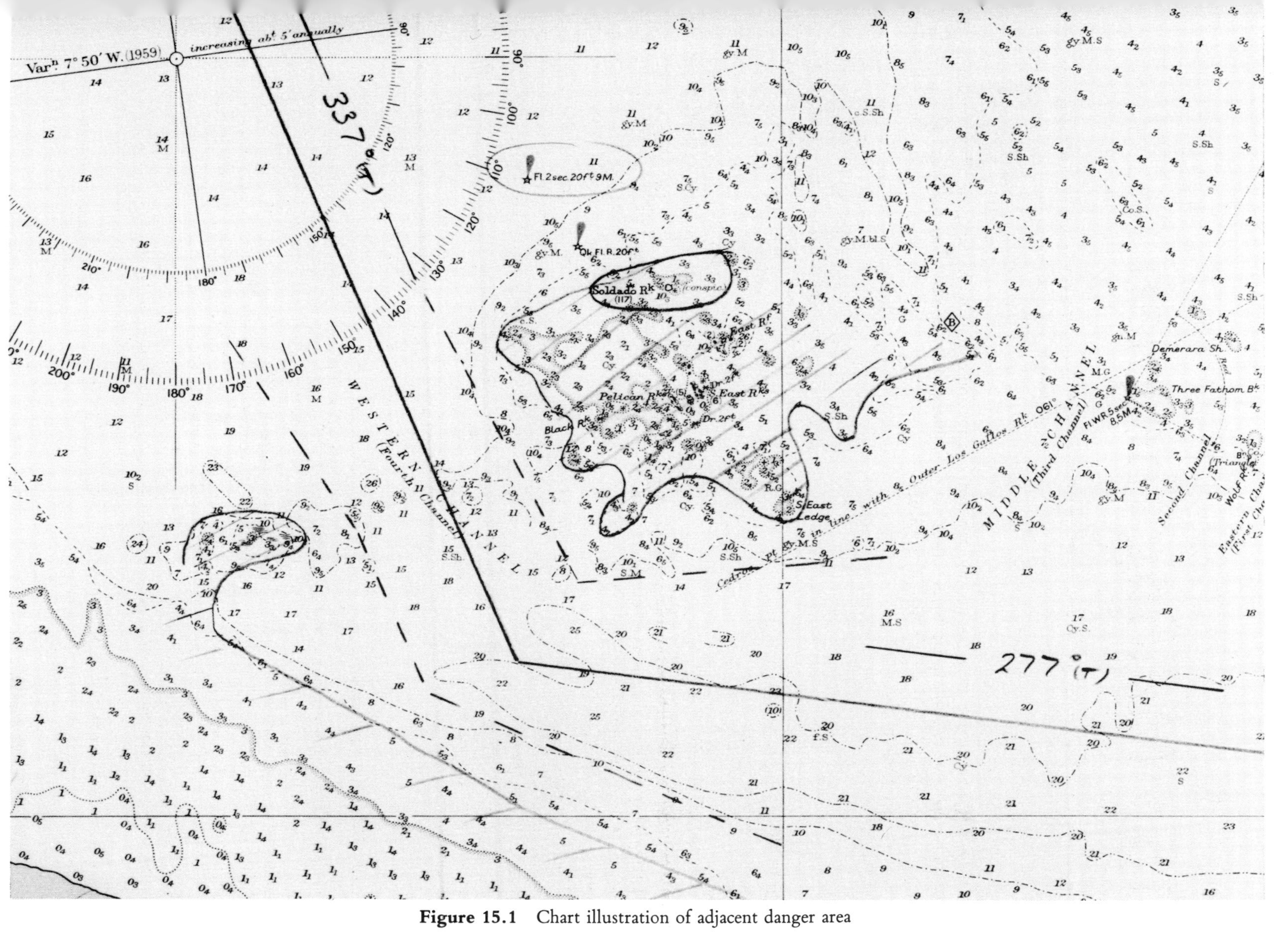

Figure 15.1 Chart illustration of adjacent danger area

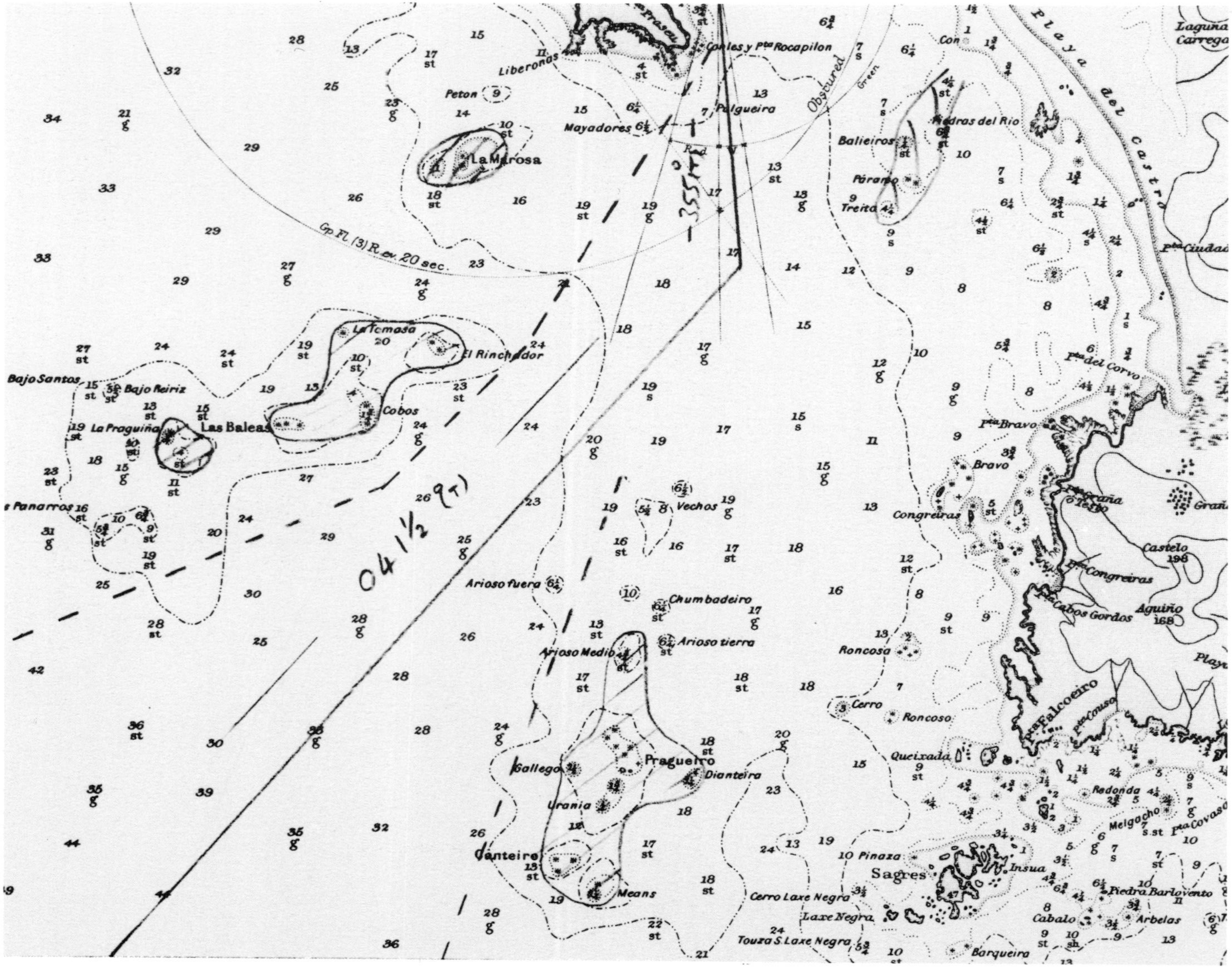

Figure 15.2 Chart illustration of margin of safety

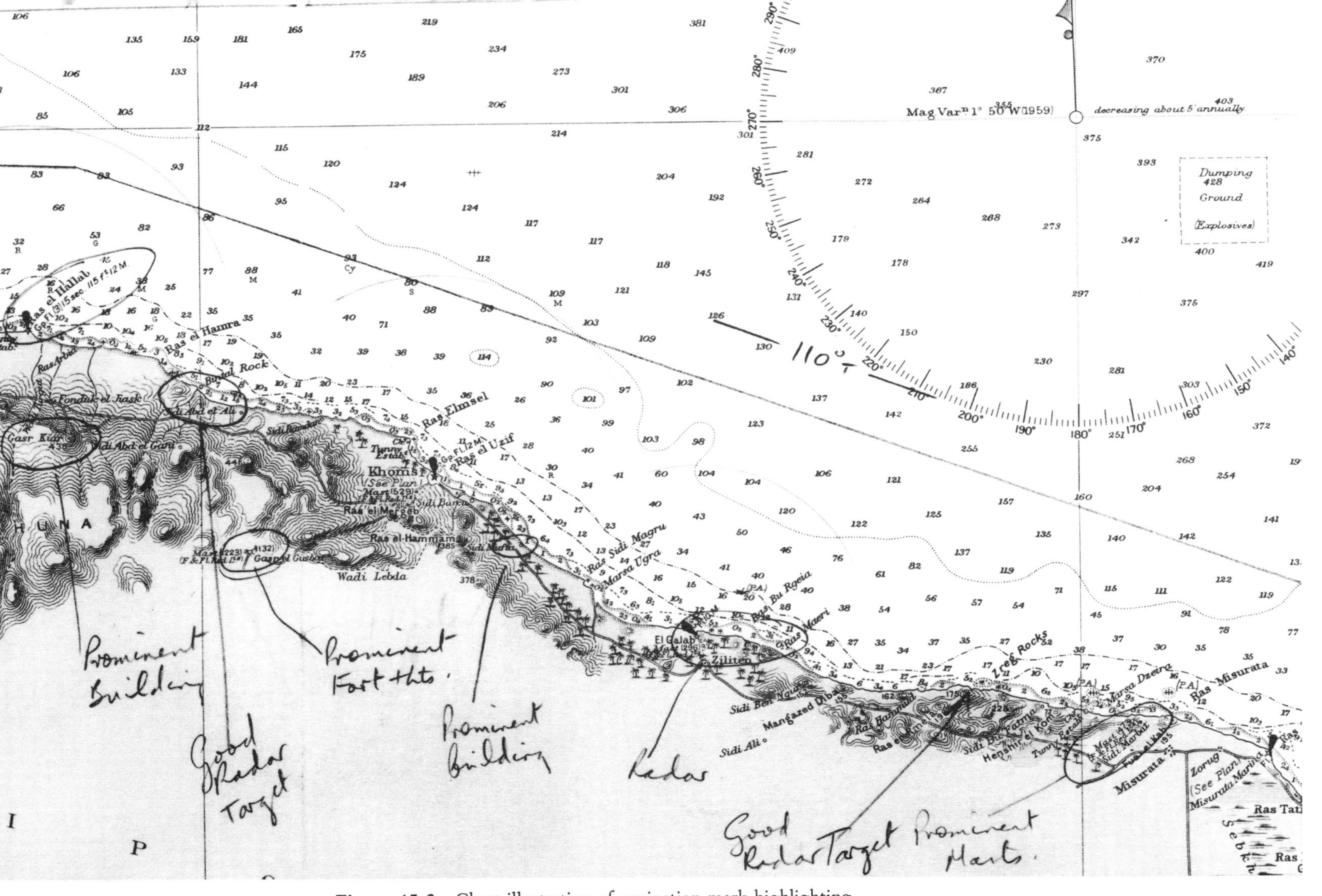

Figure 15.3 Chart illustration of navigation mark highlighting

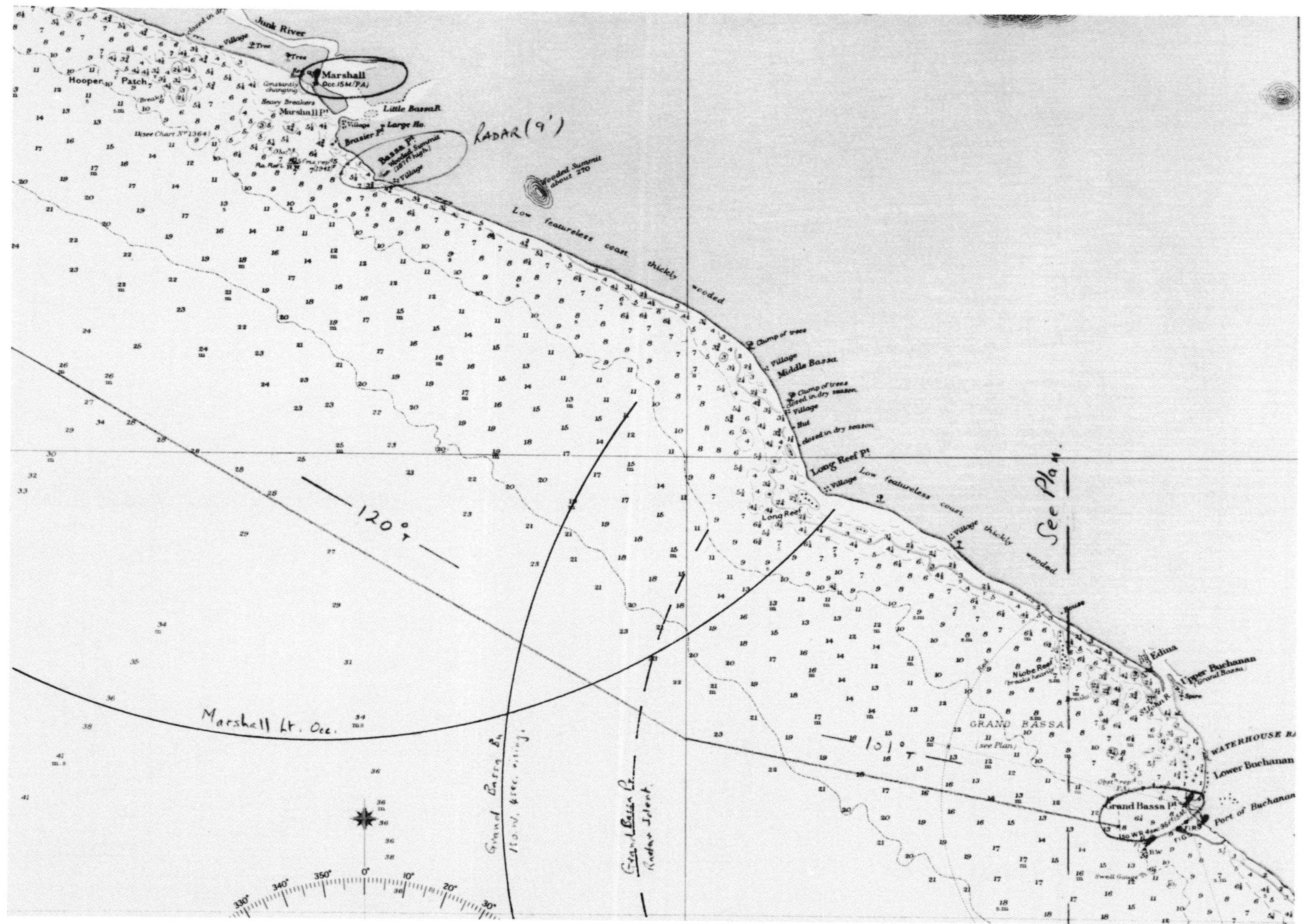

Figure 15.4 Chart illustration of navigation mark ranges

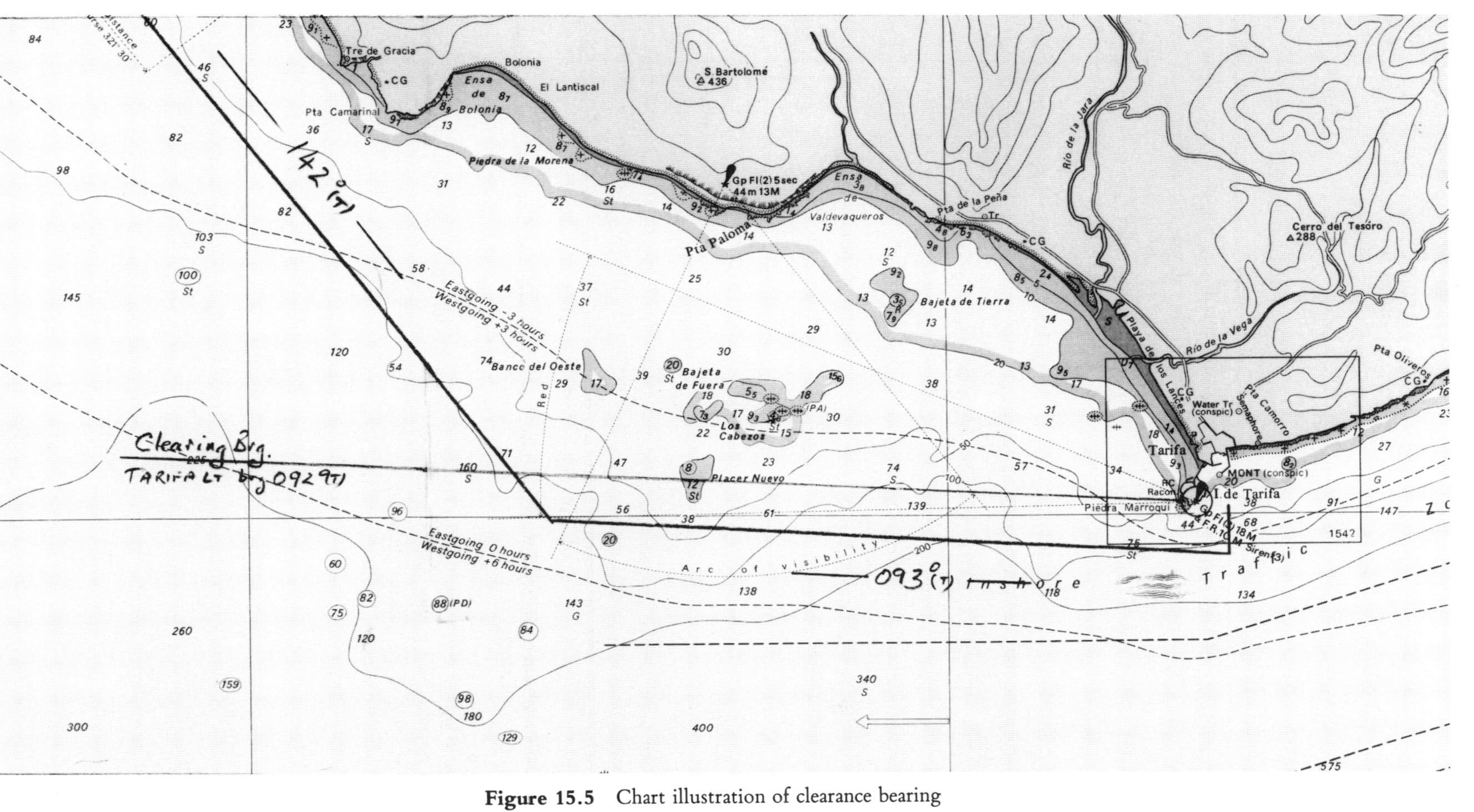

Figure 15.5 Chart illustration of clearance bearing

15.1.5 Ranges of navigation marks

On night passages too much time can be spent on fruitless searching for navigation lights that are well below the horizon. A small amount of time during passage planning will establish the visible ranges of the navigation lights, taking into account the actual height of eye (Figure 15.4). When these ranges are applied to the chart, the positions and times that the lights are expected to rise and dip are clearly seen. This process is essential when appraising whether the visual bearings are available as a means of position fixing.

If visibility is impaired then naturally these ranges will be reduced, perhaps giving the navigator his first indication of changing conditions.

Occasionally certain geographic points have maximum radar identifiable ranges quoted in the Sailing Directions. This information can be used in a similar way to that of the lights. However, the data is not nearly so precise and can be greatly affected by the radar characteristics (see Section 5.7.2).

15.1.6 Clearing bearings

Clearing bearings and transits should be marked where they will assist the OOW in assessing at a glance that the vessel is clear of a particular danger or when it is approaching an area of reduced safety margins (Figure 15.5). Transits also have value in assessing rapidly the compass error provided the objects are sufficiently well spaced.

It is better not to draw a clearing bearing as a solid line through the course line as it may, under certain circumstances, be mistaken for a new course line. It is, in fact, good practice in all passage planning to leave the area immediately adjacent to the course line as uncluttered as possible (see Section 15.1.12).

15.1.7 Under-keel clearance

Expected under-keel clearances at points on the track where they are relevant as a cross-check can be estimated and marked on the chart (Figure 15.6). This particularly applies to alter-course positions and crossing bars or other shallow patches.

15.1.8 Routine checks and changes

Mark, adjacent to the track, the points at which routine checks and changes should take place (Figure 15.7). These should include:

1. Change to manual steering
2. Crewman standing by the steering position
3. The testing of steering gear (Notice M.1471; see Section 1.6.7)
4. Advise the engine room of the impending approach to navigational dangers such as heavy traffic concentrations or areas of restricted searoom
5. The testing of engines prior to manoeuvring in confined waters
6. Upgrading the bridge manning, e.g. master to the bridge
7. Clearing anchors
8. Start parallel indexing
9. Change in prime position fixing system
10. Speed change
11. Changing charts.

There is a tendency to allocate some of these instructions to alter-course positions; this is not very practical as the OOW is, at that time, fully occupied in checking the position and altering course. A little bit of thought will decide whether the change is required before or after the alteration.

15.1.9 Wheel-over positions

In areas where margins of safety are very small, account will have to be taken of the time and space needed to bring the ship around to a new course (Figure 15.8). This is done by examining the ship's turning circle data for a moderate amount of helm (10–15°) and the intended approach speed. The point at which to begin the alteration can be marked and referenced by a bearing and/or range from a nearby radar or visual mark. The advantage of doing this is that the moment at which to begin the turn can be appraised without having to return to the chart to fix the position.

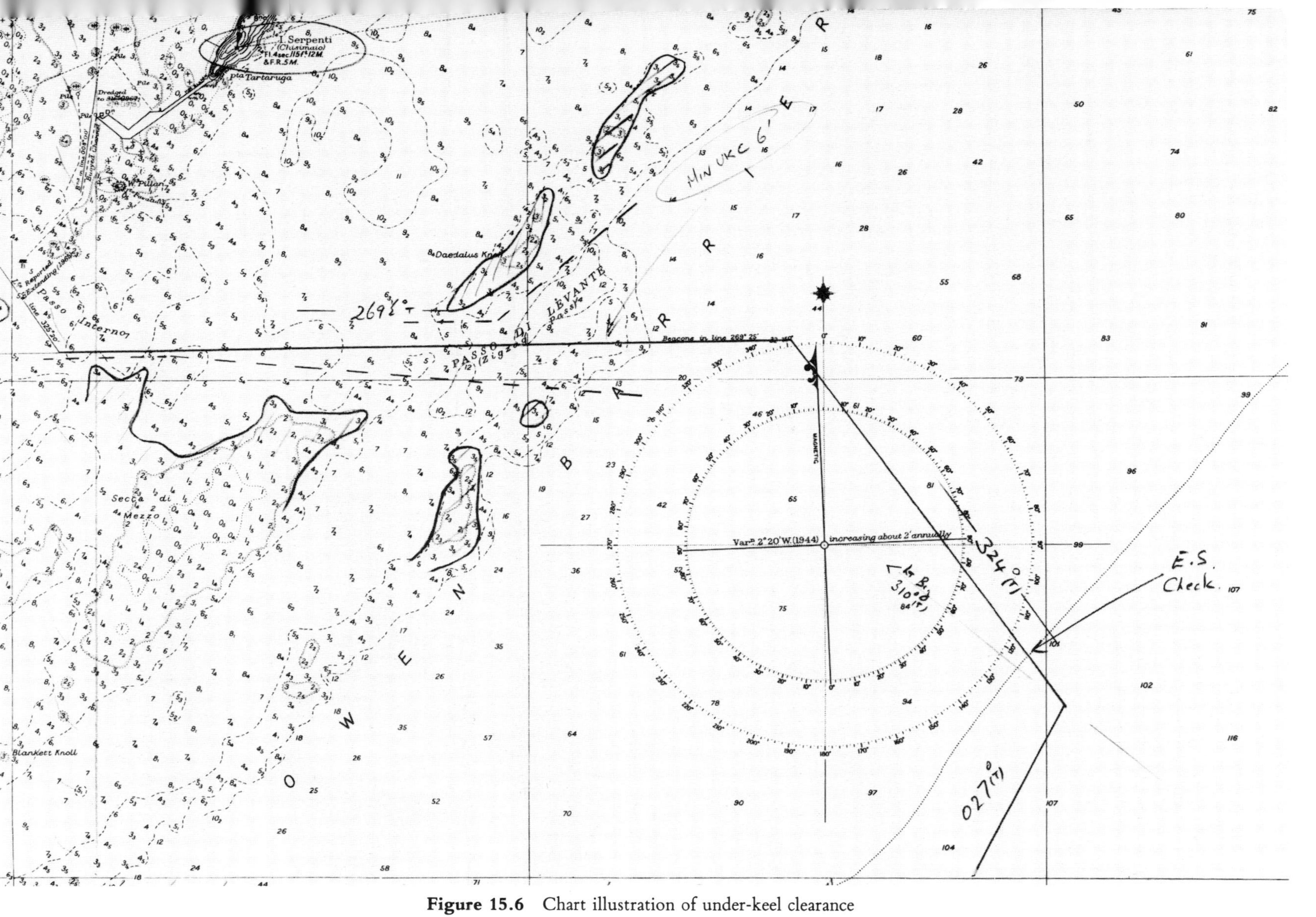

Figure 15.6 Chart illustration of under-keel clearance

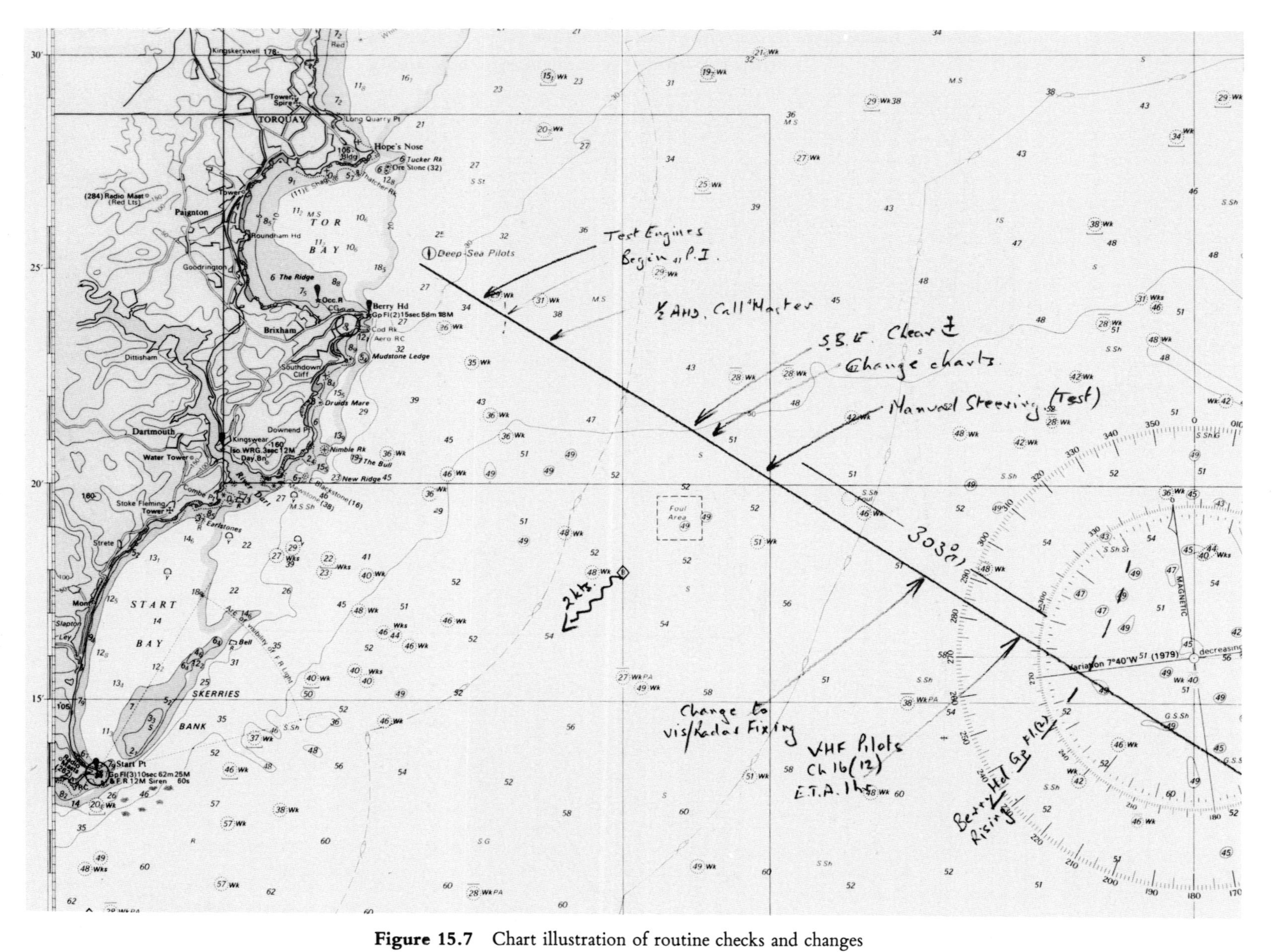

Figure 15.7 Chart illustration of routine checks and changes

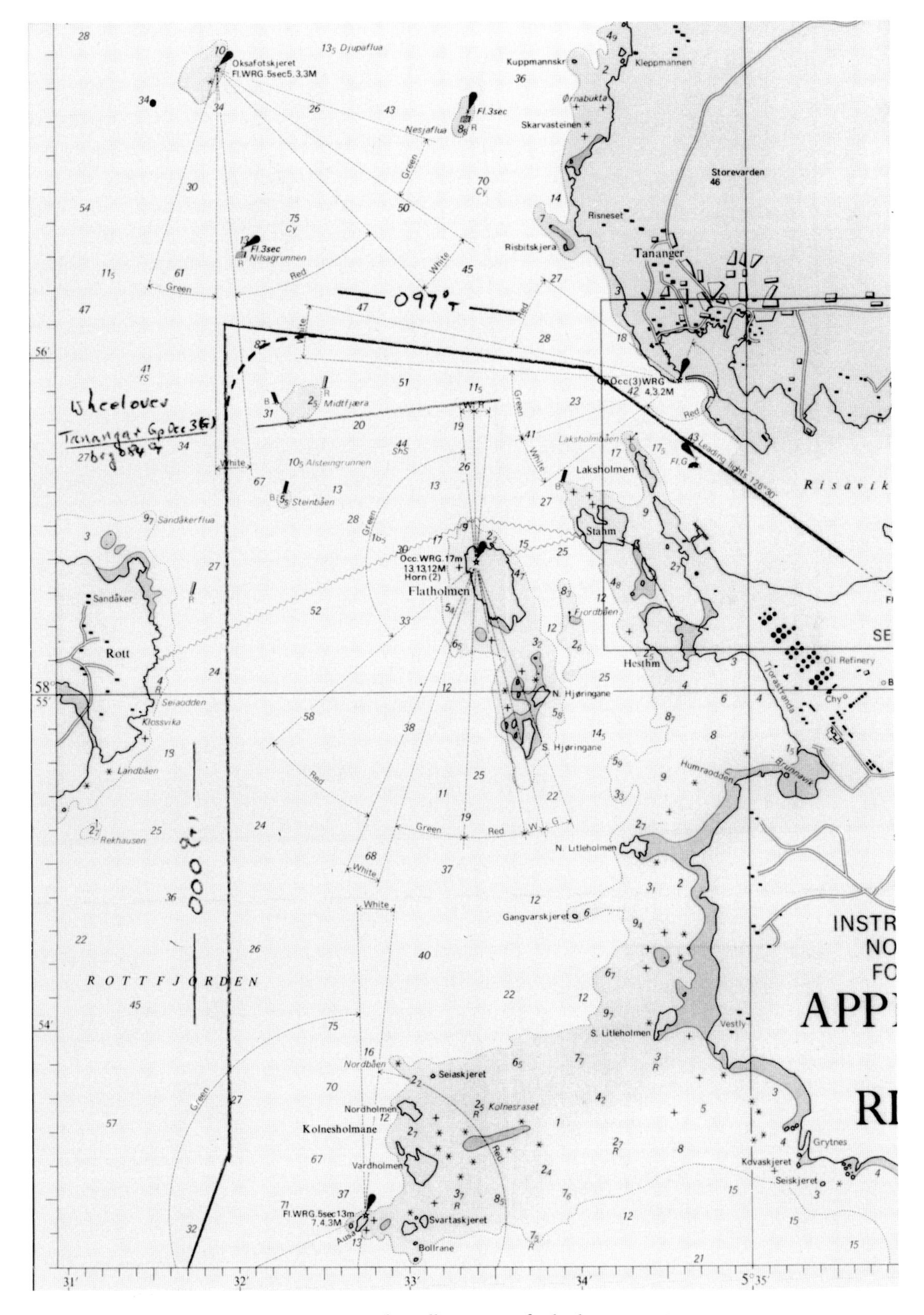

Figure 15.8 Chart illustration of wheel-over position

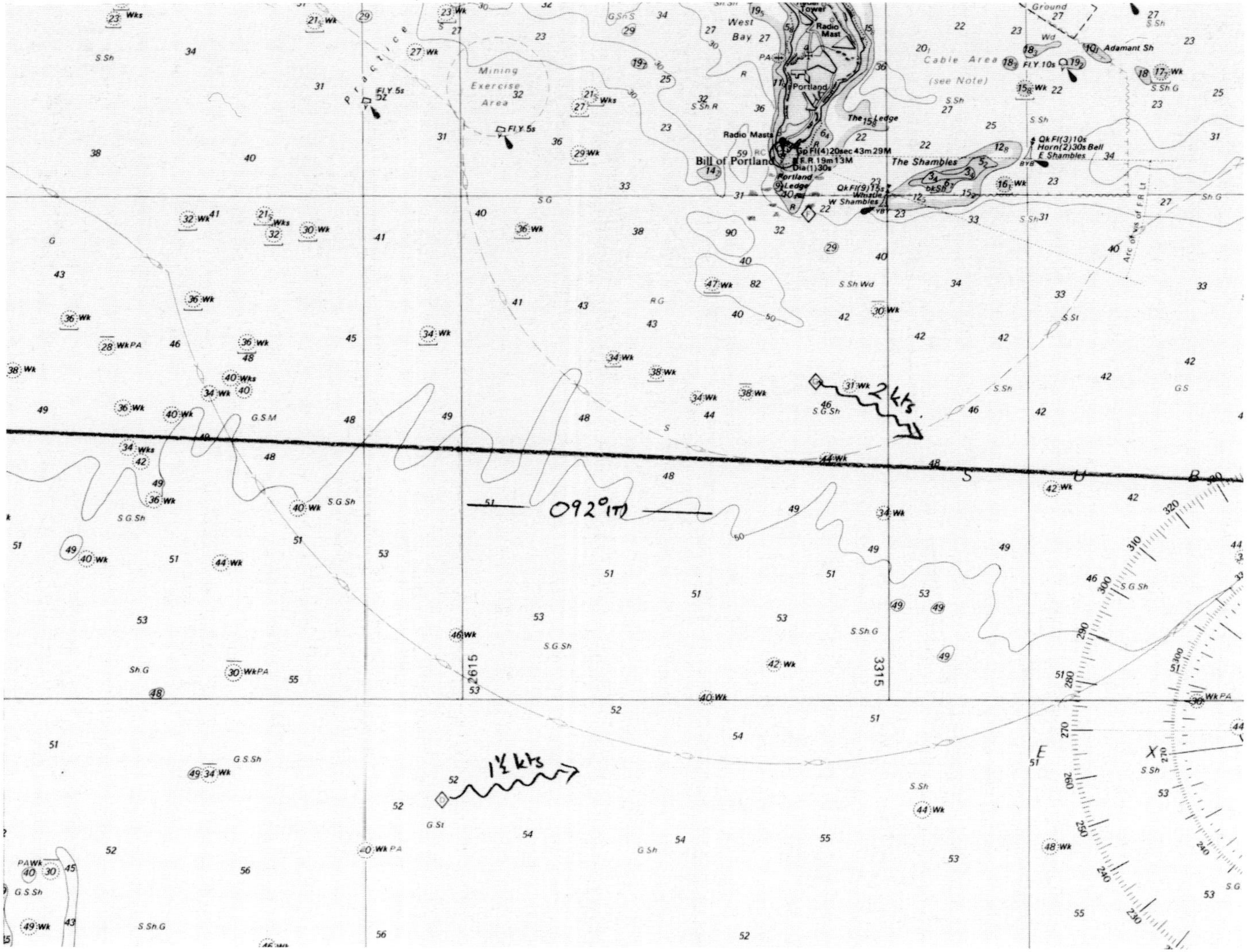

Figure 15.9 Chart illustration of tidal data

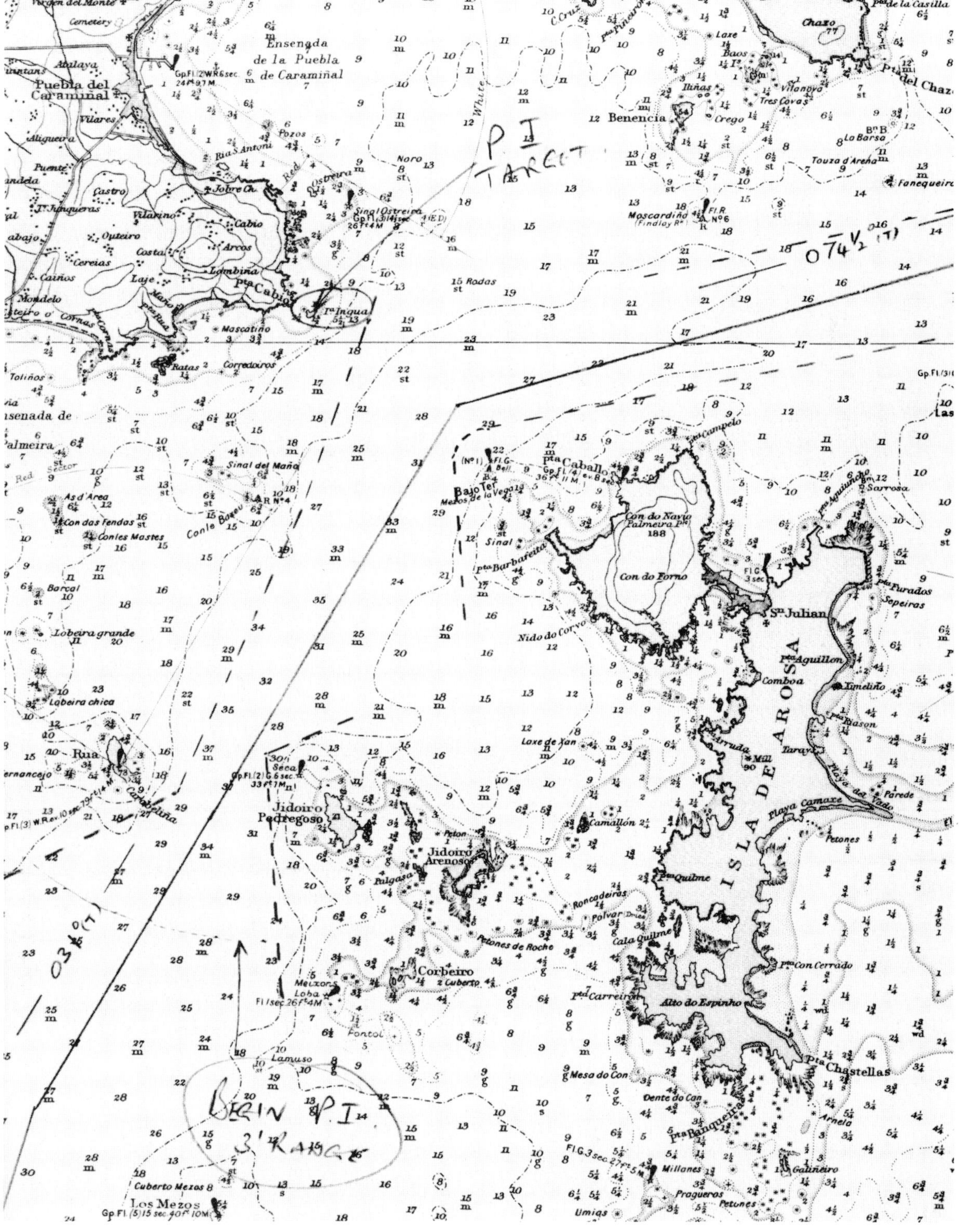

Figure 15.10 Chart illustration of parallel indexing data

15.1.10 *Tide/Current*

Estimated tidal rates and sets can be marked at the tidal diamonds and highlighted to attract the attention if the values are significant (Figure 15.9). Similarly, current values can be inserted at relevant points.

15.1.11 *Parallel indexing*

Although several construction lines may be put on to the chart during the data extraction process, the only information that need appear on the chart thereafter is an indication of where parallel indexing is to begin, a clear indication of the indexing target and the occasional clearing range (Figure 15.10). All other work related to PI should be carried out on the radar reflection plotter. (See also Sections 5.7.2.2 and 20.1.4.)

15.1.12 *Obscuring of charted data*

While inserting additional data on to the chart, the navigator must take care not to obscure important printed data. There is also the need to leave plenty of clear space either side of the plotted course line for the OOW to fix the vessel's position.

15.2 The bridge notebook

15.2.1 *As a summary of the chart*

Having been marked with all the information needed to conduct the passage safely, the charts are now ready for the passage. The bridge notebook should be a summary, derived from the charts, forming a sequential checklist for the OOW which, if necessary, he can carry with him.

All the entries should be time related and in order. To achieve this, no attempt should be made to write up the notebook until the chart is fully marked. The track on the chart can then be followed and every item noted in turn. Space can be left at appropriate places for additional data such as night orders, pilot's advice, etc.

15.2.2 *Layout*

The manner in which the information in the notebook is laid out is generally a personal choice, tailored to suit the requirements of the particular ship and company. The end result must have the information laid out clearly and unambiguously, remembering that several different people will be using it.

Tabular format lends itself well to the display of many small pieces of data, and two suggested layouts are shown (Figures 15.11, 15.12). This does not preclude the navigator from using narrative form by writing right across the page when it is necessary.

15.2.3 *Positions*

Positions for alter-course points and waypoints when on a coastal passage are often better quoted as ranges and bearings from a prominent geographical point than latitude and longitude, as there is less chance of plotting error and generally it is quicker to identify the point in question (Figure 15.13). Latitude and longitude will be necessary when the land is a long way off or if there is no suitable geographical point on the chart in use at the time. Obscure geographical points may need referencing by latitude and longitude.

15.2.4 *Abbreviations*

A certain amount of abbreviation of wording is inevitable in the notebook, but take care that this does not inadvertently lead to misinterpretation, especially with foreign names. To indicate what is being referred to, do not neglect to include: Lt; By; Lt Ho; Pt; Bk Wr; Hd; Bn; etc. When using sectored lights the light characteristic should be quoted (see Figure 15.13).

15.2.5 *Predicted tidal set*

The predicted tidal set and rate is entered in the appropriate column, particularly where adjustments may need to be made to the course steered. The OOW

Sheet ....

BRIDGE NOTE BOOK

Draft. F ....

A .... Passage From.......................... To.......................... Nominal Speed...........kts

M ....

Actual Time	Est. S.T.	A/C Positions and Way Points.	Course to make good	Dist. to next a/c	Current Set & Rate	Est. Spd over ground.	Prime Position Fixing System & Accuracy.	Position Fixing Check Systems.	Remarks.

Figure 15.11 Bridge notebook suggested layout 1

Sheet ....

BRIDGE NOTE BOOK

Passage: FROM.. Date..

TO .... Draft.. metres

Nominal speed...................................... kts

Time	Distance to next a/c	A/C Positions and Way Points	Course to make good	Tide/Current	Position Fixing Systems

Figure 15.12 Bridge notebook suggested layout 2

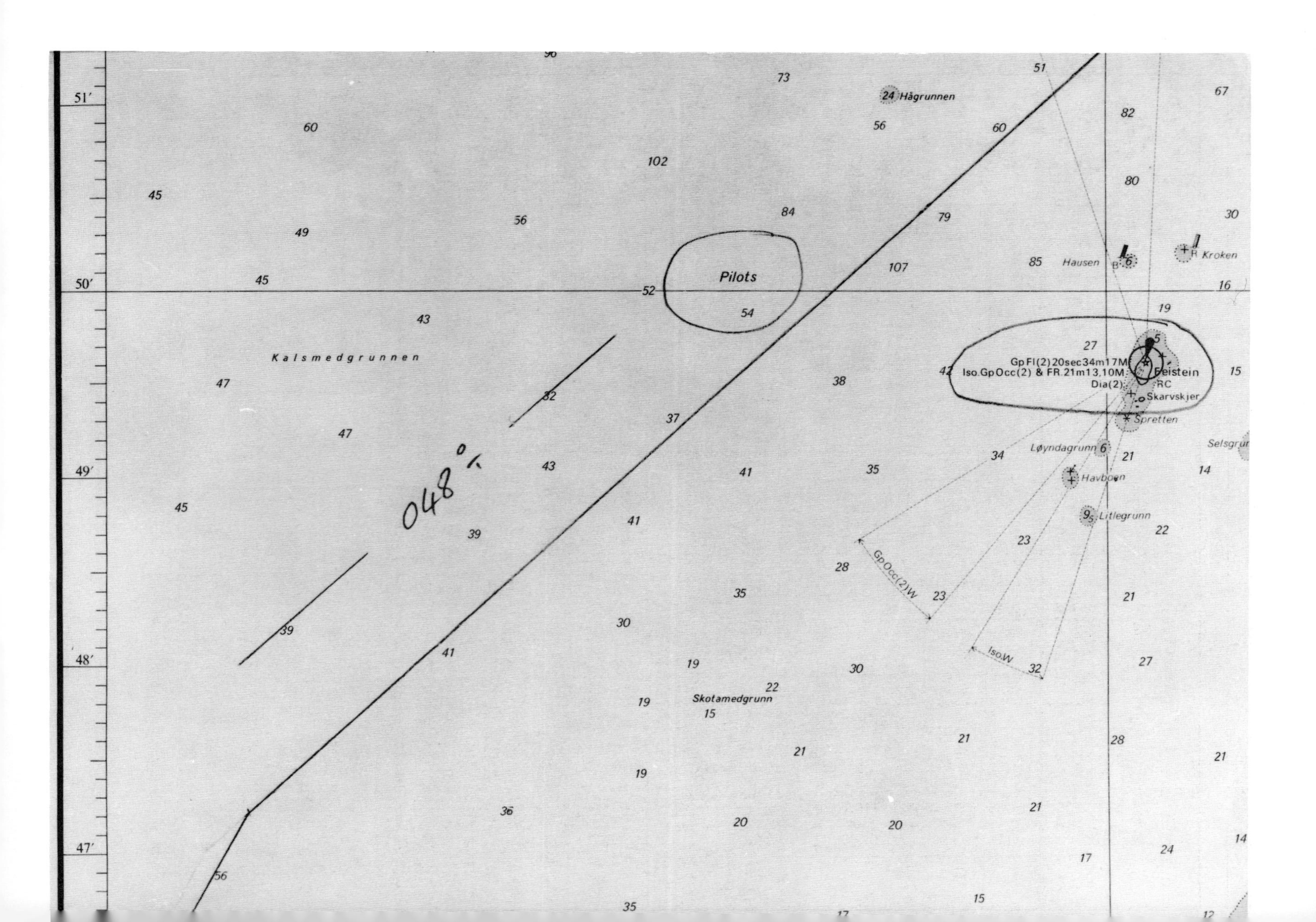
Hågrunnen
Pilots
Kalsmedgrunnen
Skotamedgrunn
Hausen
Kroken
GpFl(2)20sec34m17M
Iso.GpOcc(2) & FR.21m13,10M
Dia(2)
Feistein
RC
Skarvskjer
Spretten
Løyndagrunn
Havboen
Litlegrunn
GpOcc(2)W
IsoW
048°
51′
50′
49′
48′
47′

BRIDGE NOTE BOOK

Draft F ..X X..........

A ..X X..........

M ..X X..........

Passage From ....X X X....... To ..X X..

Actual Time	Est. S.T.	A/C Positions and Way Points	Course to make good	Dist. to next a/c	Current Set & Rate	Est. Spd. over ground	Prime P System
	0130	Feistein LT. Ho. LT. A/c [Gp. Fl. (2) 20 sec.] brg. 063 °T × 5.3M.	048 °T	7.0	X X X	X X X	X X X

Figure 15.13 Bridge notebook specimen entry of position and abbreviation

will need a clear understanding of how these adjustments are to be made, i.e. either as an instruction by the master or on his own initiative. An entry in the master's standing orders should resolve this point.

Whether these rate and set entries are made precisely, i.e. 360° notation and decimals of a knot, or more approximately as compass points and fractions of a knot, depends on personal preference. Bearing in mind the sometimes imprecise nature of these quantities, perhaps the latter method is more realistic, e.g. SW'ly × 1.5 knots instead of 238°T × 1.4 knots.

15.2.6 Speed

Enter the estimated speed over the ground, taking into account predicted current and tide effects. If possible this value should be accurate to at least 0.25 of a knot.

As mentioned in Section 14.6.1, the rate of progress along the proposed track and the corresponding ETAs at relevant points are integral parts of passage planning. How precise this speed data needs to be really depends on how rapidly the time variable data changes and on just how often the navigator is willing and able to reappraise it. The more inaccurate his estimation of progress, the more quickly the time gets out of step with the data.

In areas where the vessel is expecting to manoeuvre its engines, the speed is generally too variable to be worthy of entering in the notebook. Instead the telegraph order or demanded engine revolutions is probably more appropriate.

15.2.7 Time reference

Each entry in the notebook is sequential and it is very useful to have estimated times noted against entries. The advantage is that as the plan is executed the actual times of events can be quickly compared with the estimated times and any discrepancy noted. This could be the first indication to the navigator that there is some factor influencing progress that has been either overlooked or incorrectly estimated. Having decided on the cause of the discrepancy, he can then decide whether or not a reappraisal of data is needed for the remainder of the plan.

Some navigators prefer to include estimated times in their marking of the chart. If this is done then the estimated times on the chart should be bracketed or annotated in some way to prevent misinterpretation and confusion with the real passage times.

15.2.8 Position fixing system

The prime and secondary position fixing systems, having been identified by their suitability, will be indicated together with the time that they are expected to apply (Figure 15.14). Some indication should be made as to any significant error sources and the degree of accuracy and reliability which might be expected. For example, if the Decca Navigator is judged to be the best position fixing method for a particular area, then chain and recommended colours should be entered along with a comment on fixed and variable errors where these are significant. This does not remove the responsibility of the OOW to make proper use of the Decca data sheets. On the contrary, it directs him there in cases of large errors.

The entry for use of WT DF could include frequency and identification letters with probable accuracy, or simply a page reference to the Admiralty List of Radio Signals, vol. 2.

15.2.9 Frequency of fixing

Where the position needs to be fixed more regularly than the norm for the particular ship, then a remark to that effect should be entered. Norms for coasting should be in the master's or company's standing orders. In areas of strong variable sets or when transitting near to navigational danger, the frequency of fixing will be increased.

15.2.10 Parallel indexing

The expected time to begin parallel indexing, identification of indexing target, range scale and existence of margins of safety will be noted under 'position fixing check systems' (Figure 15.15). Note that PI is not a position fixing system. All the relevant data needed to carry out parallel indexing should have been extracted

BRIDGE NOTE BOOK

Sheet X.....

Draft. F ...XX.......
A ...XX.......
M ...XX.......

Passage From......XXX....... To.......XXX....... Nominal Speed...XX...kts

Actual Time	Est. S.T.	A/C Positions and Way Points.	Course to make good	Dist. to next a/c	Current Set & Rate	Est. Spd. over ground.	Prime Position Fixing System & Accuracy.	Position Fixing Check Systems.	Remarks.
	XX		XX	XX	XX	XX	VIS. LIZARD LT. HO. LT. x RADAR RANGES	DECCA 1B (R+G)	
	XX		XX	XX	XX	XX	DECCA 1B (R+G) (Note Fixed Errors)	RADAR RANGE & BRG.	LIZARD LT. DIPS.

Figure 15.14 Bridge notebook specimen entry of position fixing system

Sheet ..X...

BRIDGE NOTE BOOK

Draft. F ...XX......

A ...XX......

M ...XX..........

Passage From......XXX......... To.........XXX.........

Nominal Speed...XX..kts

Actual Time	Est. S.T.	A/C Positions and Way Points.	Course to make good	Dist. to next a/c	Current Set & Rate	Est. Spd. over ground.	Prime Position Fixing System & Accuracy.	Position Fixing Check Systems.	Remarks.
	XXX		029°T	3	S'ly. 1kt.	14kts.	Vis. Brgs.	BEGIN P.I. ON I? INGUA 3' RANGE SCALE x INDEX Range SECA BN 0.3n.m. (MIN 0.1n.m)	

Figure 15.15 Bridge notebook specimen entry of parallel indexing data

from the chart and stored separately in written or diagrammatic form ready for plotting directly on to the radar reflection plotter when needed. (See also Sections 5.7.2.2 and 20.1.4.)

15.2.11 Remarks

The 'remarks' section of the notebook will be used to bring to the attention of the OOW all those items not previously included; each is to be entered with the relevant time (see Section 15.1.8).

It will include times to call pilots, VHF channels to use, signals required to be shown, sunrise and sunset times and references to publications, contingency plans and checklists. Space should be left at appropriate points for night orders.

15.3 Pilotage waters

As mentioned in Section 13.4 (see also Sections 5.7.3 and 21.5), the main problem with passage planning in pilotage waters is the need to wait until the pilot boards before his expert and up-to-date knowledge of local conditions is available. However, this does not prevent the large scale charts and publications being examined in an effort to identify danger and problem areas that apply to the ship and its manoeuvring characteristics. A safe route clear of these dangers can be laid off, although it may not be followed precisely. It does, however, form a framework around which the required organizational details can be planned.

When the pilot boards, he must be immediately acquainted with the characteristics of the ship and the plan should be offered for his approval. Any amendments or additional information that he presents are then incorporated into the plan in fully written form so that at all times the OOW is briefed and can effectively monitor progress and support the pilot fully.

Two items of appraisal that should always be completed before entering the pilotage waters are parallel indexing and tidal calculations. Several notable groundings in pilotage waters could well have been avoided if parallel indexing had been employed (e.g. the grounding of the VLCC Metula in the Magellan Straits on 9 August 1974).

15.3.1 No-go areas

The process of identifying the no-go areas for the pilotage section of the passage is similar to that employed while coasting, except that the under-keel clearance safety margin will probably be considerably reduced. This involves a much more precise calculation of tidal height for the period the vessel will be in the area, and a close examination of the large scale charts. Dredged areas and buoyed channels are assessed relatively easily, but areas outside these also need appraisal with anti-collision, turning, waiting and anchoring manoeuvres in mind.

With proximity to navigational danger being the rule rather than the exception, it is also advisable to insert margin of safety lines and complete the parallel indexing preparation, especially where there are strong environmental forces or a shortage of navigation marks suitable for conning purposes.

15.3.2 Manoeuvring characteristics

The person in control of a vessel must have a good working knowledge of that vessel's manoeuvring characteristics so that he may safely carry out such operations as docking, undocking, anchoring and turning in confined water. He must have sufficient understanding of the vessel's capabilities so as to prevent the vessel reaching a dangerous position from which it is unable to escape safely.

Apart from the close manoeuvring situations generally undertaken by the pilot or master, the OOW and navigator also have occasion to use this data, e.g.:

1 When making a prediction of the effect of a reduction of speed during radar plotting for anti-collision purposes. This requires knowledge of the headreach distance and period taken to execute the manoeuvre.
2 When marking the passage plan with wheel-over positions. For this, the turning circle data for a moderate (10–15°) degree of helm is consulted.
3 When deciding on the distance from a danger area

at which to position margin of safety lines. This requires the turning circle diameter for maximum helm and engines stopped, or the headreach with crash stop and no helm, whichever is the greater. Consideration may also have to be given to the actual passing speed and any shallow water effects (see Section 15.3.3).

15.3.3 Interaction

The manoeuvring characteristics discussed in Section 15.3.2 all assumed deep water with little or no environmental effects. However, the majority of a vessel's manoeuvring is carried out in pilotage waters where these conditions are not usual. Experiments have shown that once the water depth becomes less than approximately seven times the vessel's draught, interactive forces begin to be experienced (see C. B. Barrass, *Ship Squat*, Benn, 1978). These forces have several consequences, the most important of which are

1 A reduction in under-keel clearance (squat)
2 A reduced effectiveness of the rudder
3 A reduced effectiveness of the propeller.

Outward signs that the vessel is in shallow water and hence feeling the effects of the sea-bed include an increased turbulence of the bow wave, a drop in engine revolutions, a speed drop, increased vibration and greater difficulty in maintaining the course.

The amount that the vessel squats in the water depends on several factors, such as speed, block coefficient, and the depth: draught ratio. From an operational point of view, speed is the most important, and it is worth noting that squat varies as the square of the speed. Thus a change of speed has a marked effect on the amount of squat. For example, if a vessel is experiencing severe interaction with the bottom, i.e. draught 90 per cent of water depth, at 10 knots, the squat could reduce the under-keel clearance by nearly a metre. If speed was reduced to 8 knots then squat would reduce to a little over half a metre.

A further complication can be the uneven distribution of the squat, with the bow sinking further than the stern or vice versa.

While the vessel is in shallow water there is a large volume of entrained water, which increases the vessel's inertia considerably and also changes the water flow to the propeller and rudder. As a consequence, turning circles increase in diameter, stopping distances and times increase, and the ship's head may be subject to sheer without warning. In the extreme case of interaction (depth : draught ratio of 1.1 : 1), the turning circle diameter could double for the same amount of helm.

Because shallow water effects increase with speed, the proportionate increase in stopping distance will also increase with speed. Very roughly, in shallow water, the increase in stopping distance varies as the square of the speed, whereas in deep water the relationship is linear.

Interactive effects will also show themselves when passing close to obstructions such as other ships, jetties, walls and submerged banks (see Notice M.930, Interaction between Ships), with possible catastrophic effects on the ship's handling characteristics.

In all cases where interaction is thought to be occurring, slow speeds and increased safety margins are required.

A final consideration when operating in very shallow water is the increase in draught resulting from heeling over during a turn. A VLCC of 250 000 d.w.t. could increase its draught by approximately half a metre for every degree angle of heel.

16 The contingency plan

16.1 Diversion

The circumstances that can cause a passage plan to be abandoned in favour of an alternative can be many and varied. Some of these circumstances are more predictable than others. If they are predictable, contingency plans should be prepared beforehand and kept in reserve. These alternative plans could cover periods from just a few minutes to several days (see also Chapter 21).

16.1.1 Short term alternatives

The original route chosen for the passage plan is based on an understanding of the conditions appertaining at that time. Some of these conditions, such as visibility, weather and traffic, can be very variable and may allow useful alternative routes to be considered as the passage progresses. For example, if the plan presumed difficult traffic conditions at a particular confluence point and in actuality there was very little traffic, then a different approach course might be taken and the distance shortened. Similarly the prospect of bad visibility may have resulted in an original course line being laid off a long way outside a routeing scheme, but clear visibility on the arrival of the ship in the area could allow a much shorter alternative through the scheme to be used safely.

This type of relatively short term contingency plan probably requires very little preparation and may appear only in the 'remarks' section of the bridge notebook. However, the navigator should not neglect to appraise the effect it might have on time variable data in the remainder of the plan.

16.1.2 Operational directives

It is a fairly common characteristic of the operation of certain types of vessel, such as tankers and charter vessels, that they may be diverted to a new destination whilst on passage. This should not cause any great problems if the vessel is in the open sea, provided that the required charts and publications are on board. The vessel can quickly take up its new course with little or no risk while the new plan is appraised and drawn up.

On a coastal passage or in congested waters, a major diversion can cause quite a problem for a while. An immediate diversion may not necessarily be safe or practicable if the vessel is at present in a routeing scheme or constrained in some other way. In a case like this, there is a need to look ahead along the present plan to identify a point at which the diversion may safely be carried out. This has the added advantage of allowing some time to appraise the new route. Even when an immediate diversion is possible it is not always advisable to take the new route until it has been properly appraised, especially in difficult navigational areas. The delay while a new route is appraised and planned is unlikely to exceed two hours and therefore is hardly good cause to neglect the planning.

When diversions occur regularly, the navigator probably has a fair idea of what to expect and when they are likely to occur. Under these circumstances, routes to the alternative destinations can be planned in detail from the optimum diversion points. The efficiency of this process can be greatly enhanced if the ship operators have a knowledge of the points where the alternative routes diverge and send their instructions accordingly.

16.2 Reduced progress

Any event which causes the actual rate of progress to be different from that previously estimated will affect the time related variables used in the plan. Whatever the cause of the delay, a careful watch needs to be kept on these variables, so that the point at which reappraisal and replanning becomes necessary or the time at which a contingency plan is implemented is not missed.

16.2.1 Restricted visibility

Most voyages will at some stage or other be affected by reduced visibility, be it haze, rain, mist or fog. Each occasion demands that speed be reduced to a safe level for the duration of the restriction, with consequent effect on the passage plan. The likelihood of restricted visibility is to some extent predictable, but seldom if ever with sufficient accuracy for it to be incorporated directly into the main passage plan.

There are two possible contingencies that the navigator will need to consider when preparing his plan:

1 Reduced visibility at a point where navigation is already difficult due to other causes
2 Delays due to reduced visibility resulting in changes to important time related variables at critical points.

If it is felt that either event would lead to the abandonment of the present plan, and that there was a reasonable chance of them occurring, then contingency plans for alternative routes and actions should be prepared, including an indication of the possible points of implementation.

16.2.2 Collision avoidance

In general, when deciding on a route through a possible heavy traffic area it is better to assume worst conditions and choose the route accordingly, thus minimizing possible time loss due to anti-collision manoeuvres. As mentioned in Section 16.1.1, if things turn out better than expected, perhaps a few small amendments to the route can be made and time saved.

Collision avoidance manoeuveres in open waters have a negligible effect on progress, but in confined waters, where times at various points can be important, a delay due to traffic could require the implementation of an alternative plan (see also Chapter 22).

In routeing schemes, in particular where searoom is often restricted, the anti-collision requirements are much more likely to be met by a speed change. Consequently it is sensible, when formulating the passage plan, not to assume that maximum speed will be maintained through the heavy traffic area.

16.2.3 Anchoring

When unforeseen reduced progress has resulted in the vessel missing a tidal window, there may in fact be no alternative route. In this case the vessel is obliged to wait for the next tide before proceeding. If the vessel is still in open waters, waiting presents no great navigational problems. However, in confined waters, with possible strong tidal sets and adverse weather conditions to contend with, a safe waiting or anchoring area must be sought.

Identification of these safe waiting areas should be done during the initial appraisal, together with their limitations, holding characteristics and nearest alternatives if overcrowded. Furthermore, if the passage plan includes a tidal window then a contingency plan taking the vessel to the appropriate waiting area should also be prepared.

16.3 Changed conditions

16.3.1 Abnormal tide conditions

Tidal heights seldom correspond exactly with predictions, though it is expected that 90 per cent of European predictions will be within 30 cm and

10 min. Consequently, the safety of vessels is not really affected by the inaccuracies.

In some areas the weather conditions and presence of high and low pressure systems can have a marked effect on the sea level, causing tidal surges which can alter the sea level by several metres. These discrepancies can be misleading and occasionally disconcerting, and for a vessel operating on a minimum under-keel clearance it may alter the tidal windows drastically. Negative tidal surges are obviously very important to this type of vessel, and the appraisal must always include a check for their existence in the navigation warnings.

The effect on the passage plan of these tidal aberrations is similar to the changed rate of progress but much less obvious, and consequently it is much more difficult to allow for them. If there are extreme weather systems in the area it is advisable to pay particular attention to the correlation between predicted and actual soundings.

Other causes of abnormal tidal heights are freshets near the mouths of major rivers, and subsea activity, both of which are unpredictable.

16.3.2 Unexpected traffic conditions

There is a high probability that traffic conditions in a particular area will be different from those estimated during the appraisal stage, especially for routes passing through areas of which the navigator has no experience. However, most differences are of little consequence and only occasionally warrant the plan being abandoned. In most cases minor adjustments to the original plan are all that are required.

The unexpected conditions being considered are:

1 More traffic than expected
2 Less traffic than expected
3 Congested anchorages
4 An unusual traffic flow
5 Fishing fleets
6 Naval exercises.

Condition 1 holds an increased risk; it is therefore advisable that the original plan should have assumed worst conditions. If this is done, then the second eventuality is much more likely and the possibilities of minor adjustments to the course and a shortening of the journey can be noted and included in the original plan as an alternative. A typical example would be the plan for crossing a traffic stream. This is best achieved at as near right angles as possible, and the plan would be drawn up to do this even if the approach to and departure from the area was at a narrow angle. If, on arrival in the area, traffic in the stream was observed to be negligible, then the need for the dog-leg would be removed and the vessel could take the shortest route across in safety. This is *not* an alternative, however, when crossing a routeing scheme, as movements in these areas are controlled by rule 10 of the Collision Regulations (see Section 23.6).

Congested anchorages outside port control areas are not very common and, provided all anchorage and waiting areas adjacent to the route have been previously identified, the nearest and most suitable alternative is soon found. In port control areas, expert advice is generally available and, when this is combined with the marked no-go areas on the chart, the alternative anchorage can be safely reached.

The unexpected traffic flow conditions can result in a temporary diversion from the plan, but can only be assessed and allowed for when in position. The possibilities of this condition at traffic confluence points should be noted and more attention paid to the margins of safety and the possible effects on time windows.

Known concentrations of fishing vessels are best allowed for in the original plan in a similar way to any other navigational hazard. Otherwise they must be treated in the same manner as the unexpected traffic flow mentioned above.

Naval exercise areas seldom warrant a complete avoidance, but require increased vigilance and recognition of the possibility of temporary avoiding action with consequent delay. Action will be similar to that employed with fishing fleets.

16.3.3 Navigational aids failure

The unexpected loss of navigational aid data, due to either on-board equipment failure or transmission interruption ashore, can have far reaching consequences for the passage plan. When conducting the appraisal, some consideration should have been given to the probable availability and reliability of the various systems.

If the failure involves the prime position fixing system then the secondary check system is immediately available, albeit at a lower accuracy. Areas where the operation of a particular instrument is critical should be noted and the consequences of failure foreseen.

Radar for coastal fixing generally has visual bearings, Decca and WT DF to back it up. However, in very confined waters, where a shortage or a complete lack of conning marks makes PI almost essential, radar failure inevitably leads to abandonment of the plan (see also Section 16.4.1). Two radars hopefully minimize the chances of this happening.

Decca Navigator is generally not critical to the safe operation of the vessel when other close range systems become available as hazards are approached. Loran, Omega, WT DF and satellite navigation are seldom the prime position fixing systems in areas of critical navigation, and their failure is not likely to affect progress or route except in the case of a landfall, where a reappraisal and new route could become necessary. Loss of the echo sounder in very shallow water can be upsetting but not necessarily critical, and progress will be reduced if the hand lead is to be used as an alternative.

Gyro compass failure will not affect course keeping as the magnetic compass is available as back-up, although there will probably need to be a resumption of manual steering. The fact that the radar cannot be stabilized makes the bearings less accurate and parallel indexing impracticable (see Section 5.3.4). This could, in certain circumstances, require a new prime position fixing system and possibly a new route to be assessed.

16.3.4 *Pilot's advice*

The basic passage plan for pilotage waters is agreed by the pilot and adjusted to his advice before being undertaken. However, the pilot must also respond to changed conditions such as delays for traffic and reduced visibility. The consequent new features need to be incorporated into the plan in a similar fashion to that used when the pilot first boarded.

16.3.5 *Bridge team fatigue*

Overtiredness of bridge team members has been identified as contributing towards several incidents, and regulations are pending that will require positive action to be taken to combat fatigue and ensure that watchkeepers are capable of performing their duties. Judicious planning and man management should, in most cases, alleviate this particular problem (see Section 18.1.3).

If the extent of high level manning or degree of overwork could not be foreseen, the effect of fatigue on the passage plan must be assessed and acted upon. To press on regardless of the increased risks is like making excessive speed in fog, and could seriously threaten the whole voyage.

Possible contingency plans, apart from reorganization of watchkeeping schedules, could include:

1 The reduction of speed to delay arrival at a critical point in the plan
2 The modification of the route away from a difficult area
3 The reduction of speed to allow more thinking time
4 The delay of departure from port
5 Anchoring.

16.4 Emergencies

16.4.1 *Major equipment failure*

The reliability of a vessel's main equipment (engines and steering) should be a known factor, and the implications of a breakdown at a critical point in the passage taken into account when planning.

It is worth remembering that machinery is generally more likely to fail when it is being operated under constantly varying conditions than when it is in steady state. This implies that the risk of failure increases when the engines are manoeuvred and the vessel is negotiating some difficult navigational area. Similarly, automatic control systems, being more complex, have a greater chance of failure than a simple manual control.

These risks are recognized by putting the operation of the vessel into an increased state of readiness in hazardous areas, e.g. manual control of engines and steering and the activation of duplicate and back-up systems.

From the passage planning point of view, additional personnel should be ready and anchors should be cleared ready for use. On the chart, safe waiting areas to use while equipment is repaired should be identified and the availability of outside assistance in the form of tugs noted.

16.4.2 *On-board casualty*

Because the occurrence of a casualty is totally unpredictable, a prepared contingency plan is not feasible. Depending on the degree of seriousness and the availability of outside assistance, response to the emergency could take one of four forms:

1 Medical advice by radio. In this case there need be no change to the passage plan.
2 Increased speed to destination. Time related variables in the present plan need reassessing.
3 Rendezvous with other vessel or helicopter. Probably only a small amount of reassessment is needed, this being the usual course of action if in the open sea.
4 Diversion to a new destination. Full appraisal and planning is needed for the new destination. This need not necessarily extend to the berth, as a safe outer anchorage where evacuation of the casualty can take place may suffice. Returning to the port of departure would need the same response, as the recently completed part of the plan would be, for the most part, unusable.

16.4.3 *Emergency checklists*

The response of the bridge team in the first few minutes following an emergency are often critical to both the safety of the ship and the effectiveness in meeting the emergency.

To ensure a correct response there should be immediately to hand a checklist of essential actions by the watchkeeper, with particular attention to the sequence of carrying them out. This will cover the situation for long enough to upgrade the bridge manning and allow the navigational situation to be reappraised.

The Bridge Procedures Guide includes checklists for many emergency situations (see also Section 11.17.2). These include, amongst others:

1 Main engine failure
2 Steering failure
3 Gyro compass failure
4 Collision
5 Stranding
6 Fire
7 Man overboard.

17 Distress, search and rescue

Being involved in a distress emergency, either as the recipient of aid or as the donor, can be a traumatic experience which most mariners would be happy to forego. With lives in the balance, the actions of those involved must be both prompt and correct, and towards this end the mariner should have a good working knowledge of the basic procedures to be followed. These procedures are contained in three important publications:

1 The Mersar Manual (see Section 11.15)
2 Handbook for Radio Operators (see Sections 11.18 and 19.1.5)
3 Admiralty Notices to Mariners, Annual Summary (see Section 11.4.2).

Bridge watchkeepers are advised to study all of these. The AMVER (Automated Mutual Assistance Vessel Rescue) system operated by the United States Coastguard (see Sections 23.5 and 23.9.4) is one of several voluntary and mandatory reporting schemes for merchant ships, set up to cover various parts of the world. They are designed to provide a rapid response to distress messages. Provided all ships participate by regularly reporting their position and progress, overdue vessels can be identified quickly and other vessels estimated to be nearby can be alerted accordingly.

This chapter forms a brief commentary on the contents of these publications and is not intended as a substitute. The observations have been biased towards offshore search and rescue as being perhaps more applicable to foreign-going merchant ships.

The diversion resulting from the receipt of a distress message can occur anytime and anywhere; providing it is reasonable, all ships within a hundred miles or so are likely to respond, even at some increased risk to themselves. The degree of increased risk that is acceptable is a matter of judgement, as there is no point in adding a second casualty to the first.

A prompt response is an essential feature of most rescue missions. It is also important not to overlook the need to assess the navigational problems arising from the diversion and allow for them, so as to minimize any risks. The route towards the distress should be planned, taking into account the probability of proceeding at maximum speed in less than ideal weather conditions or manoeuvring in waters not normally entered by the vessel.

17.1 The distressed vessel

17.1.1 *Radio distress*

A distressed vessel will use whatever distress frequencies are available. These might include:

1 500 kHz wireless telegraphy: the message is preceded by the automatic radiotelegraph alarm signal to alert those vessels that are not keeping a listening watch on this frequency.
2 2182 kHz radiotelephony: the message is preceded by the automatic radiotelephone alarm signal to attract attention and activate any automatic alarm on other ships; also DSC on 2187.5 kHz

4207.5 kHz, 6312 kHz, 8414.5 kHz, 1577 kHz and 16 804 kHz.
3 VHF 156.8 MHz (channel 16) and 156.525 MHz (channel 70).
4 121.5 MHz and 243 MHz aircraft distress frequencies.
5 Electronic position indicating radiobeacons (EPIRBs): these are at present on 2182 kHz, 121.5 MHz, 243 MHz and 406 MHz (see Sections 17.2.2 and 19.1.7).
6 8364 kHz from survival craft.

Nearly all coast radio stations and many ship stations monitor most or all of these frequencies (except the aeronautical ones), especially during the radio silence periods, e.g. on 500 kHz for a 3 minute period beginning at 15 minutes and 45 minutes past each hour, and on 2182 kHz for a 3 minute period beginning on the hour and 30 minutes past each hour.

The components of the distress message are very important and would include, apart from identification and position, sufficient detail to enable would-be rescuers to initiate an appropriate response. There are advantages, however, in keeping the initial message as succinct as possible. Additional data to help with the search can be included in later messages once contact has been made. (See also Section 19.1.7.)

17.1.2 *Radio direction finder use*

Information relating to the position of a distressed vessel is critical, and any data in addition to the normally transmitted estimated position is important. This can be provided by the transmission of the special signals on the distress frequencies to allow direction finding bearings to be taken by both coast stations and other ships. These individual bearings must be co-ordinated to obtain a fix. This would normally be done by the rescue co-ordination centre (RCC) in whose area the distress is located, with the assistance of the coast radio station (CRS).

17.1.3 *Aircraft casualties*

When a distressed aircraft has time, communications with air traffic control and responding ships may allow the aircraft to choose a ditching point close to potential rescuers. Merchant ships can sometimes assist this process by the transmission of homing signals, or at night by using a searchlight. Much of this is dependent on the aircraft having radio equipment capable of operating in the marine bands. This is not commonly the case.

In many cases with aircraft casualties, however, the position of ditching is only very approximate and an effective search can only be carried out with several ships and/or a specialized search and rescue aircraft. In situations like this, any information that can help to pinpoint the ditching position is vitally important. Unless the aircraft is able to transmit on a suitable frequency or releases an EPIRB there is no chance of obtaining DF bearings. Intership communications with vessels in the general area may reveal visual or audible indications of position.

17.1.4 *Cancellation of distress*

In view of the very often considerable risks and efforts taken by rescuers in answering a distress message, it is reasonable that the message is cancelled promptly when assistance is no longer required. This applies particularly to false alarms from accidentally activated EPIRBs. Once activated, it is not sufficient to simply switch them off and say nothing. The coast radio station must be advised that it was a false alarm so that the rescue services can be told to stand down.

17.2 Responding to the distress message

Under the International Convention for the Safety of Life at Sea (SOLAS) 1974, the master of a ship at sea is obliged to proceed to the assistance of persons in distress at sea (see Section 17.4.1). The Merchant Shipping (Safety Convention) Act 1949 requires the master of a British ship, on receiving distress information whilst at sea, to proceed (when circumstances are reasonable) with all speed to assist.

Most vessels are also required to monitor one or other of the radio distress frequencies, and it would be the action of a responsible master to do so, if able, regardless of regulation.

The promptness with which vessels respond to the receipt of a distress message can very often determine the success or failure of any subsequent rescue effort, especially in cold water areas.

17.2.1 *Acknowledgement and relay*

All distress messages received from whatever source or by whatever means must be acknowledged, and an attempt must be made by direction finding to confirm the position given. Where the occasion demands, the radio distress message must be relayed to ensure that all parties capable of rendering assistance are informed as soon as possible.

The relay of a distress message is particularly important in the case of a recipient whose vessel, for whatever reason, is unable to render assistance.

17.2.2 *Global maritime distress and safety system (GMDSS)*

Before the GMDSS system was introduced, search and rescue depended largely for its co-ordination on outdated and sometimes unreliable radio communication methods. Such problems as congested radio frequency bands and poor propagation conditions can lead to serious delays in instigating and executing a rescue. Recent advances in communications technology, including the Inmarsat satellites, digital selective calling, narrow band direct printing, radiotelephony and automatic distress transmitters have made it possible to provide rapid alerting and efficient co-ordination of search and rescue services (see also Sections 19.1.3.2 and 19.1.7).

The majority of the necessary infrastructure of shore-based SAR authorities is now in place, covering all sea areas. Working frequencies have been allocated and the provision of equipment to ships is proceeding to meet the fitting schedules required by SOLAS 1974, as amended 1989. With this new equipment, it is expected that in the majority of distress situations the exact position will be known immediately by satellite sensing, thus bypassing the search phase.

The system will additionally provide for urgency and safety communications as well as the dissemination of navigational and meteorological information to ships (see Section 11.4.3).

17.2.3 *Proceeding towards the distress*

Vessels answering the distress must proceed towards the distress position at their best speed, whilst reporting their action to the distressed vessel and/or CRS handling the co-ordination. The closer presence of other vessels does not relieve the master of his duty to respond. Only after the full response has been heard and co-ordinated can it be decided which vessels are needed and which are not. This decision will be made by the station co-ordinating the rescue. Meanwhile radio traffic will be monitored to establish which other vessels are responding. A plot of this response with ETAs etc. is needed as a forerunner to the search co-ordination. If communication with the distress is possible, then at this stage further relevant details can be sought of the circumstances of the distress and the facilities available or needed to deal with it.

17.2.4 *On-board preparation*

The period between first turning towards a distress and arriving on the scene is valuable preparation time needed to organize the on-board resources towards a task that may be both unfamiliar and dangerous. On most merchant ships with limited manpower and equipment, considerable organizational skill is needed to improve the chances of a successful recovery operation. Division of labour and the full briefing of key personnel will be an important feature of this preparation time.

17.3 Co-ordinating search and rescue

17.3.1 *Rescue co-ordination centres (RCC)*

In areas where these organizations exist, it is sensible to involve them because of their experience and the wide facilities available to them, including the communications necessary for an effective co-ordination of rescue efforts.

17.3.2 *Establishment of commander surface search*

Where rescue operations are taking place outside the

effective range of the RCC, its role may be filled by the on-scene unit best able to handle the communications, e.g. warship, SAR aircraft or occasionally a merchant ship. A merchant ship acting as commander surface search (CSS) will probably need to liaise fairly closely with the CRS, keeping them briefed on developments.

The task of CSS is very complex and relies heavily on good radio communication facilities and adequate manpower. On occasion, in the absence of any effective land-based help, the CSS may be self-appointed but normally only after discussions between the ships answering the distress.

In co-ordinating several ships approaching the scene of a distress, the individual ETAs and capabilities are most important. However, the first arrival should always take such action as may be necessary. Only afterwards will the pre-agreed search patterns in the designated areas be embarked upon.

7.3.3 *Control of intership communications*

When within range, VHF radio is the most con-enient form of intership communication. Channel 6 may be used for all distress traffic but it is often referable to conduct most of the organizational etails on a working channel while *still* monitoring hannel 16, as this allows urgent messages to be roadcast and received immediately. The CSS must eep control of the radio traffic. This can be done by outeing all exchanges through his station. The more nportant messages should be put into writing before oing on the air, as this will assist in making shorter, learer messages and avoiding ambiguities. Where nguage difficulties exist, 'standard phrases' should e used (see IMO Standard Vocabulary, Section 9.1.5).

7.3.4 *Assistance by aircraft/helicopters*

'aluable information concerning working with air-raft and helicopters during search and rescue (SAR) perations is contained in Annual Notice to Mariners o. 4 (see Section 17.4.2).

For the most part, limitations in communications ue to incompatibility of radio equipment means that direct involvement by civil aircraft is unlikely. Overflying civil aircraft are, however, very often the first to hear and report distress transmissions from EPIRBs (see also Section 17.1.1).

Military aircraft and helicopters may become involved, but again communications will probably still need to be routed via shore stations or naval ships. Specialized SAR aircraft are well equipped for working with ships and are sometimes able to co-ordinate the responses from merchant ships and control the search. When necessary, they can drop a radiobeacon at the distress position to assist with homing.

SAR helicopters almost invariably have VHF channel 16 and consequently communications are no problem when within range. Their main limitation is flying range and endurance. Depending on the type of helicopter and where it is based, 300 miles is at present the maximum range, and this gives very little time on scene. Under these circumstances every effort must be made to use the time efficiently.

17.3.5 *Planning the search*

When taking on the duties of CSS, the first priority is to assess the potential of each assisting vessel and aircraft including their ETAs and manoeuvring capabilities. A plot of all relevant data should be made and measured against the requirements of the distress and/or search area. The most favourable tactics can then be chosen and each assisting unit briefed. The plan may need adjusting with changing circumstances and conditions.

The characteristics of the search pattern chosen depends very much on the target and the number of assisting vessels which are available. When searching for a relatively large object of ill defined position, a radar search allows a rapid, albeit cursory, coverage of a large area. Aircraft to overfly the general search area are particularly useful in this type of search, extending the area covered appreciably.

The search patterns quoted in the Mersar Manual are basically for visual searches, to be used when the position of the distress is closely defined or where the object of the search is small. Intervals between visually searching ships is very important. Overwide spacing, allowing a person in the water to be passed undetected, may result in a delay of several hours before the same area is examined again. Too close spacing wastes time. Restricted visibility not only requires smaller spacing between searching ships but also slower search speeds and more visual/aural look-outs.

Speed of search is a compromise between the time needed to search effectively and the need to avoid unnecessary delay. The possible search area widens very rapidly with time, especially in rough weather when drift is an important variable. The Mersar Manual suggests casualty drift rates, but allowance needs to be made with an increased search area, as most floating objects will have a drift angle too. Advice from the RCC, where available, can be useful in this respect.

If it is intended to use parallel indexing to carry out a search pattern more efficiently, then an indexing target is needed. Preparations may need to be made for dropping a radar marker at the scene or alternatively asking an assisting vessel to maintain a position for the duration of the search pattern. True motion radar could be used without an indexing target but tide and current will affect geographical accuracy, not necessarily to the detriment of the search.

17.3.6 Conducting the search

Manoeuvring unwieldly merchant ships in close formation is difficult and potentially dangerous. The International Collision Regulations still apply, and consequently a good briefing of participants is essential. Any manoeuvring instructions by the CSS must be unambiguous.

Standard messages are suggested in the Mersar Manual to assist the CSS in directing the movements of searching ships. These are visual signal messages but the format can usefully be applied to the VHF. The Standard Marine Navigational Vocabulary may also prove useful in formulating unambiguous messages capable of being noted down quickly (see Section 19.1.5).

If, on conclusion of the initial search pattern, nothing is found, the situation will need to be reassessed and a fresh search instituted taking account of any changed circumstances, such as the arrival of more rescue units and the enlarged search area caused by the time lapse.

17.3.7 Rescuing survivors

The CSS will co-ordinate the rescue action directing the most suitably equipped rescue units to move in while other units stand off and assist as required.

Where survival craft are not in sight visually but are on the radar as small targets, moving in towards them can result in them being lost in the sea clutter. A vessel standing off can usefully direct under those circumstances.

When survivors are rescued, it is important that full details of the casualty are obtained quickly and passed to the CSS so that the search is not called off prematurely. Their medical needs must also be established and provision made.

17.3.8 Conclusion of search

The CSS must decide when the rescue is completed and must inform assisting ships, thus relieving them of their obligation to remain. The CSS should also inform the CRS of all relevant details.

If the search is unsuccessful and all reasonable hope of rescuing survivors has passed, the CSS, in consultation with the CRS, will call off the search and dismiss assisting units. A radio message asking all ships to keep a look-out in the area is advisable.

Before leaving the scene of a successful rescue, liferafts and lifebelts etc. should be retrieved or sunk and any remaining accumulation of debris should be the subject of a radio warning. The medical needs of survivors will have become apparent during the rescue and they will need to be transferred to land by the most appropriate route. Helicopters, where available, generally provide this route and the usual procedures when working with a helicopter need to be followed.

17.4 Extracts from official publications

Within Sections 17.4.1 and 17.4.2, the paragraph

lettering and referencing of the original publications has been retained.

17.4.1 International Convention for the Safety of Life at Sea (SOLAS) 1974. Chapter V, Regulation 10: Distress Messages – Obligations and Procedures

(a) The master of a ship at sea, on receiving a signal from any source that a ship or aircraft or survival craft thereof is in distress, is bound to proceed with all speed to the assistance of the persons in distress, informing them if possible that he is doing so. If he is unable or, in the special circumstances of the case, considers it unreasonable or unnecessary to proceed to their assistance, he must enter in the logbook the reason for failing to proceed to the assistance of the persons in distress.
(b) The master of a ship in distress, after consultation, so far as may be possible, with the masters of the ships which answer his call for assistance, has the right to requisition such one or more of those ships as he considers best able to render assistance, and it shall be the duty of the master or masters of the ship or ships requisitioned to comply with the requisition by continuing to proceed with all speed to the assistance of persons in distress.
(c) The master of a ship shall be released from the obligation imposed by paragraph (a) of this regulation when he learns that one or more ships other than his own have been requisitioned and are complying with the requisition.
(d) The master of a ship shall be released from the obligation imposed by paragraph (a) of this regulation, and if his ship has been requisitioned, from the obligation imposed by paragraph (b) of this regulation, if he is informed by the persons in distress or by the master of another ship which has reached such persons that assistance is no longer necessary.
(e) The provisions of this regulation do not prejudice the International Convention for the unification of certain rules with regard to Assistance and Salvage at Sea, signed at Brussels on 23 September 1910, particularly the obligation to render assistance imposed by article 11 of that convention.

17.4.2 Admiralty Annual Notice to Mariners no. 4, 1992: Distress and Rescue at Sea – Ships and Aircraft

Contents of Notice

Contents of Notice

Part III
Bridge Operations

18 Bridge operations and the bridge team

The manner in which bridge operations are formulated and conducted is of fundamental importance to the safety of the ship. Studies conducted over many years continually identify weaknesses in bridge organization as significantly contributing towards casualties. Statistically, for every casualty there are of the order of 400 near misses, which leads one to believe that there is considerable scope for improvement. Since every watchkeeping officer is certificated as competent, it is not major failures that are leading to the accidents but small undetected failures which compound until eventual disaster.

There is with modern shipping operations an increased complexity and stress which results in a greater incidence of small omissions and mistakes, and consequently bridge teamwork as a system of co-operative human activity is seen as the most effective way of providing against the 'one man' error and its possible fatal consequences.

This chapter considers the various components of bridge management and their relationship with the safe navigation of ships, both as an on-board process and as part of the overall operational shipping scene.

There are already several shipping Acts and M Notices (see Section 18.3.1) which lay down what is required of the bridge watchkeepers, and make recommendations as to good practice. More studies are being carried out into other aspects such as manning levels, watchkeeping schedules, workloads and fatigue, etc.

All these factors must find a place in the system that is employed on the bridge and, because of the inevitable complexity, it is also essential that some means of cross-checking is included. Individual bridge watchkeepers should not feel they are working in isolation but should be in the position of receiving active support, not only from the master but also from the other watchkeepers too; that is, they should be working as one of a team.

18.1 Bridge manning

18.1.1 Crew availability

The total number of personnel carried on the ship who are capable of carrying out a bridge function is, generally speaking, not under the control of the seafarer. Economic pressures over a number of years have resulted in most vessels being manned by a minimum number consistent with normal operation, which naturally leads to an undermanned situation in abnormal circumstances.

The master, in recognizing this situation, must therefore seek to alleviate the underlying problem by making proper use of the resources, i.e. officers, crew and equipment, that he does have. To do this successfully requires an appreciation by all concerned that every changed navigational condition will probably call for a changed bridge routine and manning level. As a consequence, personnel with a bridge function become members of a bridge team with 24 hour responsibility, and all other activities are secondary. To help decide how many persons are needed for a particular set of conditions, it is necessary to first define fairly precisely the essential bridge functions.

18.1.2 *Ship type and company requirements*

The number of essential bridge roles does not vary much between a coaster and a supertanker. The number of personnel needed to fulfil those roles is to some extent affected by the physical size of the bridge but more often by the equipment fitted, e.g. bridge control of engines, data loggers, hand held radios, internal and ship to ship/shore communications, etc. and their disposition.

Most shipping companies will include in their standing orders some direction as to manning levels for most normal circumstances, but it remains for the master to interpret these instructions intelligently and adjust them to meet any abnormal circumstances.

18.1.3 *Planning for long periods of high level manning*

All masters and shipping companies recognize the increased risks brought on by reduced visibility, congested waters and heavy weather and allow for it by instituting a high level manning routine. A problem will arise when these circumstances persist for several watches and the watchkeeping officers, and the master in particular, become overtired after spending too many hours on the bridge. This very fact will increase the risks to which the vessel is exposed.

In many cases the high level manning requirement can be foreseen during the passage planning stage and a judgement made on its likely extent. The ship routine can then be adjusted in good time and the master and watchkeepers fully rested. For example, an extended period of reduced visibility in coastal waters may require two officers on the bridge, in which case the second and third officers may share the watchkeeping between them on a '4 on – 4 off' or '6 on – 6 off' basis, while the master and chief officer take alternate watches. Alternatively, the master may arrange to be on the bridge just for the more important stages of the passage (see also Section 16.3.5).

18.1.4 *Safety before expediency*

There is a strongly held belief amongst many ships' masters and officers that their employers will frown on any action that delays the prompt execution of the voyage. Much of the fault for this belief lies with shipping company senior management, who have not stated the priorities clearly enough or ensured their adherence. Operations managers, who usually have the closest contact with the ship, do not necessarily have the same priorities. Certainly there are many commercial reasons for meeting an ETA or schedule, and any time wasting or unnecessary delay should naturally be avoided. However, these commercial pressures must never be used as an excuse to ignore regulations or take insupportable risks. The consequences of a collision or stranding are potentially enormous, as several shipowners have already discovered, and they would invariably prefer a safe but late arrival to an incident. The masters and watchkeepers should also consider the possible consequences to themselves, the least of which could be the end of their career and at worst the end of their life.

Included amongst these insupportable risks are excessive speed in reduced visibility, leaving the bridge undermanned for whatever reason, and having an overtired bridge team owing to poor man management or not planning the passage effectively.

18.1.5 *Team roles*

The effectiveness of a bridge team does not depend solely on numbers. 'Spectators' are nothing more than a distraction; when the rest of the bridge team members are not fully aware that a person is spectating, dangerous misunderstandings can result.

To operate efficiently, bridge team members need a clear understanding of their function for each level of bridge manning. The basis for this understanding will stem from company standing orders defining manning levels and roles, supplemented and/or amplified by the master's standing orders.

The defining of functions and team roles should not be allowed to develop into demarcation, as circumstances may require one member to back up another's efforts. It is essential that all team members are fully aware of the total situation and hence able to be supportive of one another without neglecting their own function.

18.1.6 Need for direct orders on manning level

If misunderstandings regarding the role of a bridge team member at any specific time are to be avoided, the bridge manning level must be made obvious to everyone at all times. Standing orders will specify what constitutes the various levels, but there is also a need for a specific routine for changing that level. Using informal verbal exchanges to change the manning level can easily lead to a misunderstanding when the pressure is on or if there are any personality differences. A written entry in the bridge notebook or the use of a display board will give the change and new level its proper emphasis.

18.1.7 Pilot on board

Faith in a pilot's competence is seldom misplaced, but this does not detract from the fact that the responsibility for the safe conduct of the vessel remains with the master and his watchkeeping officers. There is also no reason to suppose that the pilot does not need and appreciate the kind of support that the bridge team normally provides to the master when navigating in difficult circumstances. Consequently, the pilot's presence on the bridge supplements, rather than replaces, the normal bridge organization.

18.1.8 Berthing

A high proportion of deaths and injuries sustained on board ship occur during docking and undocking operations. With manpower resources stretched to the limit and with difficult communication conditions, especially if language differences exist, misunderstandings and mistakes are bound to occur. These problems can be minimized by:

1 The proper briefing of key personnel
2 The establishment of effective communications (see also Section 19.2)
3 Starting operations only when everyone is in position and ready.

18.2 Bridge procedures

18.2.1 The need for formalized procedures

It has always been recognized that when several people are working together to a common end it is necessary for all members of the team to have an understanding of the whole operation, in addition to their own role in the overall effort. In this way efficiency is improved by the elimination of misunderstandings, duplication of effort and the overlooking of important items.

On the bridge of a ship the consequences of a mistake through poor organization can be catastrophic, and many companies have instituted procedures in their standing orders designed to improve bridge teamwork. These procedures, if followed, should have the effect of:

1 Enhancing the understanding between master and pilot and between master and OOW
2 Reducing anxiety in difficult navigational situations
3 Improving discipline in emergencies.

Most of the items in these procedures are obvious to an experienced officer and under normal circumstances might be felt unnecessary. However, few if any of the items in these procedures can afford to be overlooked and it is in conditions of stress and distraction that the procedures become essential as a check. To act as a check these procedures must be written in a simple and brief form and be immediately available at all times, i.e. as a displayed checklist card rather than as a book in a bookcase.

18.2.2 Handing over the watch

By the time an officer has been through this routine a couple of hundred times, there is a great tendency to start assuming things. However, that which might be obvious to the person closely involved with the situation for the past four hours may not be the least bit obvious to the freshly awakened man who is taking over. On the basis that every item of information that is being transferred is important, to ensure the effective assimilation of this information it must be handed over in a formal manner with as much in written form as is practicable. It is the legal duty (see Section 12.4.1) of the officer handing over not to do so until he is sure that his relief is fully appraised. Similarly, the relieving officer must not accept the hand-over until he is satisfied with the situation.

The passage plan and bridge notebook provides much of this information, but the report of the current status of the ship and its tactical situation adds many more items. To ensure a complete report, a written checklist can be most useful. Attempts to hand over whilst being distracted by other events taking place, e.g. anti-collision manoeuvres, should be avoided as this greatly increases the chances of something being overlooked.

18.2.3 Handing over to master

The master has the right to take charge of the bridge as and when he sees fit. An important point that is sometimes overlooked is that he needs to be fully appraised of the current situation before taking over. A heavy responsibility remains with the OOW to pass on all relevant information even after the master has taken full charge. He must never assume anything is obvious if the safety of the ship depends on it (see 'Grounding of MV Sevillan Reefer', *Seaways*, August 1984, p. 19).

The master's presence on the bridge for whatever reason can leave the OOW in some doubt as to who is in charge unless the procedure for handing over is clearly defined. It is essential that this procedural point is made in the master's standing orders or on a procedures checklist so that the OOW is never in any doubt as to when he has full control of the vessel.

18.2.4 Handing over to pilot

The pilot is engaged for his local knowledge and experience in the manoeuvring of ships in his local area. He does not necessarily have any knowledge of the peculiarities of a particular ship, the capabilities of its crew or, for that matter, its present tactical situation. If his efforts are to be fully effective then all these relevant points must be brought to his attention as soon as he arrives on the bridge and before he begins his duties. The bridge team does in fact have a legal duty towards the pilot, not only to prepare for his arrival by planning the pilotage stage of the passage, but also during the pilotage to monitor progress and support the pilot in his con (see Section 12.4.1). The time available for the transfer of this data is generally quite short, and consequently it is important to have as much as possible prepared beforehand in simple written form. In addition to the draft passage plan, there should be the ship's manoeuvring characteristics in diagrammatic form and a standard checklist of items relating to the ship, its equipment and crew status. Only after this procedure has been completed can the pilot take up his duties in safety and only then with the full back-up of the bridge team.

18.2.5 Other procedures

There are several other routine and emergency situations for which procedures should be developed (see Section 11.17.2). Each of these procedures should have its associated checklist to assist the OOW.

When 'full away' on passage there is one almost routine procedure that is seldom addressed properly and in many cases is completely overlooked, namely that of dealing with conditions of reducing visibility. Everyone is well aware that the vessel should proceed at a safe speed in reduced visibility (see also Sections 22.1.2 and 22.2.2) but, apart from the practical difficulty of estimating the actual visibility, especially at night, there is a general reluctance of watchkeepers to act positively and in good time. There are literally dozens of excuses for doing nothing, all apparently valid at the time, such as:

1 'It probably won't get any worse!'
2 'It looks like it's clearing!'
3 'Perhaps it's not as bad as it looks.'
4 'There's no traffic about.' (I see no ships!)
5 'Will the old man agree with me?'
6 'We cannot afford to lose time.'
7 'Perhaps I can hang on until 8 bells.'

The OOW should be left in no doubt, by the provision of explicit instructions and checklists, what action he must take for any given set of visibility conditions. The master should also include in his standing orders specific visibility conditions and circumstances under which he is to be called, e.g. 'advise me immediately if you estimate visibility in any direction to have reduced to five miles or less', or 'if visibility in any direction forward of the beam is estimated to be three miles or less, institute procedure X immediately' rather than a rather vague 'call me if visibility reduces'.

The procedure checklists should specify the actions to be taken and their sequence as the vessel is placed in a higher state of readiness.

18.3 Extract from official publication

Within Section 18.3.1, the paragraph numbering of the original publication has been retained.

18.3.1 UK DTp Merchant Shipping Notice M.1102: Operational Guidance for Officers in Charge of a Navigational Watch

1 Regulation 4 of the Merchant Shipping (Certification and Watchkeeping) Regulations 1982 (SI 1982, no. 1699), which comes into operation on 28 April 1984, requires the master of any ship to which the regulations apply to give directions to the deck watchkeeping officers responsible for navigating the ship safely during their periods of duty, having particular regard to the matters set out in schedule 1 to the regulations and to the operational guidance specified by the secretary of state.

2 The operational guidance specified by the secretary of state is set out in the appendix to this notice. The guidance incorporates the substance of resolution 1 adopted by the International Conference on the Training and Certification of Seafarers 1978.

3 The provision of this guidance is not a substitute for written standing orders, which should always be provided by the master to specify his own particular requirements. Such standing orders should draw the attention of watchkeeping officers to the guidance laid down in this and other relevant Merchant Shipping Notices and emphasize those points which are of special importance to the particular ship. There is, of course, no suggestion that these written orders should reproduce the guidance laid down in this and other Merchant Shipping Notices, or that such guidance is a requirement to be complied with without due regard to circumstances.

Appendix: Operational guidance for officers in charge of a navigational watch

Introduction

1 This appendix contains operational guidance of general application for officers in charge of a navigational watch, which masters are expected to supplement as appropriate. It is essential that officers of the watch appreciate that the efficient performance of their duties is necessary in the interests of the safety of life and property at sea and the prevention of pollution of the marine environment.

General

2 The officer of the watch is the master's representative and his primary responsibility at all times is the safe navigation of the ship. He should at all times comply with the applicable regulations for preventing collisions at sea (see also paragraphs 22 and 23).

3 It is of special importance that at all times the officer of the watch ensures that an efficient look-out is maintained. In a ship with a separate chartroom the officer of the watch may visit the chartroom, when essential, for a short period for the necessary performance of his navigational duties, but he should previously satisfy himself that it is safe to do so and ensure that an efficient look-out is maintained.

4 The officer of the watch should bear in mind that the engines are at his disposal and he should not hesitate to use them in case of need. However, timely notice of intended variations of engine speed should be given where possible. He should also know the handling characteristics of his ship, including its stopping distance, and should appreciate that other ships may have different handling characteristics.

5 The officer of the watch should also bear in mind that the sound signalling apparatus is at his disposal and he should not hesitate to use it in accordance with the applicable regulations for preventing collisions at sea.

Taking over the navigational watch

6 The relieving officer of the watch should ensure that members of his watch are fully capable of performing their duties, particularly as regards their adjustment to night vision.

7 The relieving officer should not take over the watch until his vision is fully adjusted to the light conditions and he has personally satisfied himself regarding:

(a) Standing orders and other special instructions of the master relating to navigation of the ship;
(b) Position, course, speed and draught of the ship;
(c) Prevailing and predicted tides, currents, weather, visibility and the effect of these factors upon course and speed;
(d) Navigational situation, including but not limited to the following:
  (i) Operational condition of all navigational and safety equipment being used or likely to be used during the watch;
  (ii) Errors of gyro and magnetic compasses;
  (iii) Presence and movement of ships in sight or known to be in the vicinity;
  (iv) Conditions and hazards likely to be encountered during his watch;
  (v) Possible effects of heel, trim, water density and squat on under-keel clearance.

8 If at the time the officer of the watch is to be relieved a manoeuvre or other action to avoid any hazard is taking place, the relief of the officer should be deferred until such action has been completed.

Periodic checks on navigational equipment

9 Operational tests of shipboard navigational equipment should be carried out at sea as frequently as practicable and as circumstances permit, in particular when hazardous conditions affecting navigation are expected: where appropriate these tests should be recorded.

10 The officer of the watch should make regular checks to ensure that:

(a) The helmsman or the automatic pilot is steering the correct course;
(b) The standard compass error is determined at least once a watch and, when possible, after any major alteration of course; the standard and gyro compasses are frequently compared and repeaters are synchronized with their master compass;
(c) The automatic pilot is tested manually at least once a watch;
(d) The navigation and signal lights and other navigational equipment are functioning properly.

Automatic pilot

11 The officer of the watch should bear in mind the necessity to comply at all times with the requirements of regulation 19, chapter V of the International Convention for the Safety of Life at Sea, 1974. He should take into account the need to station the helmsman and to put the steering into manual control in good time to allow any potentially hazardous situation to be dealt with in a safe manner. With a ship under automatic steering it is highly dangerous to allow a situation to develop to the point where the officer of the watch is without assistance and has to break the continuity of the look-out in order to take emergency action. The change-over from automatic to manual steering and vice versa should be made by, or under the supervision of, a responsible officer.

Electronic navigational aids

12 The officer of the watch should be thoroughly familiar with the use of electronic navigational aids carried, including their capabilities and limitations.

13 The echo sounder is a valuable navigational aid and should be used whenever appropriate.

Radar

14 The officer of the watch should use the radar when appropriate and whenever restricted visibility is encountered or expected, and at all times in congested waters, having due regard to its limitations.

15 Whenever radar is in use, the officer of the

watch should select an appropriate range scale, observe the display carefully and plot effectively.

16 The officer of the watch should ensure that range scales employed are changed at sufficiently frequent intervals so that echoes are detected as early as possible.

17 It should be borne in mind that small or poor echoes may escape detection.

18 The officer of the watch should ensure that plotting or systematic analysis is commenced in ample time.

19 In clear weather, whenever possible, the officer of the watch should carry out radar practice.

Navigation in coastal waters

20 The largest scale chart on board, suitable for the area and corrected with the latest available information, should be used. Fixes should be taken at frequent intervals: whenever circumstances allow, fixing should be carried out by more than one method.

21 The officer of the watch should positively identify all relevant navigation marks.

Clear weather

22 The officer of the watch should take frequent and accurate compass bearings of approaching ships as a means of early detection of risk of collision; such risk may sometimes exist even when an appreciable bearing change is evident, particularly when approaching a very large ship or a tow or when approaching a ship at close range. He should also take early and positive action in compliance with the applicable regulations for preventing collisions at sea and subsequently check that such action is having the desired effect.

Restricted visibility

23 When restricted visibility is encountered or expected, the first responsibility of the officer of the watch is to comply with the relevant rules of the applicable regulations for preventing collisions at sea, with particular regard to the sounding of fog signals, proceeding at a safe speed and having the engines ready for immediate manoeuvres. In addition, he should:

(a) Inform the master (see paragraph 24);
(b) Post a proper look-out and helmsman and, in congested waters, revert to hand steering immediately;
(c) Exhibit navigation lights;
(d) Operate and use the radar.

Calling the master

24 The officer of the watch should notify the master immediately in the following circumstances:

(a) If restricted visibility is encountered or expected;
(b) If the traffic conditions or the movements of other ships are causing concern;
(c) If difficulty is experienced in maintaining course;
(d) On failure to sight land, a navigation mark or to obtain soundings by the expected time;
(e) If, unexpectedly, land or a navigation mark is sighted or change in sounding occurs;
(f) On the breakdown of the engines, steering gear or any essential navigational equipment;
(g) In heavy weather if in any doubt about the possibility of weather damage;
(h) If the ship meets any hazard to navigation, such as ice or derelicts;
(i) In any other emergency or situation in which he is in any doubt.

Despite the requirement to notify the master immediately in the foregoing circumstances, the officer of the watch should in addition not hesitate to take immediate action for the safety of the ship, where circumstances so require.

Navigation with pilot embarked

25 If the officer of the watch is in any doubt as to the pilot's actions or intentions, he should seek clarification from the pilot; if doubt still exists, he should notify the master immediately and take whatever action is necessary before the master arrives.

Watchkeeping personnel

26 The officer of the watch should give watchkeeping personnel all appropriate instructions and information which will ensure the keeping of a safe watch including an appropriate look-out.

Ship at anchor

27 If the master considers it necessary, a continuous navigational watch should be maintained at anchor. In all circumstances, while at anchor, the officer of the watch should:

(a) Determine and plot the ship's position on the appropriate chart as soon as practicable; when circumstances permit, check at sufficiently frequent intervals whether the ship is remaining securely at anchor by taking bearings of fixed navigation marks or readily identifiable shore objects;
(b) Ensure that an efficient look-out is maintained;
(c) Ensure that inspection rounds of the ship are made periodically;
(d) Observe meteorological and tidal conditions and the state of the sea;
(e) Notify the master and undertake all necessary measures if the ship drags anchor;
(f) Ensure that the state of readiness of the main engines and other machinery is in accordance with the master's instructions;
(g) If visibility deteriorates, notify the master and comply with the applicable regulations for preventing collisions at sea;
(h) Ensure that the ship exhibits the appropriate lights and shapes and that appropriate sound signals are made at all times, as required;
(i) Take measures to protect the environment from pollution by the ship and comply with applicable pollution regulations.

19 Communications

19.1 Intership and ship to shore communications

Although ships are still equipped to communicate by international code flags and Morse lamp, it is only in the most unusual circumstances that these methods will be employed. In general, all intership and ship to shore communications are carried out by radio.

In terrestrial methods of radio communications the radio signal travels from the transmitter through the earth's atmosphere (possibly being reflected by layers in the upper atmosphere) and subsequently arrives at the receiver. The modern approach makes use of geostationary satellites in space: the signal leaves the transmitter and travels outward through space to be received by the satellite which subsequently relays the signal back to earth. Satellite communications are attractive because the frequencies used are less affected by atmospheric conditions and the orbital height and computer technology give the potential for reliable automatic worldwide coverage. The high cost of equipment is likely to fall as usage increases. The simplicity and reliability of worldwide satellite communications have resulted in the voluntary fitting of satellite equipment on many ships. A major stimulus however is the Global Maritime Distress and Safety System (GMDSS) whose aim is to provide a mandatory unified system of distress, urgency and safety communications using the worldwide satellite coverage provided by the International Maritime Satellite Organization (INMARSAT). The system was adopted by the 1988 amendments to The International Convention for the Safety of Life at Sea (1974) and the provisions contained therein are to be phased in between 1 February 1992 and 1 February 1999 (see Section 19.3.1).

19.1.1 System outlines

Figure 19.1 shows a simple block diagram of a radio communication system. The information to be transmitted is converted into an electrical signal by some form of transducer (e.g. a microphone). The frequency of such signals is in general too low to be radiated into the atmosphere from a practical aerial. This difficulty is overcome by generating a constant signal of sufficiently high frequency known as a carrier. The low frequency information signal is used to vary some characteristic of the carrier (e.g. its amplitude, phase or frequency) by a process known as modulation. The resultant signal contains the desired information *and* is of a sufficiently high frequency to produce radiation from the aerial.

The signal is radiated by the aerial and travels by the terrestrial or satellite path and is hopefully intercepted by the aerial at the receiving station. In the receiver the information is extracted from the carrier by demodulation. After suitable amplification the information signal is used to operate a transducer (e.g. a loudspeaker or telephone earpiece) and hence produce the information in its original form (or possibly in some other form).

19.1.2 Frequency allocation

Within the radio frequency spectrum (see Section 10.1), various bands are allocated to the carrier frequencies used for intership and ship to shore communications. The allocations are made by international agrement and lie in the very high frequency

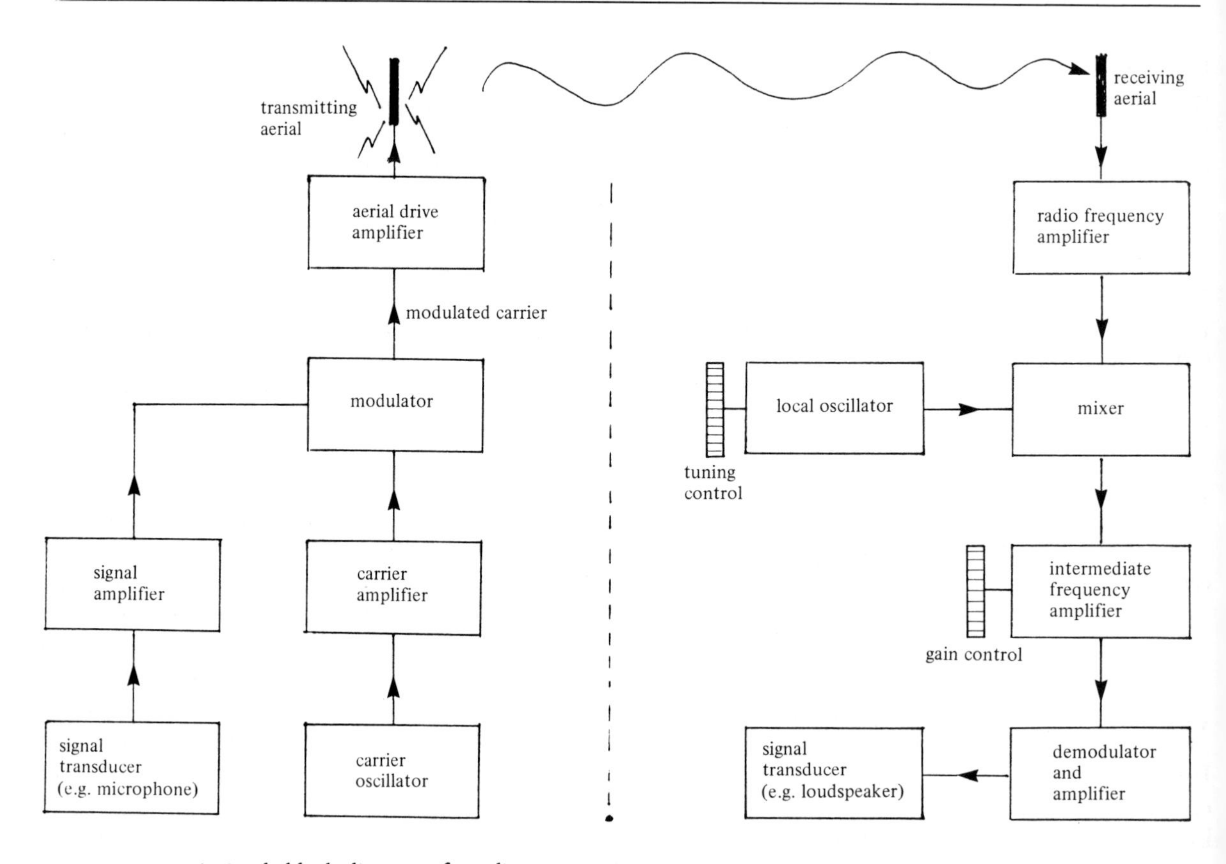

Figure 19.1 A simple block diagram of a radio communication system

(VHF), high frequency (HF), medium frequency (MF) and L band areas of the spectrum. The first three bands relate to terrestrial communications, the last to satellite use.

In terrestrial communications the signal may reach its destination by one of three possible propagation paths:

1 The direct or spacewave path, in which the signal travels directly to the receiving aerial along a 'line of sight'
2 The skywave path, in which the signal travels skywards, suffers reflection in the ionosphere and hence returns to earth in the vicinity of the receiving aerial
3 The groundwave path, in which the signal follows the curvature of the earth.

In general, for any given carrier frequency, one particular propagation path will result in least signal loss and attenuation. As a result the dominant component of the received signal will arrive by that path.

VHF transmissions follow the line of sight path and are thus employed for short range communications (up to about 30 or 40 miles). In satellite communications line of sight transmission is also used but the long ranges are achieved by redirecting the signal path at the satellite.

HF signals travel by the skywave path and these frequencies are employed for ocean coverage.

In the case of MF communications, groundwave is the dominant path. Ranges of up to 200 miles are possible in the 2 MHz region of the band, while up to 1000 miles can be achieved at 500 kHz.

The International Convention for the Safety of Life at Sea (1974) as amended to 1991 (SOLAS 1974), sets out in detail the requirements to carry radio installations and personnel. The requirements are too lengthy to reproduce in their entirety, but particular attention is drawn to the extracts presented in Section 19.3.1. These relate to the convention as amended for GMDSS but a short summary of the original requirements is provided for comparison. In the transition period from 1992 to 1999 many vessels will be able to elect to comply with either one set of requirements or the other (new vessels have to comply with GMDSS from 1995).

19.1.3 Information format

The information which is modulated on a radio carrier may take a variety of forms.

19.1.3.1 Wireless telegraphy (WT)

Historically this was the first form of radio communication used at sea. Morse characters, normally keyed by hand, are used to modulate the carrier. Calling and working frequencies exist within the HF and MF bands and the system is supported by a worldwide network of coast radio stations. It is recognized that hand-sent telegraphy does not compare favourably with the speed of connection and simplicity of operation of direct speech link by satellite. There is little doubt that year by year the volume of worldwide radio traffic using telegraphy will decrease progressively as satellite communications become more widespread and as the date for full implementation of GMDSS approaches. However, in the intervening time, a proportion of general long range radio traffic on deep sea vessels may be handled by telegraphy. Where this is the case the services of a specialist radio officer will be required. For this reason no further consideration will be given to this mode of communication. However it must be recognized that with the progressive implementation of GMDSS the navigator may well be responsible for the handling of long range radio traffic by satellite speech or data link.

19.1.3.2 Radiotelephony (RT)

In this mode of communication, the modulation takes the form of speech. Frequency allocations exist in the VHF, HF, MF and satellite bands.

VHF radio is limited in range but, where contact is established with the shore, considerable flexibility can be gained by access to international telephone networks. The improvement and expansion of VHF facilities at coast radio stations is taking place in many countries.

Facilities allowing direct calling on working channels (as opposed to channel 16) are becoming progressively more common. An automatic calling technique known as Digital Selective Calling (DSC) has also been developed. This allows a radio station to automatically establish contact with, and transfer information to, another station or group of stations. (DSC can also be used in MF and HF radiotelephony.) The technique is used in GMDSS for distress alerting of all ships in an appropriate area and will in due course facilitate direct dialling of telephone calls via a suitably equipped Coast Radio Station. The majority of all marine radio traffic is to or from vessels entering or leaving port and VHF radio is particularly appropriate in such circumstances.

HF radiotelephony has the potential for contact with distant stations but lacks the signal quality of VHF. Significant noise, fading and interference problems are experienced in the working of HF telephony traffic by way of distant coast stations in many areas. Such difficulties combined with expense and delays render this mode of communication less attractive than others.

MF radiotelephony offers effective ship to shore communication (with access to shore telephone networks) for ranges up to about 200 miles. It does not, however, offer the signal quality usually obtainable on VHF. Direct calling on a working channel (as opposed to 2182 kHz) is available in some countries.

Where shore stations and vessels are suitably equipped, satellite communications make possible worldwide direct dialling telephony from ship to ship and from ship to shore. In some countries shore to ship connection is by direct dialling while in others manual connection will be necessary. The shore-based satellite communication installation is known as a coast earth station (CES) while the shipboard equipment is known as a ship earth station (SES). The progressive implementation of GMDSS will promote the fitting of ship earth stations. Even prior to the entering into force of the GMDSS amendments certain national administrations recognized the fitting of an SES as providing the principal radio communications system and allowed dispensation in respect of specific items listed in the original SOLAS(74) carriage requirements such that the terrestrial system represented a reserve capacity. In some cases the vessels were allowed to engage in worldwide trading while others were subject to geographical limits. In the transition period from 1992 to 1999 many vessels will be able to comply with Solas(74) by virtue of having existing terrestrial communication equipment fitted to satisfy the requirements of the original convention. However some operators may elect to fit satellite equipment in anticipation of the GMDSS requirement because of the commercial benefit which may accrue from the availability of a simple and reliable worldwide communication system which can be operated by non-specialist radio communicators.

19.1.3.3 Telex

In the telex system a textual message is typed into a terminal and transmitted by automatic telegraphy to a receiving terminal where it is reproduced automatically in typewritten form. Originally the system was operated only via the shore telephone network, but from the early 1960s the improvement in terrestrial transmitters and receivers facilitated the transmission of telex over radio (TOR). Frequency allocations exist in the HF, MF and satellite bands and, while in some cases connection is manual, automation is taking place progressively.

In terms of cost resulting from tariffs, telex is cheaper than manual telegraphy and offers the obvious advantage of automatic printout of messages. This latter feature has been exploited in the development of a system known as NAVTEX (see Section 11.4.3).

The NAVTEX system uses a dedicated frequency of 518 kHz to broadcast distress and urgency alerts in addition to the normal range of navigational warnings. A printer which is integral with the receiver produces a hard copy output in English. The receiver can be programmed to print only selected types of message but important types such as distress alerts cannot be deselected. The system was inaugurated in 1982 and has since become an integral part of GMDSS. For vessels to which SOLAS(74) applies, the fitting of NAVTEX is compulsory from 1 August 1993 if they trade in areas where the service is available (see Section 19.3.1). NAVTEX, which deals with coastal warnings, and the NAVAREA system, which deals with long range warnings, together form an integral part of the World Wide Navigation Warning Service (WWNWS).

19.1.3.4 Data Communications

This is the generic term covering the spectrum of point to point data communications (in simplex or duplex mode), data broadcasts or 'groupcasts' and information retrieval from a database. Most marine systems of this kind are designed for users who are not computer experts.

Textual messages or data can be prepared using a keyboard and visual display unit (VDU). Clearly, while high speed error-free preparation of text requires keyboard skill, the microprocessor's word processing facility will allow relatively unskilled operators to prepare and edit messages in their own time. Once coded, the message can be transmitted computer to computer at a rate far in excess of that possible with any other form of communication. When the message has been prepared, it may be transmitted instantly and automatically or stored to be sent at a later time together with other messages in a batch.

To achieve the high rates of data transfer, operation must be by satellite or VHF link. The former already offers direct dialling facilities, but the latter suffers from the fact that general direct dialling facilities are not always available.

The received information is handled automatically by the computer and can be displayed on a VDU (possibly with colour and graphics) and/or printed out in typewritten form, or stored on a disc or cassette.

The shipboard computer can often be programmed to contact a database ashore on a regular automatic basis for display to the user or for automatic updating of information stored in a shipboard memory (e.g. information relating to VTS at a given port). Several databases specializing in marine information have been established by organizations such as Lloyds, BIMCO and meteorological/hydrographic bodies.

19.1.3.5 Diagrams and pictures

By optical scanning, the intensity levels of a diagram can be used to modulate a radio carrier and hence facilitate reproduction at a remote location. This technique is known as facsimile, and has been widely used for the transmission of synoptic weather charts. The reproduction time is quite slow. A typical weather map requires about 15 minutes of broadcast (see also Section 11.12.2).

In the future, it is possible that pictorial information will be transmitted by computer graphics or television as best suits a particular application. Techniques exist whereby a slow scan (freeze frame) image from a television camera can be transmitted by satellite telephone link in less than a minute. This offers a wide and exciting range of possibilities, not least of which is the possibility of improving remote medical and damage repair advice.

19.1.4 Management of communications

SOLAS(74) as amended to 1991 requires that every ship (to which the convention applies) shall carry personnel qualified for distress and safety radiocommunication purposes to the satisfaction of the Administration. In respect of management of communications (as opposed to maintenance of communication equipment) the implication of this requirement depends on whether the ship complies with the convention by virtue of the carriage of terrestrial communication equipment or the carriage of GMDSS equipment.

For many vessels fitted only with traditional terrestrial equipment, the regulations will result in the need to carry a specialized radio officer who will manage a wide range of communication channels with the probable exception of the routine use of VHF radiotelephony. In smaller vessels, the requirements may be satisfied by a navigating officer who is also qualified as a radiotelephone operator (see also Section 11.18). Where a radio officer is not carried, a navigating officer must manage all radio communications subject to the limitations of radiotelephony.

For vessels complying with GMDSS requirements, the person responsible for managing the communications must hold a GMDSS General Operator Certificate. The person need not be a specialized radio officer in the traditional sense. In most cases the role will be filled by a navigating officer. SOLAS(74) as amended to 1991 sets out the radio watch requirements under GMDSS, which are reproduced in Section 19.3.1.

19.1.5 Telephony procedure and discipline

Both at present, and in the future, the effectiveness of the watchkeeper's role as a radio communicator will depend heavily on attention to procedure and discipline in the use of radiotelephony. Although English is regularly used as an international language in marine radiotelephony, it has to be appreciated that the accents of speakers, native and otherwise, can render words difficult to distinguish. Receiver noise and radio interference can impair even perfect diction. For these reasons the use of standard words and phrases, in an established format, greatly facilitate the reception of messages in difficult conditions. In this respect, attention is drawn to the Standard Marine Navigational Vocabulary (SMNV). This was developed by the International Maritime Consultative Organization (IMCO) prior to its change of title to International Maritime Organization (IMO). It was not intended that the vocabulary should be mandatory. The hope was that, by encouraging the persistent use of the standard phrases in shore establishments and ships at sea, in due course these same phrases would become standard usage among seamen worldwide. Part I of the SMNV, which contains an introduction and basic conventions, forms an annexe to Notice M. 1252 and is reproduced in Section 19.3.2.

The remainder of the SMNV, which contains a glossary of terms and an extensive phrase vocabulary, is too lengthy to reproduce but can be obtained in the complete annexe to Notice M.1252.

The work begun by SMNV has been extended by a research project which has produced an 'Essential English for International Marine Use'. It is intended primarily for VHF radio communications and was devised by a team of professional mariners and language specialists. Known as SEASPEAK, it expands the concept of a standard vocabulary and facilitates the assembly of a very wide variety of messages. While as yet it has received no formal adoption, it is noteworthy that the project was sponsored by the United Kingdom Department of Trade and Industry and has had active support and encouragement from the IMO Secretariat.

The efficiency and quality of radiotelephony depends not only on effective speech procedure but also on disciplined use of the available frequencies. The procedure appropriate to the maritime mobile radiotelephony service is set out in detail by the Handbook for Radio Operators (see also Section 11.18), which is produced by British Telecom International. This information is too lengthy to be reproduced in a text on navigation control. It must be recognized that any navigating officer qualified to control radiotelephony equipment is expected to be familiar with these procedures and that the said publication should be available at the ship's radio station. It should be noted that the adoption of GMDSS necessitated an extensive revision of the Handbook and a revised version is due to be published in 1993. Particular concern has been expressed by the IMO in respect of discipline in the use of VHF radiotelephony channels. Disquiet over this problem has been highlighted by the publication of Notice M.1026 (see Section 19.3.3). The proper use of VHF channels is crucial to the safe and effective operation of this important area of marine communications.

19.1.6 Log book

SOLAS(74) as amended to 1991 requires the keeping of radio records to the satisfaction of the Administration. For United Kingdom ships the requirements of the Administration are set out in The Merchant Shipping (Radio Installations) Regulations 1992 (see Section 19.3.7).

For radiotelephony the log book is in a standard printed form and has two sections. The first section requires essential details of the installation and the operators, while the second section forms a complete diary of the radio service. The latter has, for each page, an original sheet and a perforated carbon copy sheet. In keeping the log book a wide variety of entries may be made but certain important entries are essential and these are set out in the extracts reproduced in Section 19.3.7. The log book is an important official document and must be maintained complete and up to date. All entries must be in strict chronological order and blank spaces must not be left. The master must inspect and sign each day's entries in the log. In cases where the master is not the radiotelephone operator, the operator is responsible for these duties but must submit the log to the master for his signature. In doing so he must draw the master's attention to any entries of importance or interest.

In the case of British foreign-going ships, the logbook has to be delivered to the superintendent of a mercantile marine office within 48 hours of the vessel's arrival at her final port of destination in the United Kingdom. Where half-yearly or other agreements apply, delivery must be made within 21 days of termination of the agreement. The carbon copies must be retained by the operator and disposed of as required by the operating company or shipowner as appropriate.

19.1.7 Distress, urgency and safety communication

In any situation involving distress or safety, effective and disciplined communications are essential (see also Chapter 17). The procedure adopted will to some extent depend on the radio equipment fitted in compliance with SOLAS(74). Apart from provisions relating to NAVTEX (see Section 19.1.3.3) and emergency position-indicating radio beacons (EPIRBs) (see Section 19.1.7.5), vessels constructed before 1 February 1995 may comply with the convention in its pre-GMDSS form until 1 February

1999. Thereafter full compliance with GMDSS provisions is required.

Vessels which elect to comply with SOLAS(74) in its pre-GMDSS form will in many cases be required to carry a specialist radio officer. In the event of distress arising on such a vessel, the distress working, and in particular the Morse telegraphy, will be managed by that officer. However a situation could arise where the radio officer was injured or incapacitated and hence unable to operate the radio station. Under these circumstances it is essential that other officers should be capable of activating an automatic distress call. Recommendations in respect of this are set out in Notice M.1235 (see Section 19.3.4). Where a specialized radio officer is not carried, the distress working will be under the control of a navigating officer who is qualified to operate radiotelephony equipment. The use of radiotelephony by a non-specialist radio officer is also one of the elements of GMDSS and it is thus expedient to consider it in that context.

A major limitation of the traditional distress system is that in oceanic areas of low traffic density and areas not serviced by effective coast station coverage, the probability of a distress call generating an alert is greatly reduced. This was a major factor in promoting the development of GMDSS. A novel feature of GMDSS requirements is that it specifies equipment carriage requirements in terms of trading area rather than vessel size. The effects of the carriage requirements prior to the 1988 GMDSS amendments were such that in most cases a vessel of 1600 tons gross, or over, was required to carry a radio telegraphy installation and a specialized radio officer whereas a vessel of less than that tonnage could comply by fitting radiotelephony operated by a navigation officer holding a restricted radiotelephony certificate. Under GMDSS, the carriage requirements are based on four sea areas. The definitive requirements for vessels to which the convention applies are set out in the extracts presented in Section 19.3.1. For ease of general understanding they can be summarized in general terms as follows:

(i) Sea Area A1 means an area within the VHF coverage of a coast station with continuous DSC alerting facility.

For Area A1 a vessel is required to carry VHF radiotelephony with DSC facility.

(ii) Sea Area A2 means an area (excluding A1) within MF coverage of a coast station with continuous DSC alerting facility.

For Area A1 + A2, a vessel is required to carry VHF and MF radiotelephony, both with DSC facility.

(iii) Sea Area A3 means an area (excluding A1 and A2) within the coverage of an INMARSAT satellite with continuous alerting facility.

For Area A1 + A2 + A3 the A1 and A2 provisions apply and additionally there is a requirement to carry *either* HF radiotelephony with DSC facility *or* an INMARSAT SES.

(iv) Sea Area A4 means an area outside the other three areas.

The requirements are the same as for A1 + A2 + A3 except that the vessel must carry HF with DSC *and* an INMARSAT SES.

Additionally provisions relating to NAVTEX, EPIRBs and radar transponders apply to all vessels subject to the convention (see Sections 19.3.1 and 19.3.9).

The distress, urgency and safety procedures for use in radio telephony are set out in the Handbook for Radio Officers and any officer holding a GMDSS General Operator Certificate must be thoroughly familiar with these procedures which are too lengthy to reproduce here. It should be noted that the adoption of GMDSS necessitated an extensive revision of the Handbook and a revised version is due to be published in 1993. In the event of a distress situation the communications should be managed by an officer who holds a GMDSS General Operator Certificate. In most cases the responsibility will fall to a navigating officer who may be required to operate terrestrial radiotelephony equipment (VHF, MF and HF) and satellite communication systems. It is expedient to consider the distress frequency arrangements in each case.

19.1.7.1 VHF radio telephony distress frequencies

Channel 16 (156.80 MHz) is designated as the international distress and calling channel. Additionally

Channel 70 (156.525 MHz) is assigned for distress and safety purposes using DSC.

19.1.7.2 MF radiotelephony distress frequencies

The international distress and calling frequency is 2182 kHz. Additionally the frequency of 2187.5 kHz is assigned for distress and safety purposes using DSC. Because 2182 kHz serves also as a routine calling frequency, silence periods have to be observed for 3 minutes after each whole hour and half hour. During this time all transmissions (other than distress) must cease and a watch must be kept to provide the maximum opportunity for any distress call to be heard.

Attention is drawn to certain essential elements of radiotelephony distress procedures which are set out in Notice M.1119 (see Section 19.3.5). This notice addresses itself to the requirement (under The Merchant Shipping (Radio Installation) Regulations 1980) whereby, on vessels compulsorily fitted with radiotelephony installations, instruction cards giving a clear summary of radiotelephony distress, urgency and safety procedures must be displayed in full view of the radiotelephony operating position. Three cards are mentioned, each of which has a distinct function. Card 1 and Card 2 are reproduced in Section 19.3.5. Card 3 displays essential information relevant to phonetic pronunciation and international code usage to facilitate effective emergency communications in cases of language difficulties. The Merchant Shipping (Radio Installation) Regulations 1980 were replaced by The Merchant Shipping (Radio Installation) Regulations 1992 on 1 February 1992. These regulations specify that a card of instructions in English giving a clear summary of the radiotelephone distress, urgency and safety procedures shall be displayed at each radio telephone operating position.

19.1.7.3 HF radiotelephony distress frequencies

Five frequencies in the HF band are designated for distress and safety communications using DSC. In units of kilohertz they are 4207.5, 6312, 8414.5, 12577 and 16804.5.

The frequency of 8364 kHz is designated as a distress frequency for telegraphy and manual speech radiotelephony and is the one traditionally allocated to the portable radio equipment for survival craft. It should be borne in mind that while a vessel in distress would normally use the designated distress frequencies, it may nonetheless, if the circumstances make it necessary, use any frequency to transmit a distress message.

19.1.7.4 INMARSAT distress alerts

The type of satellite communication systems fitted on board ship in compliance with GMDSS makes it possible, using a very straightforward procedure, to send a distress alert either by telephone or telex to the appropriate CES. Selection of the distress channel and appropriate CES is extremely simple and thereafter the message is sent in the same way as any other telephone message or telex. On receipt of the distress alert, the information will be passed to the appropriate Rescue Co-ordination Centre (RCC) (see Section 17.3). Vessels in the distress area can be alerted by Enhanced Group Calling (EGC). This is a global satellite facility which makes it possible to address messages to specific groups of vessels in specific geographical areas.

19.1.7.5 Emergency position-indicating radio beacons (EPIRBs)

Chapter IV, Regulation 7 of SOLAS(74) (see Section 19.3.1) as amended to 1991 specifies certain requirements for the carriage of satellite emergency position-indicating radio beacons (EPIRBs). The requirements enter into force on 1 August 1993. The beacons have to be capable of transmitting a distress alert either through the polar orbiting satellite service (COSPAS-SARSAT) operating in the 406 MHz band or, if the ship is engaged on voyages only within INMARSAT coverage, through the INMARSAT geostationary satellite service operating in the L band. The beacons are required to be installed in an easily accessible position, ready for manual release, and must be able to be carried to a survival craft by one person. The beacon must be capable of floating free if the ship sinks, capable of automatic activation when afloat and also of manual activation.

When activated the beacons which operate with polar orbiting satellites transmit a distress signal which is relayed to an earth station known as a Local User Terminal (LUT). From the distress signal it is possible to identify the vessel from a code contained within the signal and determine the position from Doppler shift measurement (see also Section 9.1). Delays in the reception of the alert may occur until the beacon falls within the coverage of the orbiting satellite and until a ground station lies within the satellite coverage.

The L band beacon when activated transmits a distress signal which will be received by a geostationary satellite if the beacon lies within its coverage. The signal is relayed to the appropriate CES which can decode the vessel's identity and position from the signal transmitted by the beacon. The position information stored in the beacon memory can be updated by direct connection to the ship's navigation interface.

Vessels engaged on voyages exclusively in Sea Area A1 may in lieu of a satellite EPIRB carry an EPIRB which can transmit an alert using DSC on Channel 70 (VHF) and provide for location by means of a radar transponder operating in the 9 GHz Band (see Section 19.1.7.6).

Although they are not part of GMDSS, there are some EPIRB'S which operate on 2182 kHz, the MF distress and calling frequency. Others operate on 121.5 and 243 MHz which are aeronautical distress frequencies. These beacons comply with SOLAS(74) as amended to 1983 (i.e. prior to the 1988 amendments).

19.1.7.7 Radar transponders

Chapters III of SOLAS(74) as amended to 1991 (see Section 19.1.9), which deals with life-saving appliances, specifies carriage requirements for radar transponders according to vessel type and size. Chapter IV, Regulation 7 of SOLAS(74) as amended to 1991 (which deals with radio equipment and the implementation of GMDSS) requires the carriage of one radar transponder (see Section 19.3.1). This transponder may be one of those required under Chapter III requirements.

The transponders are sometimes referred to as Search and Rescue Transponders (SARTs). The transponder is required to operate in the 9 GHz band, i.e. it uses a wavelength of approximately 3 centimetres (X band) (see also Section 5.1.1, 5.2.9.4, 5.2.12.2 and Table 5.4). The transponder is a small Racon beacon which is designed to produce a signal on the radar screens of vessels engaged in locating a casualty. It can only be detected by vessels operating X band radar equipment. In this connection it should be noted that prior to the 1988 amendments to SOLAS(74), Chapter V, Regulation 12 of the convention, which dealt with shipborne navigational equipment, set out carriage requirements for radar but made no specification regarding operating frequency. In practice the situation was that almost all ships probably had an X band radar and some would also have S band equipment. Whatever the reality, the 1988 amendments to the regulation (see Section 10.3.1 at paragraphs (g) and (h)) have the effect of *requiring* most vessels to have at least one X band radar. The requirement must be met by 1 February 1995.

The detailed principles underlying the operation and detection of radar beacons are set out in the Radar and ARPA Manual.

19.1.8 Radio communications in collision avoidance

VHF has the potential to make a positive contribution to collision avoidance in certain circumstances but sadly, if misused, it may well lead to the collision which it was sought to avoid. In this section the potential contribution of VHF will be discussed, the need for discipline in its use underlined, and the dangers of its misuse highlighted. Notice M.845 draws particular attention to these dangers, and is reproduced in Section 19.3.6.

Both rules 5 and 7 of the Collision Regulations (see Section 22.3.1) require the mariner to use all means available to keep a look-out (by sight *and* by hearing) and to determine if risk of collision exists. There can be no doubt that in many encounters information relating to the movement or intention of a positively identified vessel must assist the mariner in making a full appraisal of the situation. It can be said that radar plotting will reveal what a vessel has done until the last observation, and visual sighting will show what the vessel is doing now, but neither has the ability to indicate with certainty what the vessel will do next, although the provisions of the Collision Avoidance Regulations might provide some indication as to what might or might not be expected. VHF radio can make the unique contribution of allowing one vessel to make other vessels aware of its intentions. Other vessels must take such information into account when planning collision avoidance strategy and, while they must always comply with the collision regulations, they may be better able to choose a manoeuvre which complements rather than cancels that of the other vessel. The potential of the contribution is well illustrated by the special case of the disciplined use of VHF made in pilotage and VTS situations. In such situations, the knowledge that another vessel has an ETA at a critical point (e.g. a bridge, a bend, etc.) at a certain time and has specified intentions thereafter is clearly invaluable.

With regard to more general encounters, under rule 19(e) (see Section 22.3.1) a vessel which hears the fog signal of another vessel apparently forward of her beam, must reduce her speed to steerage way, *except* in cases where it has been determined that risk of collision does not exist. As this rule applies only to restricted visibility, it would be dangerous to assume that the fog signal was necessarily associated with a specific radar echo. The use of related VHF speech and fog signal sounding could resolve this ambiguity. However, like *all* VHF assistance in collision avoidance techniques, it can only be used if both vessels have *positively* identified the radar echo (or visual presence) of the vessel with whom they are in VHF contact.

The cornerstone of the disciplined use of VHF to provide assistance in collision avoidance is such positive identification. Each vessel must positively identify the other vessel(s) involved by carefully cross-checking the position of such vessel(s) as observed visually, or by radar, against that received by VHF from a vessel which has given its name or call sign and the time for which the position was specified. If it is intended to use VHF information, identification must be achieved well before the development of a close-quarters situation. In a developing encounter no discussion of possible manoeuvres should be attempted if positive identification has not been achieved.

A major danger in the use of VHF in collision avoidance is misunderstandings arising from language difficulties. Although SMNV and SEASPEAK are well documented there is no guarantee that other vessels will use them. Where serious language difficulties arise it is safer to forego the potential of VHF communications.

If positive identification can be achieved, and clear communications can be established, then where it is considered that an exchange of spoken information may remove uncertainty in an encounter, VHF communication should be used providing that this does not prevent orthodox compliance with the Collision Avoidance Regulations.

However, no apology is made for restating the fact that positive identification is absolutely essential. In addition, the following dangers should be borne in mind:

1 It must not be assumed that a message broadcast has been received by all other vessels involved.
2 Although a vessel has broadcast an intention to carry out certain manoeuvres, all means available must none the less be used to monitor the movements of such a vessel.
3 The ability to speak directly to another vessel in an encounter may well encourage a smaller alteration than might otherwise be considered appropriate. In this respect the ability of other vessels to detect such an alteration must be given due consideration (see Section 22.2.6).
4 Where a close-quarters situation is already developing, valuable time may be wasted in attempting to make VHF contact.
5 It is dangerous to make agreements by VHF to carry out manoeuvres which contravene the Collision Regulations.

19.1.9 Routine navigation communications

Radio communications provide the best method of keeping up to date with day-to-day changes in navigation, weather and safety information. The effectiveness of the system clearly depends on the shore-based organization and network in the area in which the vessel is trading. However, whatever the service, the ship has a responsibility to make arrangements to monitor the appropriate listening frequencies.

Where a specialist radio officer is carried, such monitoring will be an integral part of telegraphy and telephony duties, but navigating officers will normally be expected to maintain a listening watch on VHF channel 16 (see Section 19.3.1).

Where no specialist radio officer is carried, the navigating officer must service the requirement on such frequencies as are appropriate to the area (see Section 19.3.1). In this connection, the following points should be given due consideration:

1 Silence periods should be strictly observed and logged. At all times the watchkeeper must be alert to the possible reception of distress and urgency messages. Where appropriate DSC and satellite equipment should be checked and operational.
2 A general listening watch must be maintained for navigation and weather warnings, but particular attention must be paid to times scheduled for the broadcast of navigation warnings and weather forecasts. Details are furnished by Admiralty List of Radio Signals, vol. 3 (see also Sections 11.6, 11.11 and 11.12). Where appropriate NAVTEX or other WWNWS direct printing equipment should be checked and operational.
3 A general listening watch must be kept for communications addressed to the observing vessel, but particular attention must be paid to those times scheduled for the broadcast of traffic lists (these are given in Admiralty List of Radio Signals, vol. 1). In a traffic list, the coast station broadcasts the names and/or call signs of the vessels for which it is holding messages. Early reception of commercial messages may have considerable bearing on contingency plans in the event of diversion (see Section 16.1).
4 Where the vessel is in an area to which a ship position reporting scheme applies, care should be taken to make the necessary or voluntary reports (including defects if appropriate). In oceanic systems, the effective reporting of movements by all vessels ensures the maximum speed and effectiveness of any SAR operation that may have to be implemented (see Chapter 17). Details of reporting systems are provided in Admiralty List of Radio Signals, vol. 1. In VTS areas it makes an essential contribution to the safe, efficient and free flow of traffic (see Section 23.5).

19.2 On-board communications

On-board communications may take the form of writing, direct speech, speech by line telephony and speech by short range radiotelephony. These are practices which should be part of the everyday experience of those involved in navigation control. A brief discussion of each of these areas will be given in the following sections.

19.2.1 Written communications

These are relevant to items such as master's night orders and the bridge notebook, the operational aspects of which have been discussed in Sections 11.25 and 15.2.

Written communications should be clear, concise and above all free from ambiguity. Where the communication is to be read at a later time in the absence of the author (e.g. master's night orders) it is good practice to have it read by at least one other officer at the time of writing. This will, hopefully, reveal any possible alternative interpretation or ambiguity not foreseen by the writer. Given the flexibility of many languages there is always the danger that two different people will interpret the same arrangement of words in different ways. Particular care is necessary where the reader, or the writer, are not native speakers of the language used. In such cases the regular use of standard phrases is particularly helpful.

19.2.2 Direct speech

This will be used on the bridge for the passing of information, the issue and acknowledgement of orders. It is important to differentiate between the above functional communications and conversational asides which may spontaneously arise. This is best done by establishing a procedure whereby orders are acknowledged by repetition and the understanding of information is confirmed in a similar way.

In spoken communication, diction, intonation and vocabulary are extremely important. Particular care is necessary where non-native speakers or distinctive accents are involved. Orders are most effective when given crisply using well measured, standard phrases. While adequate volume is clearly necessary in the passing of information, shouting over short distances is frequently counter-productive in achieving quick, reliable and effective execution of orders.

19.2.3 Speech by line telephony

This method will be used to pass information or orders from the bridge to personnel in other parts of the vessel. To avoid misunderstandings in the use of shipboard telephones, the proper calling and answering procedure should be used. At the commencement of any call both parties should identify themselves and the station from which they speak.

Speech is composed of a very wide variety of sound frequencies blended together in widely differing proportions to give the infinite range of colourful characteristics associated with the human voice. All audio systems can only pass a limited range of frequencies known as the bandwidth, and the extent of this range limits the fidelity with which any sound can be reproduced. Telephone systems frequently have fairly narrow bandwidths, as a result of which some tones are excluded while others are made more obvious. This may well exaggerate distinctive accents. It is thus important to pronounce words clearly and to stress weak syllables so that they are not lost. As in direct speech, volume is also important. In general, normal conversational level should be adequate, though in some situations it may be necessary to raise the voice to compete with background noise. Shouting will be counter-productive because the telephone is designed to receive speech at normal levels. Excess volume merely saturates the circuitry and causes distortion. Where persistent or excessive background noise is a problem, the design and siting of the equipment used should be investigated.

19.2.4 Short range radiotelephony

Chapter III of SOLAS(74) as amended to 1991, which deals with lifesaving appliances, sets out carriage requirements in respect of two-way VHF radiotelephony apparatus. The requirements, which vary according to the type and size of vessel, are set out in the extract reproduced in Section 19.3.9.

Hand held short range radio transmitters and receivers can be used for internal communications on board ship or between ship and its lifeboats or life-rafts. Due care and attention should be given to the speech techniques already mentioned in Sections 19.2.2 and 19.2.3.

Proper procedure must be followed in station identification before the passing of messages. There is considerable potential for misunderstandings where several hand held radios on the same ship are operating on the same frequency. While the operational range of hand held radios is fairly short, the possibilities of the signals being received on another ship close by cannot be ignored. At least one accident has occurred where an order intended for one hand held station was executed by another. To avoid any possible confusion a procedure in line with that set out in the Handbook for Radio Operators should be followed. This is reproduced in Section 19.3.8.

19.3 Extracts from official publications

Within Sections 19.3.1–19.3.9, the paragraph numbering of the original publications has been retained.

19.3.1 International Convention For The Safety of Life at Sea (1974) as Amended to 1991 – Chapter IV: Radiocommunications

Introduction

This does not form part of the extracts.

Chapter IV of SOLAS(74) was completely replaced by the 1988 amendments to the convention which introduce the Global Maritime Distress and Safety System. The provisions will be implemented over the period 1 February 1991 to 1 February 1999. During this transitional period some vessels will be able to comply with the convention in its pre-GMDSS form and for reference purposes some major provisions are summarized in the paragraph below.

Under the pre-GMDSS convention passenger vessels irrespective of size and cargo ships of 1600 tons gross tonnage and upwards, unless exempted, were required to be fitted with a radio telegraphy station and carry specialist radio personnel. Passenger vessels irrespective of size and cargo ships of 300 tons gross tonnage and upwards were required to be fitted with a VHF radiotelephone installation. Cargo ships of 300 tons gross tonnage but less than 1600 tons gross tonnage were required, if not fitted with a telegraphy station, to be fitted with a radiotelephone station. This could be operated by non specialist radio personnel, e.g. deck officers holding restricted radiotelephony operator's certificates.

Important extracts from the convention as amended for GMDSS are set out in the remainder of this section.

Regulation 1 Application

1 This chapter applies to all ships to which the present regulations apply and to cargo ships of 300 tons gross tonnage and upwards.

4 Every ship shall comply with regulations 7.1.4 (NAVTEX) and 7.1.6 (satellite EPIRB) not later than 1 August 1993.

5 Subject to the provisions of paragraph 4, the Administration shall ensure that every ship constructed before 1 February 1995:

.1 during the period between 1 February 1992 and 1 February 1999:

.1.1 either complies with all applicable requirements of this chapter; or

.1.2 complies with all applicable requirements of chapter IV of the International Convention for the Safety of Life at Sea, 1974 in force prior to 1 February 1992; and

.2 after 1 February 1999, complies with all the applicable requirements of this chapter.

6 Every ship constructed on or after 1 February 1995 shall comply with all the applicable requirements of this chapter.

7 No provision in this chapter shall prevent the use by any ship, survival craft or person in distress, of any means at their disposal to attract attention, make known their position and obtain help.

Regulation 7 Radio equipment – General

1 Every ship shall be provided with:

.1 a VHF radio installation capable of transmitting and receiving:

.1.1 DSC* on the frequency 156.525 MHz (channel 70). It shall be possible to initiate the transmission of distress alerts on channel 70 from the position from which the ship is normally navigated;** and

.1.2 radiotelephony on the frequencies 156.300 MHz (channel 6), 156.650 MHz (channel 13) and 156.800 MHz (channel 16);

.2 a radio installation capable of maintaining a continuous DSC watch on VHF channel 70 which may be separate from, or combined with, that required by subparagraph .1.1;**

.3 a radar transponder capable of operating in the 9 GHz band, which:

.3.1 shall be so stowed that it can be easily utilized; and

.3.2 may be one of those required by regulation III/6.2.2 for a survival craft;

.4 a receiver capable of receiving international NAVTEX service broadcasts if the ship is engaged on voyages in any area in which an international NAVTEX service is provided;

.5 a radio facility for reception of maritime safety information by the INMARSAT enhanced

* Digital selective calling (DSC) for all ships and HF direct-printing telegraphy (NBDP) carriage requirements for ships of 300 tons gross tonnage and over but less than 1,600 tons gross tonnage are subject to review in accordance with resolution A.606(15) – Review and evaluation of the GMDSS. Unless otherwise specified this footnote applies to all DSC and NBDP requirements prescribed in the Convention.

** Certain ships may be exempted from this requirement (see regulation 9.4).

group calling system if the ship is engaged on voyages in any area of INMARSAT coverage but in which an international NAVTEX service is not provided. However, ships engaged exclusively on voyages in areas where an HF direct-printing telegraphy* maritime safety information service is provided and fitted with equipment capable of receiving such service, may be exempt from this requirement;**

.6 subject to the provisions of regulation 8.3, a satellite emergency position-indicating radio beacon (satellite EPIRB) which shall be:

.6.1 capable of transmitting a distress alert either through the polar orbiting satellite service operating in the 406 MHz band or, if the ship is engaged only on voyages within INMARSAT coverage, through the INMARSAT geostationary satellite service operating in the 1.6 GHz band;***

.6.2 installed in an easily accessible position;

.6.3 ready to be manually released and capable of being carried by one person into a survival craft;

.6.4 capable of floating free if the ship sinks and of being automatically activated when afloat; and

.6.5 capable of being activated manually.

2 Until 1 February 1999 or until such other date as may be determined by the Maritime Safety Committee, every ship shall, in addition, be fitted with a radio installation consisting of a radiotelephone distress frequency watch receiver capable of operating on 2182 kHz.

3 Until 1 February 1999, every ship shall, unless the ship is engaged on voyages in sea area A1 only, be fitted with a device for generating the radiotelephone alarm signal on the frequency 2182 kHz.

4 The Administration may exempt ships constructed on or after 1 February 1997 from the requirements prescribed by paragraphs 2 and 3.

*Digital selective calling (DSC) for all ships and HF direct-printing telegraphy (NBDP) carriage requirements for ships of 300 tons gross tonnage and over but less than 1,600 tons gross tonnage are subject to review in accordance with resolution A.606(15) – Review and evaluation of the GMDSS. Unless otherwise specified this footnote applies to all DSC and NBDP requirements prescribed in the Convention.

** Reference is made to the recommendation on promulgation of maritime safety information, to be developed by the Organization (see MSC 55/25, annex 8).

*** Subject to the availability of appropriate receiving and processing ground facilities for each ocean region covered by INMARSAT satellites.

Regulation 8 Radio equipment – Sea area A1

1 In addition to meeting the requirements of regulation 7, every ship engaged on voyages exclusively in sea area A1 shall be provided with a radio installation capable of initiating the tansmission of ship-to-shore distress alerts from the position from which the ship is normally navigated, operating either:

.1 on VHF using DSC; this requirement may be fulfilled by the EPIRB prescribed by paragraph 3, either by installing the EPIRB close to, or by remote activation from, the position from which the ship is normally navigated; or

.2 through the polar orbiting satellite service on 406 MHz; this requirement may be fulfilled by the satellite EPIRB, required by regulation 7.1.6, either by installing the satellite EPIRB close to, or by remote activation from, the position from which the ship is normally navigated; or

.3 if the ship is engaged on voyages within coverage of MF coast stations equipped with DSC, on MF using DSC; or

.4 on HF using DSC; or

.5 through the INMARSAT geostationary satellite service; this requirement may be fulfilled by:

.5.1 an INMARSAT ship earth station;* or

.5.2 the satellite EPIRB, required by regulation 7.1.6, either by installing the satellite EPIRB close to, or by remote activation from, the position from which the ship is normally navigated.

*This requirement can be met by INMARSAT ship earth stations capable of two-way communications, such as Standard -A or Standard-C ship earth stations. Unless otherwise specified, this footnote applies to all requirements for an INMARSAT ship earth station prescribed by this chapter.

2 The VHF radio installation, required by regulation 7.1.1, shall also be capable of transmitting and receiving general radiocommunications using radiotelephony.

3 Ships engaged on voyages exclusively in sea area A1 may carry, in lieu of the satellite EPIRB required by regulation 7.1.6, an EPIRB which shall be:

.1 capable of transmitting a distress alert using DSC on VHF channel 70 and providing for locating by means of a radar transponder operating in the 9 GHz band;
.2 installed in an easily accessible position;
.3 ready to be manually released and capable of being carried by one person into a survival craft;
.4 capable of floating free if the ship sinks and being automatically activated when afloat; and
.5 capable of being activated manually.

Regulation 9 Radio equipment – Sea areas A1 and A2

1 In addition to meeting the requirements of regulation 7, every ship engaged on voyages beyond sea area A1, but remaining within sea area A2, shall be provided with:

.1 an MF radio installation capable of transmitting and receiving, for distress and safety purposes, on the frequencies:
.1.1 2,187.5 kHz using DSC; and
.1.2 2,182 kHz using radiotelephony;
.2 a radio installation capable of maintaining a continuous DSC watch on the frequency 2187.5 kHz which may be separate from, or combined with, that required by subparagraph .1.1; and
.3 means of initiating the transmission of ship-to-shore distress alerts by a radio service other than MF operating either:
.3.1 through the polar orbiting satellite service on 406 MHz; this requirement may be fulfilled by the satellite EPIRB, required by regulation 7.1.6, either by installing the satellite EPIRB close to, or by remote activation from, the position from which the ship is normally navigated; or
.3.2 on HF using DSC; or
.3.3 through the INMARSAT geostationary satellite service; this requirement may be fulfilled by:
.3.3.1 the equipment specified in paragraph 3.2; or
.3.3.2 the satellite EPIRB, required by regulation 7.1.6, either by installing the satellite EPIRB close to, or by remote activation from, the position from which the ship is normally navigated.

2 It shall be possible to initiate transmission of distress alerts by the radio installations specified in paragraphs 1.1 and 1.3 from the position from which the ship is normally navigated.

3 The ship shall, in addition, be capable of transmitting and receiving general radiocommunications using radiotelephony or direct-printing telegraphy by either:

.1 a radio installation operating on working frequencies in the bands between 1605 kHz and 4000 kHz or between 4000 kHz and 27 500 kHz. This requirement may be fulfilled by the addition of this capability in the equipment required by paragraph 1.1; or
.2 an INMARSAT ship earth station.

4 The Administration may exempt ships constructed before 1 February 1997, which are engaged exclusively on voyages within sea area A2, from the requirements of regulations 7.1.1.1 and 7.1.2 provided such ships maintain, when practicable, a continuous listening watch on VHF channel 16. This watch shall be kept at the position from which the ship is normally navigated.

Regulation 10 Radio equipment – Sea areas A1, A2 and A3

1 In addition to meeting the requirements of regulation 7, every ship engaged on voyages beyond sea areas A1 and A2, but remaining within sea area A3, shall, if it does not comply with the requirements of paragraph 2, be provided with:

.1 an INMARSAT ship earth station capable of:
.1.1 transmitting and receiving distress and safety

communications using direct-printing telegraphy;

.1.2 initiating and receiving distress priority calls;

.1.3 maintaining watch for shore-to-ship distress alerts, including those directed to specifically defined geographical areas;

.1.4 transmitting and receiving general radiocommunications, using either radiotelephony or direct-printing telegraphy; and

.2 an MF radio installation capable of transmitting and receiving, for distress and safety purposes, on the frequencies:

.2.1 2187.5 kHz using DSC; and

.2.2 2182 kHz using radiotelephony; and

.3 a radio installation capable of maintaining a continuous DSC watch on the frequency 2187.5 kHz which may be separate from or combined with that required by subparagraph .2.1; and

.4 means of initiating the transmission of ship-to-shore distress alerts by a radio service operating either:

.4.1 through the polar orbiting satellite service on 406 MHz; this requirement may be fulfilled by the satellite EPIRB, required by regulation 7.1.6, either by installing the satellite EPIRB close to, or by remote activation from, the position from which the ship is normally navigated; or

.4.2 on HF using DSC; or

.4.3 through the INMARSAT geostationary satellite service, by an additional ship earth station or by the satelite EPIRB required by regulation 7.1.6, either by installing the satellite EPIRB close to, or by remote activation from, the position from which the ship is normally navigated;

2 In addition to meeting the requirements of regulation 7, every ship engaged on voyages beyond sea areas A1 and A2, but remaining within sea area A3, shall, if it does not comply with the requirements of paragraph 1, be provided with:

.1 an MF/HF radio installation capable of transmitting and receiving, for distress and safety purposes, on all distress and safety frequencies in the bands between 1605 kHz and 4000 kHz and between 4000 kHz and 27 500 kHz:

.1.1 using DSC;

.1.2 using radiotelephony; and

.1.3 using direct-printing telegraphy; and

.2 equipment capable of maintaining DSC watch on 2187.5 kHz, 8414.5 kHz and on at least one of the distress and safety DSC frequencies 4207.5 kHz, 6312 kHz, 12 577 kHz or 16 804.5 kHz; at any time, it shall be possible to select any of these DSC distress and safety frequencies. This equipment may be separate from, or combined with, the equipment required by subparagraph .1; and

.3 means of initiating the transmission of ship-to-shore distress alerts by a radiocommunication service other than HF operating either:

.3.1 through the polar orbiting satellite service on 406 MHz; this requirement may be fulfilled by the satellite EPIRB, required by regulation 7.1.6, either by installing the satellite EPIRB close to, or by remote activation from, the position from which the ship is normally navigated; or

.3.2 through the INMARSAT geostationary satellite service; this requirement may be fulfilled by:

.3.2.1 an INMARSAT ship earth station; or

.3.2.2 the satellite EPIRB, required by regulation 7.1.6, either by installing the satellite EPIRB close to, or by remote activation from, the position from which the ship is normally navigated; and

.4 in addition, ships shall be capable of transmitting and receiving general radiocommunications using radiotelephony or direct-printing telegraphy by an MF/HF radio installation operating on working frequencies in the bands between 1605 kHz and 4000 kHz and between 4000 kHz and 27 500 kHz. This requirement may be fulfilled by the addition of this capability in the equipment required by subparagraph .1.

3 It shall be possible to initiate transmission of distress alerts by the radio installations specified in subpara-

graphs 1.1, 1.2, 1.4, 2.1 and 2.3 from the position from which the ship is normally navigated.

4 The Administration may exempt ships constructed before 1 February 1997, and engaged exclusively on voyages within sea areas A2 and A3, from the requirements of regulations 7.1.1.1 and 7.1.2 provided such ships maintain, when practicable, a continuous listening watch on VHF channel 16. This watch shall be kept at the position from which the ship is normally navigated.

Regulation 11 Radio equipment – Sea areas A1, A2, A3 and A4

1 In addition to meeting the requirements of regulation 7, ships engaged on voyages in all sea areas shall be provided with the radio installations and equipment required by regulation 10.2, except that the equipment required by regulation 10.2.3.2 shall not be accepted as an alternative to that required by regulation 10.2.3.1, which shall always be provided. In addition, ships engaged on voyages in all sea areas shall comply with the requirements of regulation 10.3.

2 The Administration may exempt ships constructed before 1 February 1997, and engaged exclusively on voyages within sea areas A2, A3 and A4, from the requirements of regulations 7.1.1.1 and 7.1.2 provided such ships maintain, when practicable, a continuous listening watch on VHF channel 16. This watch shall be kept at the position from which the ship is normally navigated.

Regulation 12 Watches

1 Every ship, while at sea, shall maintain a continuous watch:

.1 on VHF DSC channel 70, if the ship, in accordance with the requirements of regulation 7.1.2, is fitted with a VHF radio installation;

.2 on the distress and safety DSC frequency 2187.5 kHz, if the ship, in accordance with the requirements of regulation 9.1.2 or 10.1.3, is fitted with an MF radio installation;

.3 on the distress and safety DSC frequencies 2187.5 kHz and 8414.5 kHz and also on at least one of the distress and safety DSC frequencies 4207.5 kHz, 6312 kHz, 12 577 kHz or 16 804.5 kHz, appropriate to the time of day and the geographical position of the ship, if the ship, in accordance with the requirements of regulation 10.2.2 or 11.1, is fitted with an MF/HF radio installation. This watch may be kept by means of a scanning receiver;

.4 for satellite shore-to-ship distress alerts, if the ship, in accordance with the requirements of regulation 10.1.1, is fitted with an INMARSAT ship earth station.

2 Every ship, while at sea, shall maintain a radio watch for broadcasts of maritime safety information on the appropriate frequency or frequencies on which such information is broadcast for the area in which the ship is navigating.

3 Until 1 February 1999 or until such other date as may be determined by the Maritime Safety Committee, every ship while at sea shall maintain, when practicable, a continuous listening watch on VHF channel 16. This watch shall be kept at the position from which the ship is normally navigated.

4 Until 1 February 1999 or until such other date as may be determined by the Maritime Safety Committee, every ship required to carry a radiotelephone watch receiver shall maintain, while at sea, a continuous watch on the radiotelephone distress frequency 2182 kHz. This watch shall be kept at the position from which the ship is normally navigated.

Regulation 16 Radio personnel

Every ship shall carry personnel qualified for distress and safety radiocommunication purposes to the satisfaction of the Administration. The personnel shall be holders of certificates specified in the Radio Regulations as appropriate, any one of whom shall be designated to have primary responsibility for radiocommunications during distress incidents.

Regulation 17 Radio records

A record shall be kept, to the satisfaction of the Administration and as required by the Radio Regulations, of all incidents connected with the radiocommunication service which appear to be of importance to safety of life at sea.

19.3.2 UK DTp Merchant Shipping Notice M1252 Standard Marine Navigational Vocabulary

Introduction

This vocabulary has been compiled:

- to assist in the greater safety of nagivation and of the conduct of ships.
- to standardize the language used in communication for navigation at sea, in port-approaches, in waterways and harbours.

These phrases are not intended to supplant or contradict the International Regulations for Preventing Collisions at Sea or special local rules or recommendations made by IMO concerning ships' routeing. Neither are they intended to supersede the International Code of Signals nor to supplant normal radiotelephone practice as set out in the ITU Regulations.

It is not intended that use of the vocabulary shall be mandatory, but rather that through constant repetition in ships and in training establishments ashore, the phrases and terms used will become those normally accepted and commonplace among seamen. Use of the contents of the vocabulary should be made as often as possible in preference to other wording of similar meaning.

In this way it is intended to become an acceptable 'language', using the English tongue, for the interchange of intelligence between individuals of all maritime nations on the many and varied occasions when precise meanings and translations are in doubt, increasingly evident under modern conditions at sea.

The typographical conventions used throughout most of this vocabulary are as follows:

( ) brackets indicate that the part of the message enclosed within the brackets may be added where it is relevant.

/ oblique stroke indicates that the items on either side of the stroke are alternatives.

... dots indicate that the relevant information is to be filled in where the dots occur.

Standard Marine Navigational Vocabulary

When spelling is necessary, only the letter spelling table contained in the International Code of Signals, Chapter X, and in the Radio Regulations should be used.

Part 1 General

1 Procedure/message markers

When it is necessary to indicate that phrases in this vocabulary are to be used, the following messages may be sent:

'Please use the standard Marine Navigational Vocabulary.'

'I will use the Standard Marine Navigational Vocabulary.'

If necessary, external communication messages may be preceded by the following message markers:

QUESTION	indicates that the following message is of interrogative character
ANSWER	indicates that the following message is the reply to a previous question
REQUEST	indicates that the contents of the following message are asking for action from others with respect to the ship
INFORMATION	indicates that the following message is restricted to observed facts
INTENTION	indicates that the following message informs others about immediate navigational actions intended to be taken
WARNING	indicates that the following message informs other traffic participants about dangers
ADVICE	indicates that the following message implies the intention of the sender to influence the recipient(s) by a recommendation

INSTRUCTION	indicates that the following message implies the intention of the sender to influence the recipient(s) by a regulation.

2 Standard verbs

Where possible, sentences should be introduced by one of the following verb forms:

IMPERATIVE

Always to be used when mandatory orders are being given

You must	Do not	Must I?
Indicative	*Negative*	*Interrogative*
I require	I do not require	Do I require?
I am	I am not	Am I?
You are	You are not	Are you?
I have	I do not have	Do you have?
I can	I cannot	Can I? / Can you? *is it possible?*
I wish to	I do not wish to	Do you wish to?
I will – *future*	I will not – *future*	
You may	You need not	May I? – *permission*
Advise	Advise not	
There is	There is not	Is there?
		What is/are?
		Where is/are?
		When is/are?

Note: See section 1 – Message markers.

3 Responses

Where the answer to a question is in the affirmative, say:

'yes ...' – followed by the appropriate phrase in full.

Where the answer to a question is in the negative, say:

'no ...' – followed by the appropriate phrase in full.

Where the information is not immediately available but soon will be, say:

'Stand by'.

Where the information cannot be obtained, say:

'No information'.

Where a message is not properly heard, say:

'Say again'.

Where a message is not understood, say:

'Message not understood'.

4 Distress/urgency/safety messages

MAYDAY (repeated three times)	is to be used to announce a distress message
PAN PAN (repeated three times)	is to be used to announce an urgency message
SECURITE (repeated three times)	is to be used to announce a safety message

5 Miscellaneous phrases

5.1 What is your name (and call sign)?

5.2 How do you read me?

5.3 I read you ... with signal strength ...

(bad/1)	(1/barely perceptible)
(poor/2)	(2/weak)
(fair/3)	(3/fairly good)
(good/4)	(4/good)
(excellent/5).	(5/very good).

5.4 Stand by on channel ...

5.5 Change to channel ...

5.6 I cannot read you.
(Pass your message through vessel ...).
(Advise try channel ...).

5.7 I cannot understand you.
Please use the ...
(Standard Marine Navigational Vocabulary).
(International Code of Signals).

5.8 I am passing a message for vessel ...

5.9 Correction ...

5.10 I am ready to receive your message.

5.11 I am not ready to receive your message.

5.12 I do not have channel ... Please use channel ...

6 Repetition

If any parts of the message are considered sufficiently important to need safeguarding, use the word 're-peat'.

Examples:
'You will load 163 repeat 163 tons bunkers.'
'Do not repeat not overtake.'

7 Position
When latitude and longitude are used, these shall be expressed in degrees and minutes (and decimals of a minute is necessary), north or south of the Equator and east or west of Greenwich.

When the position is related to a mark, the mark shall be a well-defined charted object. The bearing shall be in the 360 degree notation from true north and shall be that of the position *from* the mark.

Examples:
'There are salvage operations in position 15 degrees 34 minutes north 61 degrees 29 minutes west.'
'Your position is 137 degrees from Barr Head lighthouse distance two decimal four miles.'

8 Courses
Always to be expressed in 360 degree notation from north (true north unless otherwise stated). Whether this is *to* or *from* a mark can be stated.

9 Bearings
The bearing of the mark or vessel concerned, is the bearing in the 360 degree notation from north (true north unless otherwise stated), except in the case of relative bearings. Bearings may be either *from* the mark or *from* the vessel.

Examples:
'The pilot boat is bearing 215° from you.'
'Your bearing is 127° from the signal station.'
Note: Vessels reporting their position should always quote their bearing *from* the mark, as described in paragraph 7.

Relative bearings
Relative bearings can be expressed in degrees relative to the vessel's head or bow. More frequently this is in relation to the port or starboard bow.

Example:
'The buoy is 030° on your port bow.'
Relative D/F bearings are more commonly expressed in the 360 degree notation.

10 Distances
Preferably to be expressed in nautical miles or cables (tenths of a mile) otherwise in kilometres or metres, the unit always to be stated.

11 Speed
To be expressed in knots:

(a) without further notation meaning speed through the water; or
(b) 'ground speed' meaning speed over the ground.

12 Numbers
Numbers are to be spoken:

'One-five-zero' for 150.
'Two point five' for 2.5.

13 Geographical names
Place names used should be those on the chart or Sailing Directions in use. Should these not be understood, latitude and longitude should be given.

14 Time
Times should be expressed in the 24 hour notation indicating whether UTC, zone time or local shore time is being used.

Note: In cases not covered by the above phraseology normal radiotelephone practice will prevail.

19.3.3 UK DTp Merchant Shipping Notice M.1026: Proper Use of the VHF Channels at Sea

1 VHF communication technique

1.1 Preparation
Before transmitting, think about the subjects which have to be communicated and, if necessary, prepare

written notes to avoid unnecessary interruptions and ensure that no valuable time is wasted on a busy channel.

1.2 Listening

Listen before commencing to transmit to make certain that the channel is not already in use. This will avoid unnecessary and irritating interference.

1.3 Discipline

VHF equipment should be used correctly and in accordance with the Radio Regulations. The following in particular should be avoided:

.1 Calling on channel 16 for purposes other than distress, urgency and very brief safety communications when another calling channel is available
.2 Communications not related to safety and navigation on port operation channels
.3 Non-essential transmissions, e.g. needless and superfluous signals and correspondence
.4 Transmitting without correct identification
.5 Occupation of one particular channel under poor conditions
.6 Use of offensive language.

1.4 Repetition

Repetition of words and phrases should be avoided unless specifically requested by the receiving station.

1.5 Power reduction

When possible, the lowest transmitter power necessary for satisfactory communication should be used.

1.6 Communications with shore stations

1.6.1 Instructions given on communication matters by shore stations should be obeyed.

1.6.2 Communications should be carried out on the channel indicated by the shore station. When a change of channel is requested, this should be acknowledged by the ship.

1.6.3 On receiving instructions from a shore station to stop transmitting, no further communications should be made until otherwise notified (the shore station may be receiving distress or safety messages and any other transmissions could cause interference).

1.7 Communications with other ships

1.7.1 During ship to ship communications the ship called should indicate the channel on which further transmissions should take place. The calling ship should acknowledge acceptance before changing channel.

1.7.2 The listening procedure outlined in paragraph 1.2 should be followed before communications are commenced on the chosen channel.

1.8 Distress communications

1.8.1 Distress calls/messages have absolute priority over all other communications. When hearing them all other transmissions should cease and a listening watch should be kept.

1.8.2 Any distress call/message should be recorded in the ship's log and passed to the master.

1.8.3 On receipt of a distress message, if in the vicinity, immediately acknowledge receipt. If not in the vicinity, allow a short interval of time to elapse before acknowledging receipt of the message in order to permit ships nearer to the distress to do so.

1.9 Calling

1.9.1 Whenever possible, a working frequency should be used. If a working frequency is not available, channel 16 may be used, provided it is not occupied by a distress call/message.

1.9.2 In case of difficulty to establish contact with a ship or shore station, allow adequate time before repeating the call. Do not occupy the channel unnecessarily and try another channel.

1.10 Changing channels

If communications on a channel are unsatisfactory, indicate change of channel and await confirmation.

1.11 Spelling

If spelling becomes necessary (e.g. descriptive names, call signs, words which could be misunderstood) use the spelling table contained in the International Code of Signals and the Radio Regulations.

1.12 Addressing
The words 'I' and 'You' should be used prudently. Indicate to whom they refer.

Example
Seaship, this is Port Radar, Port Radar, do you have a pilot?
Port Radar, this is Seaship, I do have a pilot.

1.13 Watchkeeping
1.13.1 Ships fitted only with VHF equipment should maintain watch on channel 16 when at sea.

1.13.2 Other ships should, where practicable, keep watch on channel 16 when within the service area of a shore station capable of operating on that channel.

1.13.3 In certain cases governments may require ships to keep a watch on other channels.

2 VHF communication procedure

2.1 Calling
When calling a shore station or another ship, say the name of that shore station or ship once (twice if considered necessary in heavy radio traffic conditions) followed by the phrase 'this is' and the ship's name twice, indicating the channel in use.

Example
Port City, this is Seastar, Seastar, on channel 14.

2.2 Exchange of messages
2.2.1 When communicating with a ship whose name is unknown but whose position is known, that position may be used. In this case the call is addressed to all ships.

Example
Hello all ships, this is Pastoria, Pastoria. Ship approaching number four buoy, I am passing Belinda Bank Light.

2.2.2 Where a message is received and only acknowledgement of receipt is needed, say 'received'. Where a message is received and acknowledgement of the correct message is required, say 'received, understood', and repeat message if considered necessary.

Example
Message: Your berth will be clear at 0830 hours.
Reply: Received, understood. Berth clear at 0830 hours.

2.2.3 During exchange of messages, a ship should invite a reply by saying 'over'.

2.2.4 Where appropriate, the following message should be sent:

Please use/I will use the Standard Marine Navigational Vocabulary.

When language difficulties exist which cannot be resolved by the use of the Vocabulary, the International Code of Signals should be used. In this case the word INTERCO should precede the groups of the International Code of Signals.

Example
Please use/I will use the International Code of Signals.

2.2.5 Where a message contains instructions or advice, the substance should be repeated.

Example
Message: Advise you pass astern of me.
Reply: I will pass astern of you.

2.2.6 If a message is not properly received, ask for it to be repeated by saying 'Say again'.

2.2.7 If a message is received but not understood, say 'message not understood'.

2.2.8 If it is necessary to change to a different channel, say 'change to channel . . .' and wait for acknowledgement before carrying out the change.

2.2.9 The end of a communication is indicated by the word 'out'.

3 Standard messages

3.1 Since most ship to shore communications are exchanges of information, it is advisable to use standard messages which will reduce transmission time.

19.3.4 UK DTp Merchant Shipping Notice M.1235: Use of the MF Radio Installation by Unskilled Persons On Board Radiotelegraph Ships in an Emergency

1 On most radiotelegraph ships only one radio officer is carried. If by chance he were to be incapacitated through an accident, illness or other serious mishap whilst his ship was at sea, it might well be that there would be no one else on board capable of operating that radio equipment installed in the radio room. Bearing in mind the possibility of a distress incident involving either the ship itself or another in its vicinity it is clearly desirable that some provision should be made for one or more other officers on such ships to be capable of operating the radio equipment for distress purposes.

2 All ships which are fitted with a radiotelegraph installation in compliance with the Merchant Shipping (Radio Installation) Regulations 1980, as amended, are provided with a radiotelephone distress frequency watchkeeping receiver, a radiotelephone transmitter and a radiotelephone alarm signal generating device. In many cases the transmitter and alarm signal generating device, together with a receiver capable of use on 2182 kHz, are installed in, or operated from, the radiotelegraph operating room, either as part of the main or the reserve radiotelegraph equipment.

3 It is strongly recommended to owners and masters that, in radiotelegraph ships which carry only one radio officer, at least one deck officer, who should wherever possible be the holder of a radiotelephone operator's certificate, be instructed on the method of bringing the radiotelephone equipment into use. The instruction should be adequate to enable the equipment to be provided with electrical energy, connected to an antenna, set up on the radiotelephone distress frequency and operated properly, including the transmission of a two-tone alarm signal.

4 To enable the maximum possible use to be made of the radio installation the instruction should also be adequate to enable the radiotelegraph automatic keying device to be brought into operation in the event of the radiotelephone transmitter being out of use, or if there is no response to the two-tone alarm.

5 A set of simple guidelines for the emergency operation of the equipment should be displayed in the radio room. It is essential that all the relevant switches and controls on the equipment can be identified and it is recommended that in cases where any confusion may arise they should be numbered using coloured labels.

6 The conditions which might involve the use of the radio equipment by a deck officer might also necessitate the use of the emergency lighting in the radiotelegraph operation room. It is essential therefore that deck officers are familiar with the locations of the switches for the emergency lamp.

19.3.5 UK DTp Merchant Shipping Notice M.1119: Radiotelephone Distress Procedure

CARD 1

NAME OF SHIP .... **CALL SIGN**............

DISTRESS TRANSMITTING PROCEDURES
(For use only when **IMMEDIATE ASSISTANCE** required)

1. Ensure transmitter is switched to 2182 kHz.

2. If possible **transmit two-tone ALARM SIGNAL for $\frac{1}{2}$ to 1 minute.**

3. Then say:
MAYDAY, MAYDAY, MAYDAY
THIS IS......**(Ship's name or call sign 3 times)**......**MAYDAY**
followed by ship's name or call sign

POSITION ....

Nature of Distress ....

Aid Required .... **OVER.**

4. Listen for a reply and if none heard **repeat** above procedure, particularly during the 3-minute silence period commencing at each hour and half-hour.

EXAMPLE—If possible ALARM SIGNAL followed by:

"MAYDAY, MAYDAY, MAYDAY,
This is NONSUCH, NONSUCH, NONSUCH,
MAYDAY, NONSUCH,
Position 54 25 North 016 33 West,
I am on fire and require immediate assistance, OVER."

NOTE–If language difficulties arise, use CARD 3.

CARD 2
RECEPTION OF SAFETY MESSAGES

Any message which you hear prefixed by one of the following words concerns SAFETY—

MAYDAY **PAN–PAN** **SECURITE**
(pronounced SAY–CURE–E–TAY)

If you hear these words, pay particular attention to the message and call the master or the officer on watch

MAYDAY (Distress) Indicates that a ship, aircraft or other vehicle is threatened by grave and imminent danger and requests immediate assistance.

PAN–PAN (Urgency) Indicates that the calling station has a very urgent message to transmit concerning the safety of a ship, aircraft or other vehicle, or of a person.

SECURITE (Safety) Indicates that the station is about to transmit a message concerning the safety of navigation or giving important meteorological warnings.

19.3.6 UK DTp Merchant Shipping Notice M.845: Dangers in the Use of VHF Radio in Collision Avoidance

1 The Department wishes to draw the attention of all concerned to the risks involved when VHF radio is used as a collision avoidance aid. In an increasing number of cases it has been found that at some stage before the collision VHF radio was being used by one or both parties in an attempt to avoid collision. The use of VHF radio in this role is not always helpful and may even prove dangerous.

2 Uncertainties can arise over the identification of vessels and the interpretation of messages received. At night, in restricted visibility or when there are more than two vessels in the vicinity the need for positive identification of the two vessels is essential but this can rarely be guaranteed. Even where positive identification has been achieved there is still the possibility of a misunderstanding between the parties concerned due to language difficulties, however fluent they are in the language being used. An imprecise, or ambiguously expressed, message can have serious consequences.

3 Valuable time can be wasted while mariners on vessels approaching each other try to make contact on VHF radio instead of complying with the requirements of the Collision Regulations. There is the further danger that if contact has been established, identification has been achieved and no language or message difficulty exists, a course of action is chosen which does not comply with the Collision Regulations. This can lead to the collision it was intended to avoid.

4 The Department of Trade recognizes that most ships are now fitted with VHF radio facilities and are capable of bridge to bridge communication. Although the practice of using VHF radio as a collision avoidance aid may be resorted to on occasion, especially in pilotage waters, the risks described in this notice should be clearly understood and the Collision Regulations complied with.

19.3.7 The Merchant Shipping (Radio Installations) Regulations 1992

Regulation 17(1) Schedule 3: Radio log

The following shall be recorded in the radio log as they occur, together with the time of their occurrence–

(a) a summary of communications relating to distress, urgency and safety traffic;
(b) a record of important incidents connected with the radio service;
(c) where appropriate, the position of the ship at least once a day.

Regulation 36(2) Schedule 5: Radio log – Radiotelephone ship

Part A

The radio log book, the form of which is at Part B below, is compiled in two sections which shall be completed in accordance with the following–

Section A – Particulars of the radiotelephone operators on board.

Section B – Diary of the radio service.
(a) the name of the radiotelephone operator and the times at which the watch commences and ends;
(b) the times at which radio watch is for any reason discontinued, together with the reason and the time at which radio watch is resumed;
(c) a summary of communications exchanged between the ship station and coast stations or other ship stations, including the serial numbers and the dates of any messages passed;
(d) a summary of all communications relating to distress, urgency and safety traffic;
(e) a record of all incidents connected with the radio service, including the radiotelephone installation and the VHF radiotelephone installations, which occur during the watch and appear to be of importance to safety of life at sea;
(f) details of the tests and checks required by regulation 21(1);
(g) if the ship's rules permit, the position of the ship at least once a day.

19.3.8 Handbook for Radio Operators

On-board communications

Calls for internal communications on board ship when in territorial waters consist of:

(a) From the master station:
The name of the ship followed by a single letter (ALFA, BRAVO, CHARLIE, etc., indicating the substation), not more than three times;
The words THIS IS;
The name of the ship followed by the word CONTROL.
(b) From the substation:
The name of the ship followed by the word CONTROL, not more than three times;
The words THIS IS;
The name of the ship followed by a single letter (ALFA, BRAVO, CHARLIE, etc., indicating the substation).

19.3.9 International Convention for the Safety of Life at Sea (1974) as amended to 1991 – Chapter III: Life-saving Appliances and Arrangements

2.1 Two-way VHF radiotelephone apparatus

2.1.1 At least three two-way VHF radiotelephone apparatus shall be provided on every passenger ship and on every cargo ship of 500 tons gross tonnage and upwards. At least two two-way VHF radiotelephone apparatus shall be provided on every cargo ship of 300 tons gross tonnage and upwards but less than 500 tons gross tonnage. Such apparatus shall conform to performance standards not inferior to those adopted by the Organization. If a fixed two-way VHF radiotelephone apparatus is fitted in a survival craft it shall conform to performance standards not inferior to those adopted by the Organization.
2.1.2 Two-way VHF radiotelephone apparatus provided on board ships prior to 1 February 1992 and not complying fully with the performance standards adopted by the Organization may be accepted by the Administration until 1 February 1999 provided the Administration is satisfied that they are compatible with approved two-way VHF radiotelphone apparatus.

2.2 *Radar transponders*

At least one radar transponder shall be carried on each side of every passenger ship and of every cargo ship of 500 tons gross tonnage and upwards. At least one radar transponder shall be carried on every cargo ship of 300 tons gross tonnage and upwards but less than 500 tons gross tonnage. Such radar transponders shall conform to performance standards not inferior to those adopted by the Organization. The radar transponders shall be stowed in such locations that they can be rapidly placed in any survival craft other than the liferaft or liferafts required by regulation 26.1.4. Alternatively one radar transponder shall be stowed in each survival craft other than those required by regulation 26.1.4.

Part IV
Navigation Control

20 Execution of the passage plan

20.1 Monitoring progress

20.1.1 *Dead reckoning (DR)*

Assuming that the position of the ship is known at some particular time and that the course and speed/distance made good are also accurately known, then the new position of the ship can be determined. Unfortunately, the course and distance measuring sensors are insufficiently accurate to allow implicit reliance to be placed on this new position. Bearing in mind that the initial position could also be in doubt, this method of navigation should be used only as a back-up and check on more precise position determination or in the absence of other systems.

Over short distances (<600 nm), plane trigonometry may be used. This may be referred to as *plane sailing*. Over longer distances, Mercator sailing is essential.

Example

From a departure position in lat. 45° N, long. 15° West, a vessel steers 225° (T), a distance of 600 nm. Find the DR position using:

1 Plane sailing
2 Mercator sailing.

1 Departure position:	lat.	45° 00′ N	long.	15° 00′ W
	d.lat.	7° 04.3′ S	d.long.	9° 26.5′ W
Position arrived at:	lat.	37° 55.7′ N	long.	24° 26.5′ W

Using traverse tables: course 225°, distance 600 nm

d.lat. = 424.3′ = 7° 04.3
departure = 424.3
mean lat. = 41.5°N
∴ d.long. = 566.5′ = 9° 26.5′ W

2 Departure position:	lat.	45° 00′ N	long.	15° 00′ W	Mer. parts 3013.4
	d.lat.	7° 04.3′ S	d.long.	9° 25′ W	
Position arrived at:	lat.	37° 55.7′ N	long.	24° 25′ W	Mer. parts 2448.4

d.lat. = 600 cos 45°
= 424.3′ S
= 7° 04.3′ S

d.long = 565 tan 45°
= 565′
= 9° 25′ W

DMP = 565.0

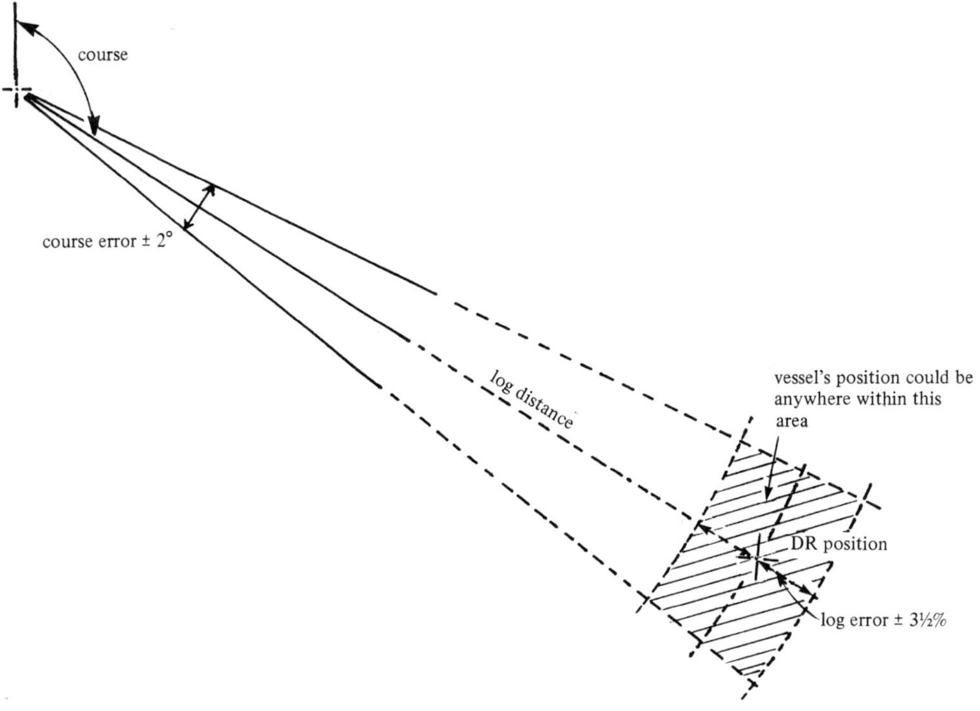

Figure 20.1 The dead reckoning position

Although we have calculated the DR position, this could also have been worked on the chart. Ignoring any errors which might exist in the data, it can be seen that even the method of 'running up' the position can introduce differences.

In any event, the basic idea is correct and it is mainly the imprecision of the data inputs which results in the uncertainty of the DR position (see also Section 1.5). Of course, the longer the uncertainty exists, the greater will be the cumulative error, and therefore the larger the error circle around the DR position (Figure 20.1).

20.1.2 The estimated position (EP)

In the previous section, it was assumed that errors in course and distance made good were due primarily to sensor errors, i.e. the effects of wind and tide (or current) were assumed to be zero. If an estimate of these two factors is made and included in the calculation then a better estimate of the ship's position will be achieved. It is important to realize that any uncertainty in knowledge of current or leeway will have to be added to the error circle. Thus the error around the estimated position may be larger than that around the DR position. A failure to take into account the effect of existing current or leeway can render the DR position virtually useless.

Example

From a departure position lat. 45° N, long. 15° W, a vessel steers a course of 225° (T) at a speed of 20 knots. Find the vessel's estimated position after 30 hours steaming if the set and rate of the current are estimated at 030° (T) and 1 knot and the estimated leeway, resulting from a NW'ly gale is 3° throughout (Figure 20.2).

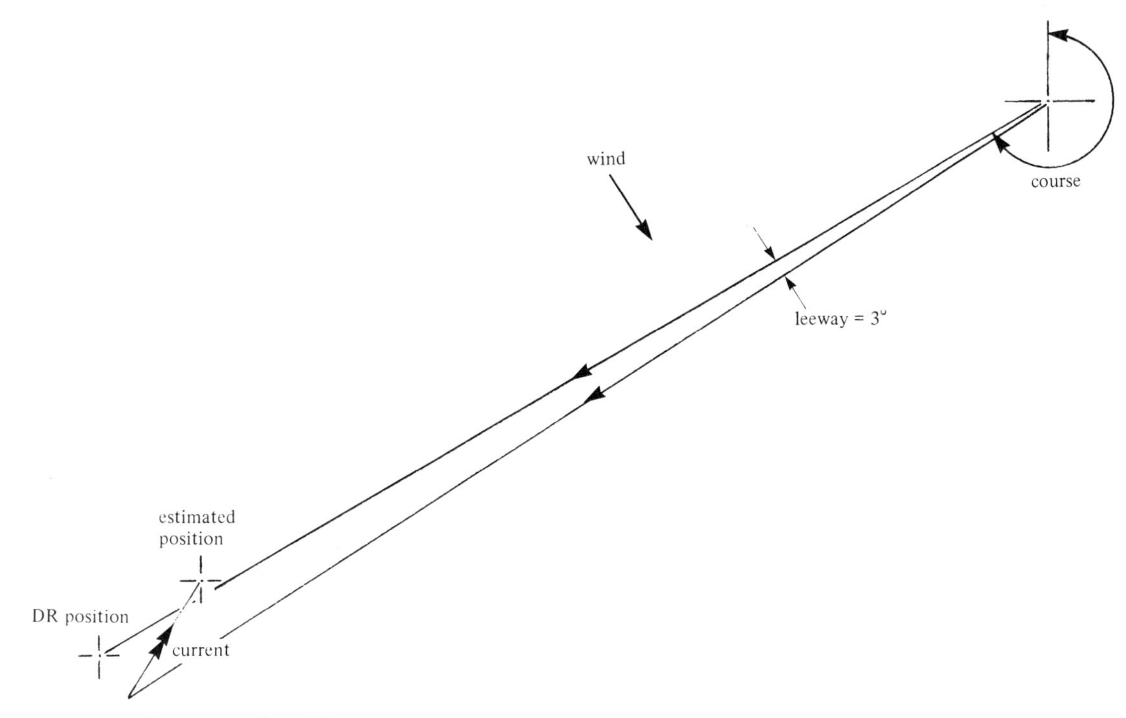

Figure 20.2 The estimated position

Departure position:	lat.	45° 00′ N	long.	15° 00′ W
	d.lat.	7° 00′ S	d.long.	8° 36′ W
Position arrived at:	lat.	38° 00′ N	long.	23° 36′ W (EP)

Course made good through the water = 222°(T), distance 600 nm

The effect of tide setting: 030°(T), drift 30 nm

d.lat.	446′ S	dep 401.5′ W
Tide d.lat.	26′ N	dep 15.0′ E
d.lat.	420′	dep 386.5′ W mean lat. = $41\frac{1}{2}$°′ N
	= 7° 00′ S	∴ d.long. = 516 = 8° 36′ W

It can be seen by comparison with the previous example that, if the leeway and current really existed but were ignored, the knowledge of the vessel's position could be very much in error, even allowing for the slightly increased error circle which must be drawn around the EP.

20.1.3 Position fixing principles

When there exists some other means of fixing position, it should be used in preference to DR and EP methods, but the reliability of this other data must always be borne in mind.

A position line is a line on a chart on which, for whatever reason, the vessel's position is believed to lie. Position lines may be obtained from many sources and in principle can be used to define the ship's position, e.g. a position line from an observation of the sun might be crossed with a bearing obtained from a radio direction finder with possibly a third position line obtained from one of the hyperbolic position fixing systems.

The vessel's position is eventually 'fixed' at the intersection of a number of position lines which may be derived from:

1 Visual sources: compass bearings, transit bearings (no problems with compass error), horizontal sextant angles, vertical sextant angles
2 Astronomical sources: meridian altitude, giving latitude; 'Marcq St Hilaire' and 'longitude by chronometer'
3 Radar: bearings, ranges
4 WT/DF: bearings
5 Hyperbolic systems: Decca, Loran C and Omega
6 Satellite navigation systems. In this case there is a direct readout of position, which is derived mathematically within the equipment by integrating a series of position lines.

Whatever the source of the position line, certain basic principles apply:

1 There should always be at least three independent position lines, the third being necessary as a check against blunders in either of the other two.
2 The angle of cut between any two position lines should never be less than about 30°.
3 The source upon which the position line is based should be established beyond question. This is most important when fixing position by radar where the plan appearance of the coastline might differ from the chart (see Sections 5.7.1 and 5.7.2). Visually, navigation marks should be identified by their colour, shape and, especially at night, by the characteristic of their lights.

While the optimum is not always possible, it is the assessment of the shortfall which forms the basis for safe navigation. It is essential that the navigator is aware of any limitations and the possible consequences thereof, and reacts accordingly. Also he must act to minimize any known deficiencies in the systems he is using, in spite of the possible extra effort required. Two examples of this are interchain fixing with Decca, and differential Omega.

In spite of the best laid plans, the time will arrive when position fixing as such is not possible. However, this does not mean that the vessel cannot proceed safely. Examples are:

1 The use of transits and leading lines (Figure 20.3(a))
2 The use of clearing lines (Figure 20.3(b))
3 The use of a single position line (Figure 20.3(c))
4 Transferring a position line as in a running fix, which may be based on terrestrial or celestial position lines (Figure 20.3(d)).

20.1.4 Parallel indexing and radar maps

The frequency with which the ship's position is fixed depends upon many factors, e.g. proximity to danger, wind and current/tide conditions, available means of fixing position, etc., and intervals of 15 to 30 minutes between fixes are usually quoted. In many cases this can be far too long. However, it is not so much of the ship's position that is important but how the position is changing, and this needs to be assessed on a minute-to-minute basis. It is just such a monitoring service that parallel indexing (PI) provides (see also Sections 5.7.2.2, 5.9.3, 14.6.3).

20.1.4.1 Parallel indexing

The advantages of PI are:

1 That it confirms that the vessel is safe from moment to moment;
2 That any trends become obvious very quickly;
3 That it reduces the manual workload at a crucial time.

An indirect advantage is that the work needs to be prepared in advance. This ensures that the navigator does study the problems prior to meeting them – a task which is easily overlooked under the weight of other immediate and more pressing chores (see Notice M.1158, in Section 5.11.5).

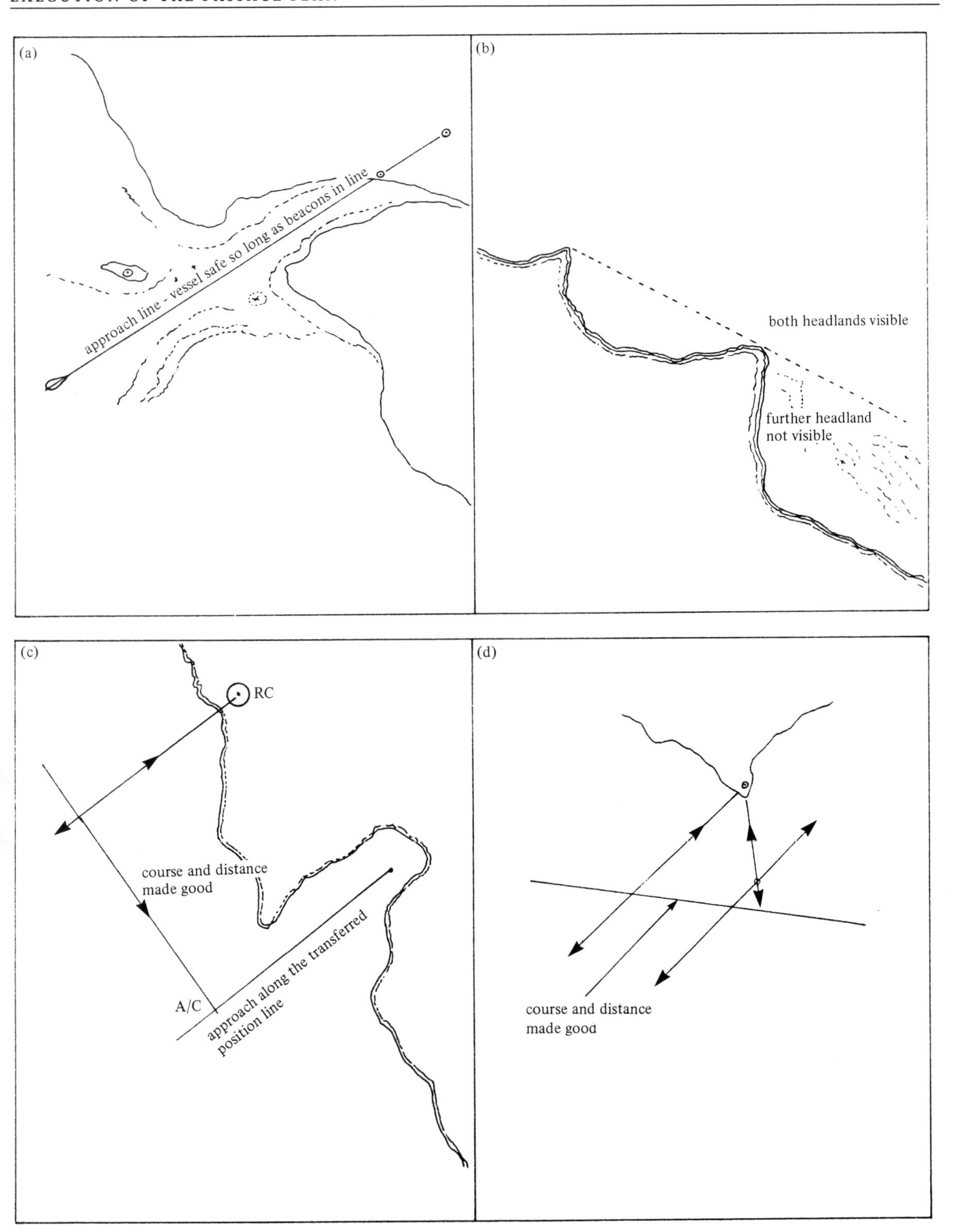

Figure 20.3(a) Leading mark approach. **(b)** Clearing line. **(c)** Use of a single position line. **(d)** Running fix

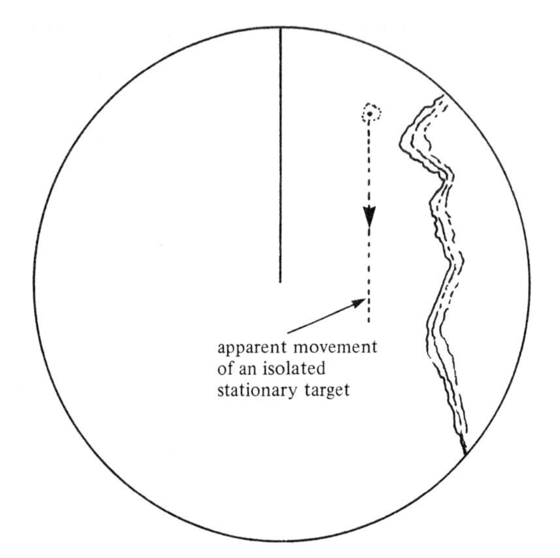

Figure 20.4 Parallel indexing principle

The basic principle (Figure 20.4) as applied to a relative motion radar display is as follows. If an isolated land target can be identified (e.g. a light vessel with a Racon), and if a line is drawn through it on the reflection plotter, parallel with the heading marker, then the target should move along that line. Any tendency for the target to move off the line is an indication that the vessel is being set off her intended track, and an early correction can be applied to prevent any further set, or to return the vessel to the required track.

This basic principle can be extended to prepare much more sophisticated PI plans (Figure 20.5).

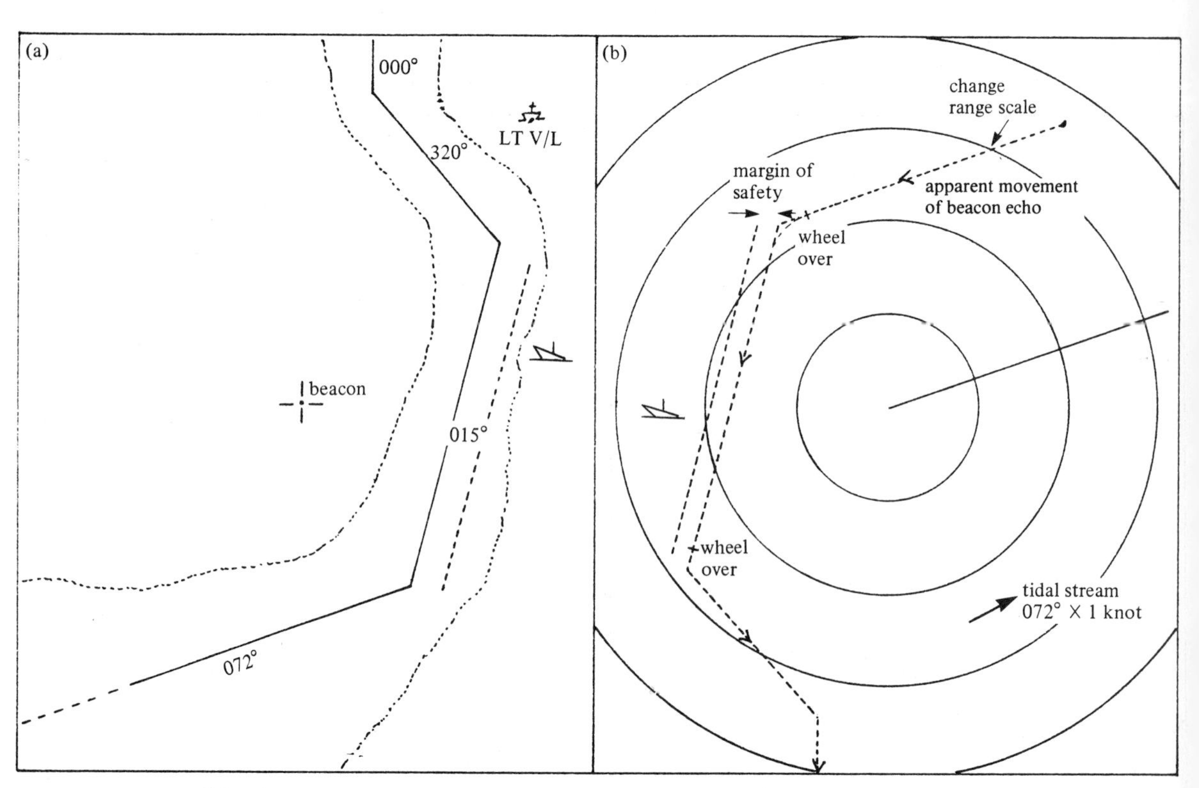

Figure 20.5 Parallel indexing. **(a)** The chart. **(b)** Relative motion display and reflection plotter

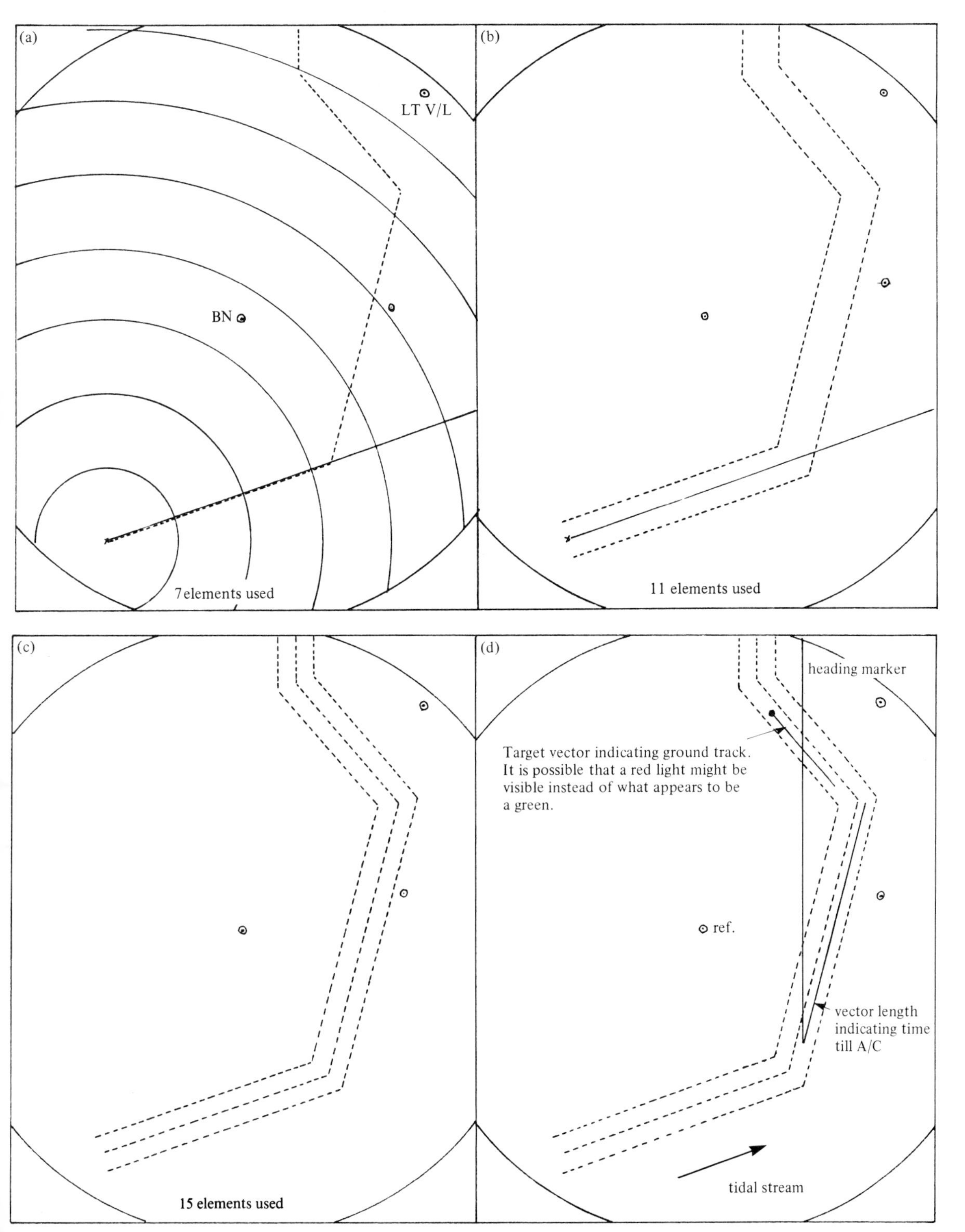

Figure 20.6 Radar Maps. **(a)** Tracklines and registration marks. **(b)** Tram lines and registration marks. **(c)** Tram lines, track lines and registration marks. **(d)** Ground locked and experiencing tide

20.1.4.2 Radar maps

As the facilities on ARPAs have increased, it has become possible to provide some limited form of 'mapping' facility. This should not be confused with the eventual possibility of an integrated electronic chart. A typical 'mapping' facility will make available some 20 lines and/or dots which can be placed in such a way as to delineate some navigational features which the navigator considers important (see Section 5.9.3).

It is possible to produce the electronic lines on the radar display (Figure 20.6). These may merely take the form of course lines which it is intended to follow, or alternatively traffic separation scheme outlines. Whatever method is preferred, two conditions are essential:

1 They must be prepared in advance of the intended passage.
2 They must be capable of being 'registered' with *real* features being displayed on the radar screen. This facility, when used with automatic ground stabilization of a true motion ARPA display, can be of considerable value, but it does require a lot of preparation (see also Section 5.9.2).

It is possible (having once prepared the map) on some ARPAs to store up to sixteen maps in the memory for subsequent use.

20.1.5 Single system position fixing

In most areas of the world there will be more than one independent means of fixing position, but occasions will arise when it is overcast in mid-Atlantic, say, when the Omega is the only available means of fixing position. In such cases it is essential that positions obtained are monitored and any inconsistency investigated.

It is important to remember that all of the systems have some inherent flaw (see Part I), e.g.: hyperbolic systems – lane slip; satellite systems – DR update between fixes; radar – ambiguous identification, etc. Although the navigator may be using radar as his primary position fixing system, he must maintain a second system (if at all possible) in readiness:

1 As a means of cross-checking positions obtained from the primary system; and
2 As back-up for the primary system, in the event of its failure (see also Section 14.4).

Equipment is becoming available (e.g. the Decca MNS 2000) which incorporates a number of navigation system receivers – Decca, Loran C, Omega, Transit (and NAVSTAR) which display the vessel's position as derived from the system which it has been programmed to recognize as the most suitable for that area. When using this equipment, the navigator must know which system is being used to derive the displayed position, and also cross-check it with the other available systems for discrepancies.

20.1.6 System integration

With the advances being made in microprocessors, it is possible to obtain positions from a number of sources, and by using information known about the system operating under the existing conditions, to integrate all of the data and indicate the vessel's most probable position (Figure 20.7).

The main advantages of system integration are:

1 The best (angle of cut) position lines can be used.
2 There is a natural cross-checking taking place.
3 The systems are not usually susceptible to the same errors.

Partially integrated systems have been introduced in which, for example, Omega is used to update the vessel's position between satellite fixes while the satellite fix is used as a check for lane slip in the Omega system. Here, the systems are working in a complementary way.

True system integration involves much more than is indicated in Figure 20.7 and so is essentially something that must be done electronically, but it is important to appreciate what is happening within the machine. The method and degree of integration varies from one manufacturer to another, but most employ some form of 'filter' which in effect searches for and removes consistent errors from each system being integrated. In addition, there is invariably an alarm system which is activated when sudden differences are detected, e.g. lane slip, or where one

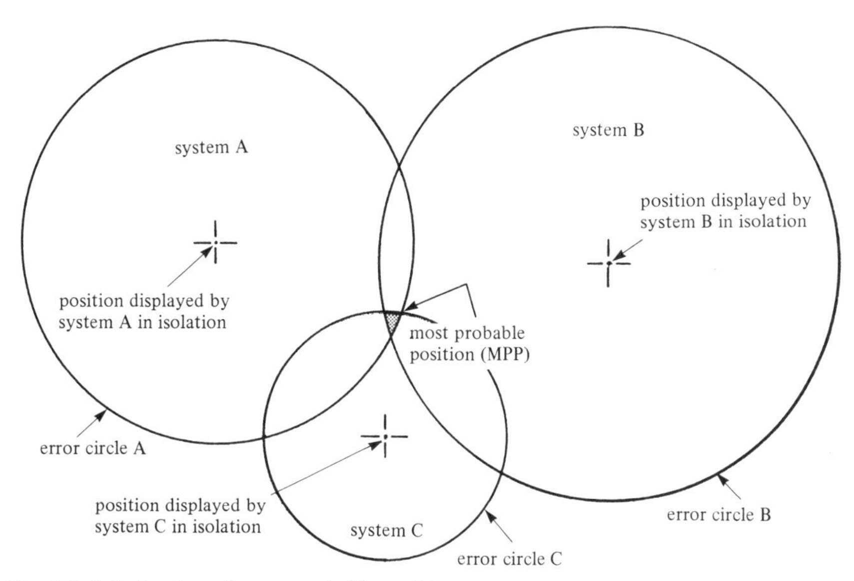

Figure 20.7 Simplified derivation of most probable position

system is markedly or consistently different from the others.

20.1.7 Errors and accuracy

Position lines, irrespective of source, are subject to errors of one form or another, and these need to be borne in mind when assessing what reliance can be placed on the indicated position.

20.1.7.1 The effect of errors in bearing

These are usually given in angular form, i.e. bearing 217° ± 2°, and here it can be seen that, as range increases, the positional uncertainty increases (Figure 20.8).

20.1.7.2 The effect of errors in range

These are usually quoted in the form of a percentage of range (e.g. with radar, as a percentage of the maximum range of the scale in use at the time) (see Section 5.10.1). Once again, positional uncertainty increases with range but in discrete steps. In both cases it is obvious that ranges and/or bearings should be taken from the nearest navigation marks.

20.1.7.3 The effect of a constant range error on position

Consider the variable range marker on a radar that is consistently reading say 0.2 nm high (or low), and this amount is not applied to readings before plotting on the chart (Figure 20.9).

It can be seen here that when using navigation marks on opposite sides of the vessel, along-track displacements of the plotted position result, while those from marks on the same side produce cross-track displacements. Where it is intended to clear some offlying danger, this can result in misplaced confidence in the clearing distance.

20.1.7.4 The cocked hat

Invariably when using three position lines, they will not intersect at a single point but will produce a 'cocked hat'. Much has been written about the vessel's most probable position in this event. Providing the resulting triangle is small, the traditional method of assuming that the vessel is at the centre is probably as good as any other, but the navigator should always be conscious of additional factors which might relate to a

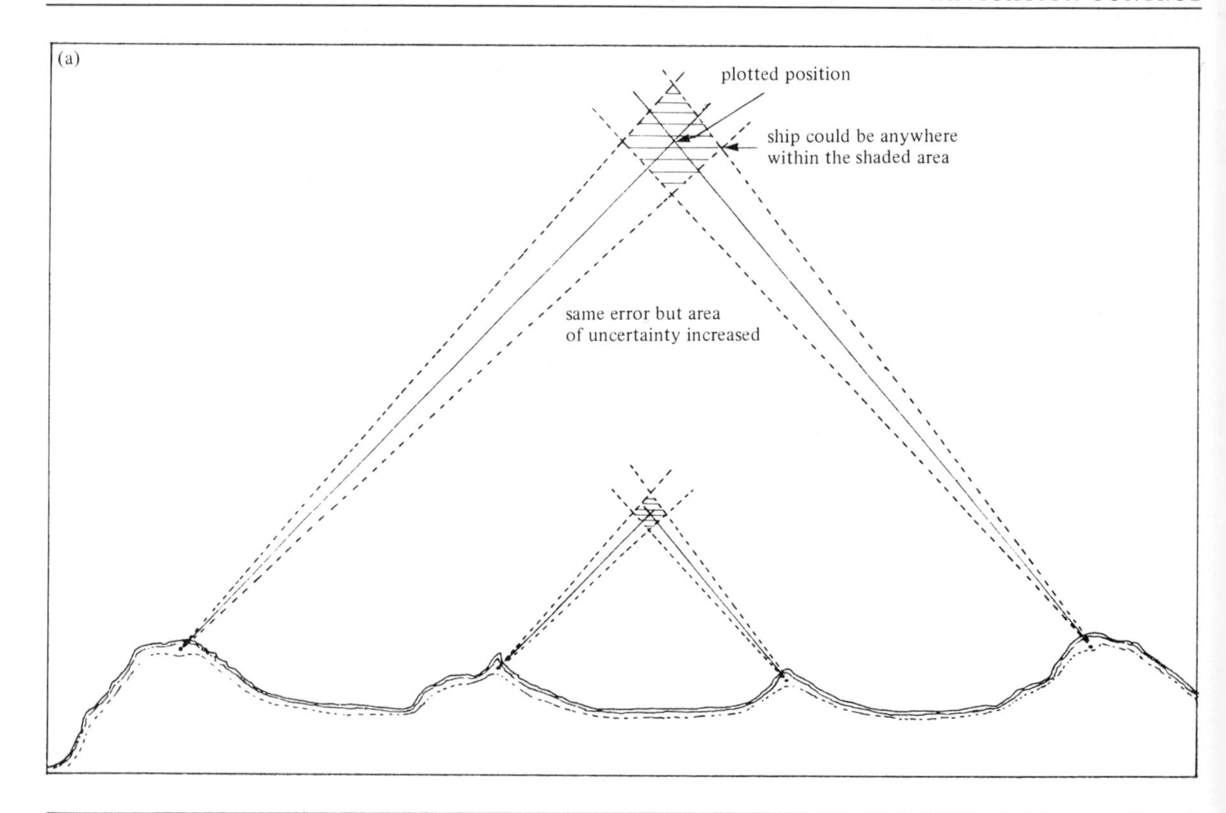

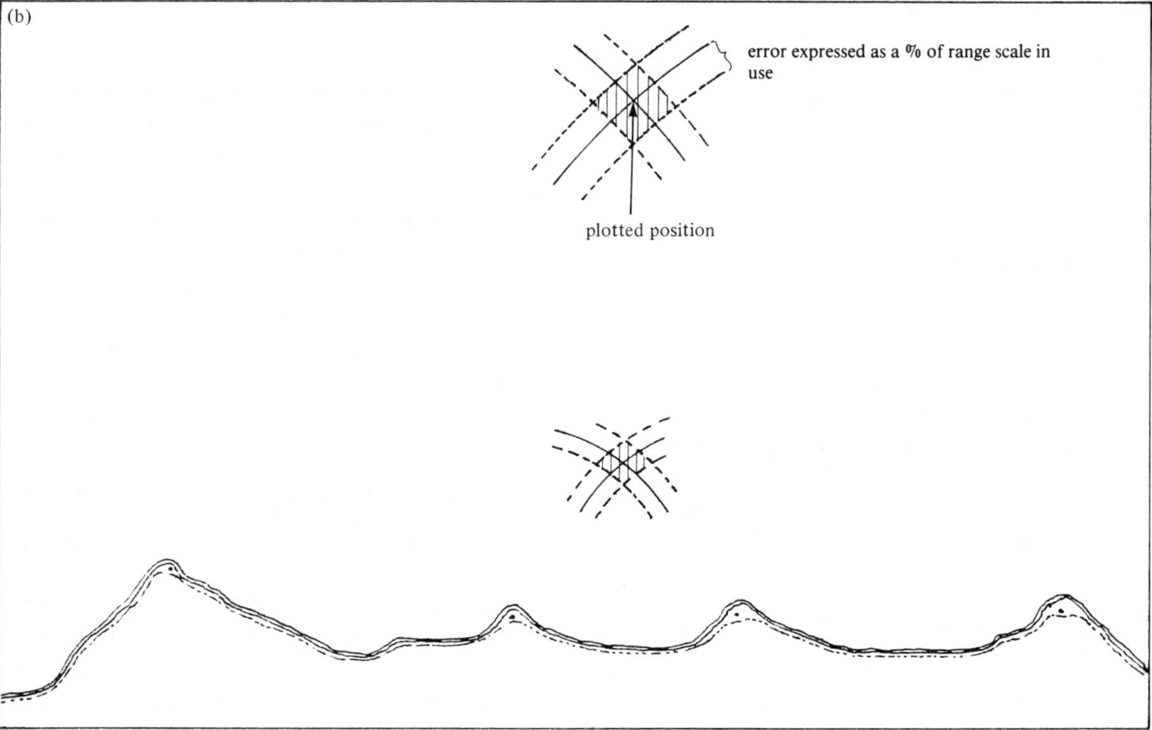

Figure 20.8(a) The effect of range on the area of uncertainty when using bearings. **(b)** The effect of range on the area of uncertainty when using ranges

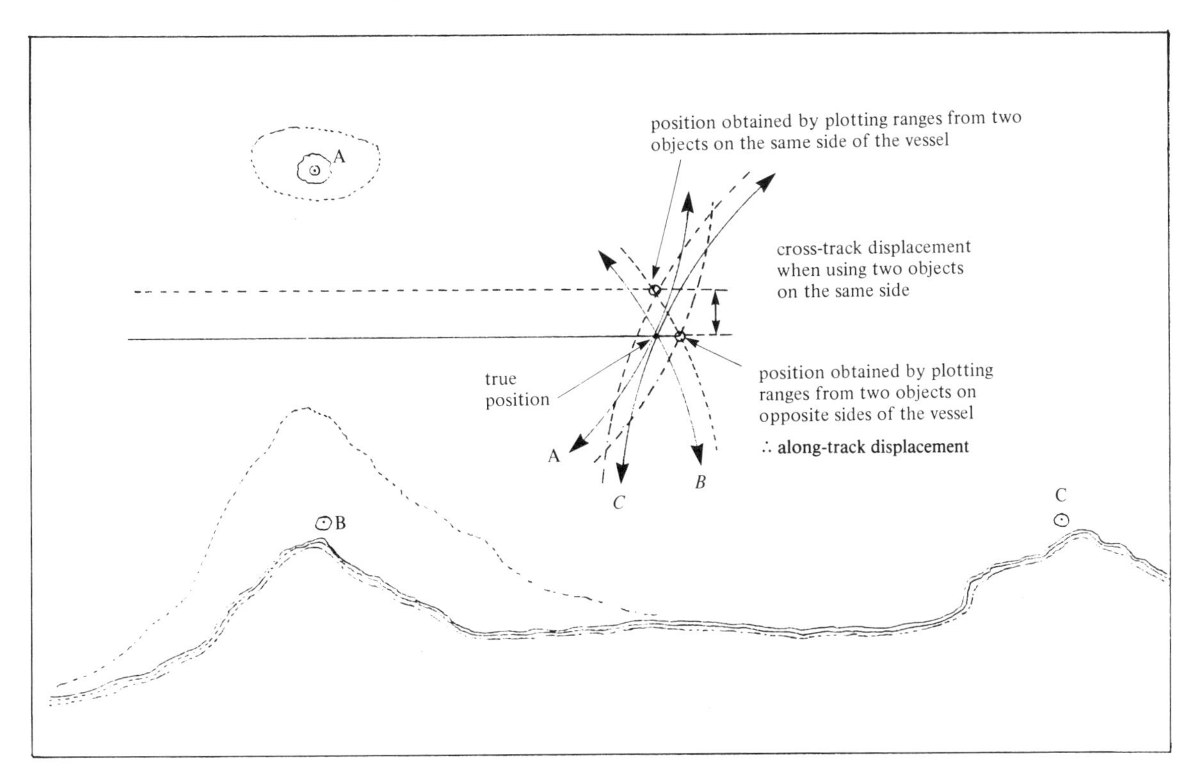

Figure 20.9 The effect of constant range error when fixing position

particular position line. For example, when measuring the sextant altitudes of stars, if the horizon was poor in the east (or west), then the mental confidence factor related to the position lines derived from stars in the east should be lower. Decca data sheets go some way towards providing this kind of information by indicating the two preferred colours for a fix.

This mental confidence factor should not be confused with the mental desire for the ship to be where one would like it to be, and many a navigator has been tempted to discard that position line which did not agree with his preconceived idea of where the ship should be. A large cocked hat should be a warning to check *all* the position lines, and not an excuse to discard the most embarrassing.

20.1.7.5 Errors in hyperbolic position lines

Fixed errors are usually quoted in hundredths of a lane and should be applied to the instrument readings before plotting (see Section 6.3), but the effects of variable errors are published in various ways. Hyperbolic system errors have been extensively investigated and, although not fully understood, where sufficiently reliable the information is published in the form of correction tables or data sheets etc. Unfortunately, most of the errors relating to hyperbolic systems stem from changing conditions in the ionosphere, which vary with time of day, time of year (which are somewhat predictable) and also with the sunspot cycle and solar flares (which are not).

20.1.7.6 Diamond, circle and ellipse of error

As a result of investigations into the position displayed by a system at a place with an accurately known position, it is possible to statistically analyse the results and indicate the area within which the true position is most likely to lie. For example, if a receiver is placed at some position and the readings noted and

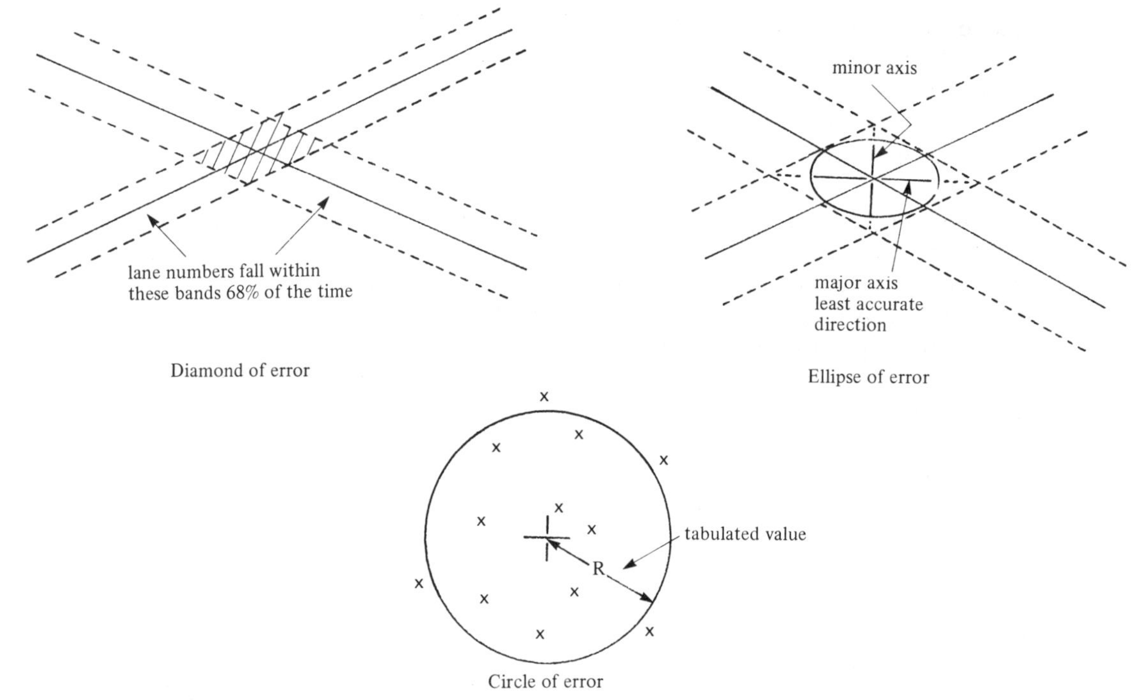

Figure 20.10 Diamond, circle and ellipse of error

plotted, then the occasions on which the receiver value will lie within certain values can be predicted.

If the scatter diagram is analysed, the information may be presented in one of a number of ways (Figure 20.10).

It is normal to quote a single accuracy level at 68 per cent (i.e. statistically, one standard deviation) which means that the vessel's true position will lie within the defined area on two out of three occasions. Implicit in this statement is the 95 per cent (i.e. nineteen out of twenty occasions) probability level, which would mean doubling the tabulated dimensions, i.e. if a variable error of say 3 cables is quoted for a Decca position, then at 95 per cent probability the circle to be drawn around the plotted position would have a radius of 6 cables.

20.1.7.7 Accuracy – absolute and repeatable

Until the advent of higher precision position fixing systems, this problem was really only faced by surveyors. Now the question arises much more frequently. Basically the question posed is: if one places, say, a Decca Navigator receiver and a satellite receiver at precisely the same position and have them both indicate the position in latitude and longitude, why do the readings differ?

Absolute accuracy assumes the existence of some 'standard' co-ordinate system to which all positions – however obtained – can be referred. This in turn assumes a common perception and knowledge of the shape of the earth. In fact, this is a long way from the truth, and those once staunch pillars of navigation – latitude and longitude – are found to crumble in the face of maritime 'relativity'. Accuracy quoted in terms of latitude and longitude is proving unsatisfactory, and so when one reads of a system having a particular accuracy, it is generally understood to mean repeatable accuracy.

Repeatable accuracy is a measure of the ability to return to the same position, the position being quoted in system co-ordinates. If the Decca co-ordinates are noted when a vessel is in a particular position, and then at some later date the vessel is positioned such that the co-ordinates are the same as those originally noted, how far will the vessel be from the original position? It is this value which is the basis of the accuracy quotation, and is a very much more meaningful value.

When using a position fixing system that displays the vessel's position in latitude and longitude, it is important to know to which grid co-ordinate system they refer, e.g. WGS72, Hayford, etc. (see Section 9.4). Although it is extremely convenient to work in latitude and longitude, and in most cases the accuracy will be adequate, for best accuracy (i.e. to achieve the quoted system accuracy) it is necessary to work in system co-ordinates. For how long this will continue to be possible is uncertain, as with the advent of equipment reading out latitude and longitude there has been a marked decline in the demand for lattice charts. There is no doubt that in the near future the hydrographic offices will have to look at the viability of continuing to produce such charts.

20.1.8 Checking for discrepancies

Despite the self-testing routines, warnings and alarms fitted to present-day navigational equipment, one of the prime duties of the officer of the watch is to monitor the proper working of the equipment. Every opportunity should be taken to apply critical tests, e.g. the availability of a transit bearing to check the error of the compass should not be allowed to pass unused, a time signal to check the chronometer error, etc. Checking positions from one system against those from another might not indicate which is in error, but should prompt further checks when a discrepancy is found.

At no time should inconsistent information be ignored or available data not used, simply because that system is out of favour.

20.2 Minor adjustments to the plan

It is important not only to know where the vessel is at every instant, but also to appreciate what effect the

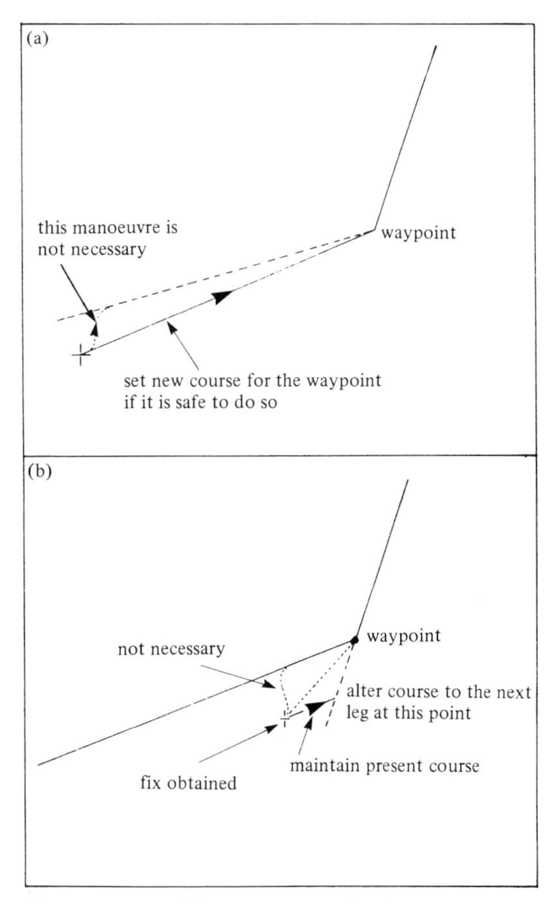

Figure 20.11 Manoeuvre to make the next waypoint

forces acting on her are likely to have, and to apply the appropriate corrections when necessary (see also Section 16.1.1).

When the effects of tide, leeway, etc. are different from those anticipated in the plan, the vessel will depart from the intended track, and it is then necessary for the navigator to act. In the majority of cases this will mean making some adjustment to the course steered, but a slavish attempt to maintain track can be unnecessary and in some cases undesirable. For example if, after some time when it was not possible to obtain a position, a fix indicates that the vessel is off track, all that may be necessary is to set a new course (making the appropriate allowances) for the next waypoint, and not to return to the original track. A vessel which is off track in the vicinity of a waypoint may find it more prudent to adjust the time and/or position of the course alteration than to try to regain the original track (Figure 20.11).

Parallel indexing and radar maps (see Section 20.1.4) can be extremely helpful in giving early warning of when minor adjustments to the plan are necessary.

21 Reacting to changed circumstances

From time to time events will occur which will demand a total reassessment of the passage plan (see Chapter 16), such as:

1 Bad weather
2 Restricted visibility and other factors causing delay
3 Changed orders
4 Distress (see also Chapter 17), emergencies (fire/accidents on board etc.).

21.1 Implementing the contingency plan

When the passage was originally planned, the likelihood of bad weather and restricted visibility should have been borne in mind and, based on this possibility, contingency plans prepared. For example, when making a winter passage in the North Sea, bays and anchorages with good holding ground which could provide shelter in the event of bad weather should have been noted in the original plan.

The same preparation should have been made for restricted visibility. Where it had been intended to use visual clearing lines and transits, alternative marks should have been noted for use with radar.

There should be a smooth transition from normal plan to contingency plan, once the decision to transfer has been made. There should be no need for an undue amount of extra work.

21.2 Reappraisal

Where a contingency plan is put into effect, the duration of departure from the original plan may be such that the original plan may no longer be appropriate. In this case, the situation will need to be reappraised with the new circumstances and conditions in mind.

21.3 Replanning

When, for example, orders have been changed and the vessel has to be diverted to another destination, quite a lot of work may have to be done (and decisions made quickly) to prepare a new plan for immediate implementation. Because of the somewhat 'urgent' nature of the replanning exercise, the following checklist might prove helpful:

1 Distance to the destination via alternative routes.
2 Tidal streams to be met on each route.
3 Traffic/shipping concentrations.
4 Are the necessary charts, light lists, etc. on board, and are the corrections up to date?
5 Is there sufficient depth of water on arrival, or will it be necessary to anchor?
6 If there is insufficient water on arrival, should speed be reduced to achieve a suitable arrival time?
7 Is there any urgency in arriving, e.g. medical assistance essential?

8 Are navigation marks and position fixing systems adequate, e.g. for night-time arrival, or is a daylight approach necessary?
9 Do the sailing directions advise any necessary precautions?
10 What is the port approach/entry procedure with regard to documentation, customs, immigration, health clearance, etc.?
11 What is the procedure for obtaining a pilot, e.g. advice time, position of pilot station, VHF calling and working channels, boarding procedures and methods?
12 Are there any circumstances which might have an effect on the plan, e.g. port closed during NW gales, vessels should stand out to sea; entry for vessels over 100 000 tons not permitted at night, etc.?

Having prepared the new plan, it is essential that all key personnel involved in its implementation – both aboard and ashore – are informed of the changes. These might include:

1 The chief officer – to warn the crew and prepare the ship to meet the changed requirements.
2 The chief engineer – to have the engines ready to manoeuvre when required.
3 The purser – to prepare the necessary documentation and advise passengers, stewards, cooks, etc.
4 The radio officer – to be ready to communicate with owners, agents, port authority and pilots as required.
5 The officer of the watch.

Because of the limited time for all of this work, it is important that the various elements are understood, and no aspect overlooked.

21.4 Upgrading the bridge team

On occasions during the voyage, the work demanded of the officer of the watch will increase to a point where it is necessary for the bridge manning to be increased (see also Section 18.1). Such occasions might include:

1 Restricted visibility
2 Emergencies, e.g. fire, damage to the vessel or cargo, injuries aboard due to accidents, distress calls
3 Port approach.

The personnel used to upgrade the bridge team will vary, depending on the circumstances. For example, the master and OOW may be all that are needed for port approach, whereas in a distress situation in fog an officer employed full-time on the radar and the radio officer conducting the communications may be required, in addition to the officer of the watch and master.

The period for which the upgrading may be required must also be borne in mind. Where the requirement for additional manning is likely to be of short duration but possibly very demanding, it is advisable to have extra personnel briefed and standing by. Where it is likely to be of longer duration, care must be taken to ensure that officers are not unnecessarily overcommitted and that the upgrading can be sustained, say, for several days in restricted visibility.

Two important points when upgrading the bridge team are:

1 There should be proper hand-over and briefing procedures (the ICS Bridge Procedures Guide provides valuable advice on this matter; see Section 11.17).
2 All members of the bridge team should fully understand the role which they are expected to perform. Where there is any doubt, the point should be clarified, and it should never be assumed that someone else will automatically be attending to it.

21.5 Port approach procedures

This will vary somewhat and will depend on both the specific features of the vessel and the port, but some general points can be made (see also Sections 13.4 and 15.3).

Details of the approach should have been planned in advance and personnel involved should have been briefed as to what will be required of them (see Section 21.3). The officer of the watch should have clear instructions when to take particular actions, e.g.: warn the engine room when 10 miles from pilot station; call the master at position marked on the chart; call Mersey Radio on channel 16 at 0330 hours and advise ETA at the pilot; etc. Other activities such as calling the crew, preparing the pilot ladder, clearing the anchors and testing equipment should be spelled out.

When the master comes to the bridge, he should make it quite clear when he is taking the con, and there should be a proper hand-over of information. Here again the ICS Bridge Procedures Guide provides valuable advice. These procedures should really be second nature, inculcated from the first occasion an officer ever goes on the bridge. They may be summarized as follows:

1 Navigational status – course, speed, position, traffic, tide, visibility, etc.
2 Port/pilot information – ETA, pilot requirements/boarding arrangements, etc.
3 Ship status – engineers warned, anchors cleared, etc.
4 Special information – e.g. buoy out of position etc.

On picking up the pilot, there should again be a proper exchange of information, with the pilot informing the master what arrangements have been made for his vessel at the port, i.e. which berth, locking arrangements, tugs, arrangements for fuelling and water, etc. The master in turn should hand-over

1 Navigational status – course, speed, telegraph setting, draft, etc.
2 Ship status – preparedness for docking etc.
3 Special information – critical speeds, engine peculiarities, etc.

Again the ICS Bridge Procedures Guide provides valuable advice and also a proforma for more detailed data about the ship, which may be handed to the pilot for subsequent reference as required. There is a tendency to regale the pilot with the vessel's history at a time when it is not necessary and when in any case he is unable to assimilate it. This should be avoided at all costs.

A common attitude which exists is that once the pilot is aboard, he can do the rest. This is neither correct nor proper. The ultimate responsibility rests aboard the ship and, therefore, it is the duty of the ship's personnel to monitor the vessel's progress right up to the berth and to be prepared to question and, if necessary, countermand orders given by the pilot. While one has come to expect a high standard of professionalism and competence among pilots, this is not necessarily the case worldwide.

22 Collision avoidance

In general, a feature of good navigation control is that a passage is planned in such a way that, except in the case of unforeseen circumstances, the critical phases of navigation will arise at a time which has been substantially anticipated. By contrast, the need to take action to avoid collision may occur at any stage of the passage and cannot be anticipated except in the most general terms of known areas of high traffic density. Any action or procedure associated with collision avoidance must be based on compliance with the International Regulations for Preventing Collisions at Sea (see also Section 11.24). This text is based on the 1972 Regulations as amended to the 1989 Amendment that came into force on 19 April 1991, and is particularly concerned with the conduct of vessels as governed by the Steering and Sailing Rules (part B of the regulations). The titles of these rules are listed in Section 22.3.1. While part B, section I applies to all conditions of visibility, Section II applies only to vessels in sight of one another, and section III applies to vessels in restricted visibility. It is thus expedient to consider clear visibility and restricted visibility separately.

22.1 Clear visibility

In the clear visibility situation, every vessel must comply with the general rules of conduct set out in section I of part B (see Section 22.3.1). In the event of risk of collision with other vessels which can be sighted visually, a vessel must give way or stand on as required by whichever rule in section II deals with the specific type of encounter. The application of the clear weather rules can in itself provide sufficient material for an entire book. It would be inappropriate in a volume devoted to navigation control to attempt to emulate, in half a chapter, the several eminent works already available which deal extensively and solely with rule of the road (see Section 22.4). In this section, it is merely intended to discuss the general practice of clear weather collision avoidance in the context of navigation control. In Section 22.2 somewhat more detailed consideration will be given to the case of restricted visibility because of its extensive relevance to radar and ARPA.

22.1.1 Look-out

Rule 5 requires a proper look-out to be kept by sight and hearing as well as by 'all means available'. There can be no doubt that the ability of radar to detect targets and provide collision avoidance data will in clear weather (as well as in restricted visibility) assist the mariner in making the required full appraisal of the situation and of risk of collision. This is particularly relevant to conditions of darkness and heavy traffic. In the keeping of a look-out, whether visual or radar, due attention must be given to the existence of blind or shadow sectors. For reference purposes, the full text of part I of the Steering and Sailing Rules (which apply in *all* conditions of visibility) is presented in Section 22.3.1.

22.1.2 Safe speed

Rule 6 requires that every vessel shall, at all times, proceed at a safe speed so that she can take proper and effective action to avoid collision and be stopped within a distance appropriate to the prevailing circumstances and conditions (see Section 22.3.1). The rule then goes on to list factors to be considered in determining a safe speed, and in so doing devotes a complete section to those factors which should be taken into account by vessels with operational radar. It is important to appreciate that the instruction to consider radar factors applies in *all* conditions of visibility.

The question of what is a 'safe speed' has been the subject of learned legal discussion. Attempts have been made to produce a formula from which a value could be calculated.

It would be inappropriate in a volume of this sort to enter into such academic debate. The number of knots which represents a safe speed is a matter of judgement in any given circumstance (courts frequently criticize vessels for excessive speed but seldom, if ever, say what a safe speed would have been). Thus, the mariner must, in a seamanlike manner, evaluate the factors listed in the rule in relation to the circumstances in which he finds himself and, using his professional judgement, adjust his speed accordingly. If he does so properly and continually reviews the position, then he has done all he can to ensure that a collision does not result because of *his* excessive speed. Should a collision result for other reasons (e.g. irrational behaviour on the part of the other vessel) then his speed will be eminently capable of defence in terms of the seamanlike way in which he reached his decision. Although the case of excessive speed is frequently debated, the possibility that a vessel's speed might be too slow to be considered safe should not be overlooked.

22.1.3 Risk of collision

Rule 7 (see Section 22.3.1) requires that every vessel shall use all available means appropriate to the prevailing circumstances and conditions to determine if risk of collision exists. In clear weather, the traditional method of establishing risk of collision is to carefully watch the compass bearing of an approaching vessel, and such method is given due prominence in the rule. However, it is important to note that rule 7(d) makes it quite clear that this is only one of the considerations to be taken into account, thus reinforcing the phrase 'all available means' appearing in rule 7(a).

Rule 7(b) specifically requires the proper use of radar equipment if fitted and operational. It must be stressed that this applies to all conditions of visibility. In clear weather there will be many occasions on which proper use means continuous operation of the equipment, together with observation of the display and extraction of data to complement that available by visual means. The ability of the radar to make CPA and TCPA readily available clearly makes a major contribution to the determination of risk of collision in any state of visibility. This is of particular relevance in conditions of darkness and heavy traffic (see Notice M.1102 in Section 18.3.1; and Notice M.1158 in 5.11.5).

Historically there has been a school of thought among some shipmasters which felt that the radar should not be switched on without good reason. The 1972 Regulations make it quite clear that you need a very good reason to switch it *off*.

22.1.4 Avoidance manoeuvres

Any manoeuvre to avoid collision in clear weather must comply with the general requirements set out in rule 8 (which applies in all conditions of visibility) and the specific requirements of the appropriate rule(s) in part II of the Steering and Sailing Rules.

22.1.4.1 General requirements

Rule 8 sets out general instructions for all action to avoid collision in all circumstances. In its application to the clear weather situation the rule is quite straightforward and self-explanatory. Some further discussions of the rule will be presented in its application to the case of restricted visibility (see Section 22.2.6).

22.1.4.2 Specific requirements

Where risk of collision exists in a specific encounter the observing vessel must, in accordance with the nature of the encounter, give way or stand on as required by the appropriate rule in section II (see Section 22.3.1) which deals with that specific type of encounter. Further, her action in giving way or standing on must be in accordance with the general instructions of rule 16 (if giving way) or rule 17 (if standing on).

Where only two ships are involved, in the open sea, the application of part II of the Steering and Sailing Rules is, in general, simple and straightforward. In congested waters, very complex situations can develop. For their solution, a thorough knowledge of the rules allied with the ability to plan and execute effective strategy is required. For detailed consideration of the tactics and strategy of manoeuvres associated with part II of the Steering and Sailing Rules, the reader is referred to existing specialist texts dealing extensively with this subject (see Section 22.4).

22.2 Restricted visibility

In restricted visibility every vessel is required to comply with the general rules of conduct set out in section I of the Steering and Sailing Rules (Section 22.3.1) and with the particular rules of conduct for vessels in restricted visibility set out in section III of the same rules. For ease of reference, rule 19, which is the only rule in that section, is reproduced in Section 23.3.1. It should be noted that it applies to vessels not in sight of one another when navigating *in or near* an area of restricted visibility.

22.2.1 *Look-out*

Rule 5, which has been discussed in Section 22.1.1, applies in all conditions of visibility. However, due regard must be paid to the prevailing circumstances and conditions of restricted visibility when complying with this rule (and all the rules of section I – see rule 19(c)). While stressing that the need to maintain visual and aural look-out remains, it is necessary to give detailed consideration to the circumstances in which the presence of another vessel is detected by radar alone.

22.2.2 *Safe speed*

Rule 6 requires vessels to proceed at a safe speed in all conditions of visibility, and this has been discussed in Section 22.1.2. Additionally, rule 19(b) instructs that such speed should be adapted to the prevailing circumstances and conditions of restricted visibility, and that a power-driven vessel shall have her engines ready for immediate manoeuvre.

In the circumstances in which other vessels are detected by radar alone, it is of particular importance that the speed chosen must take into account the rate at which collision avoidance data (see Sections 5.6 and 5.8.3) can be extracted and processed. The vessel which does not have time to carry out plotting, be it manual or automatic, is not proceeding at a safe speed.

22.2.3 *Risk of collision*

The obligation to use all available means appropriate to determine if risk of collision exists (rule 7(a)) applies in all conditions of visibility, as does the requirement to make proper use of radar equipment if fitted and operational (rule 7(b)). Rule 7 makes it clear that proper use includes long range scanning to obtain early warning of risk of collision, and radar plotting or equivalent systematic observation of detected objects. However, in restricted visibility, radar plotting (or its equivalent) becomes the prime, and in almost all cases, the sole source of collision avoidance data. Direct radio communication between positively identified vessels may make a contribution to collision avoidance, as discussed in Section 19.1.8 (see also Section 23.5).

22.2.4 *Radar plotting*

The Collision Regulations do not define 'radar plotting'. However, careful consideration of the instructions and cautions given in the rules, particularly those which specifically relate to radar, reveal that an effective procedure of collision avoidance data extraction is necessary in order to comply with the letter and the spirit of the regulations. Attempts have been made in some quarters to put forward a definition of the term radar plotting. The quest for such a definition is sometimes justified by the desire to anticipate some possible future legal interpretation of the term in an inquiry or court. It is not intended to pursue this path because there is ample evidence that the term does not lend itself to simple, precise definition. It is considered more helpful to recognize that radar plotting is a practical procedure which is inseparably entwined with the minute by minute application of the Collision Regulations in any encounter, and the extent to which the procedure must be followed is dependent on the way in which the encounter develops. For example, that which constitutes plotting in the case of another vessel whose range is 10 miles and increasing would be quite different to that required in the case of the target vessel whose range was 6 miles and decreasing. Accordingly, the remainder of this chapter (including Table 22.1 and Figure 22.1) gives an exposition of the practice of collision avoidance in restricted visibility, into which is woven the appropriate requirement to extract collision avoidance data. If this is followed, the navigator should not find himself in the position of requiring a legal definition of radar plotting. Should he, despite this, find himself in an enquiry due to the action of others, his general procedure of data extraction should be eminently capable of defence in so far as it constituted radar plotting, or equivalent systematic observation of detected objects.

The logical and systematic nature of the procedure is summarized, in the context of two widely differing plotting techniques, by Figures 22.1 and 22.2. Figure 22.1 is based on the assumption that collision avoidance data is obtained by manual plotting using a single relative motion presentation. By contrast, Figure 22.2 illustrates the same procedure where computer assistance is available. For completeness. Table 22.1 cross-references all elements of the procedure against the various plotting techniques described earlier in this volume (see Sections 5.6, 5.8 and 5.9).

Table 22.1 Systematic observation – summary of techniques

Requirement	*Considerations*	*Possible sources of information*
Proper look-out	Visibility, heavy traffic, darkness, clutter	Radar, ARPA (in addition to eye and ear)
To determine if a close-quarters situation is developing	CPA and TCPA (these must be considered in relation to position, manoeuvrability, visibility, traffic, speed and plotting facilities)	Relative plot, completed true plot, zero speed technique, appraisal aid, relative vector, data, PAD
To analyse the situation and consider possible manoeuvres	Aspect (and speed) of other vessels Recent target manoeuvres Manoeuvre must be seamanlike and in accordance with rule 19(d)(i), (ii)	True plot, true motion, completed relative plot, appraisal aid, true vector, data, manual plots, history
To determine the best manoeuvre	Forecast CPA and TCPA Manoeuvre must result in a safe passing distance. It must be positive, early and obvious to others	Forecast plot, dual displays, appraisal aid, trial manoeuvre, PAD
To ensure the effectiveness of the manoeuvre	Has the forecast been fulfilled? Has a target manoeuvred?	Forecast and continued relative plot, dual displays, appraisal aid, true and relative vectors
Resumption	Are all targets past and clear? Will they pass at a safe distance on resumption?	Forecast, dual displays, appraisal aid, trial manoeuvre, PAD

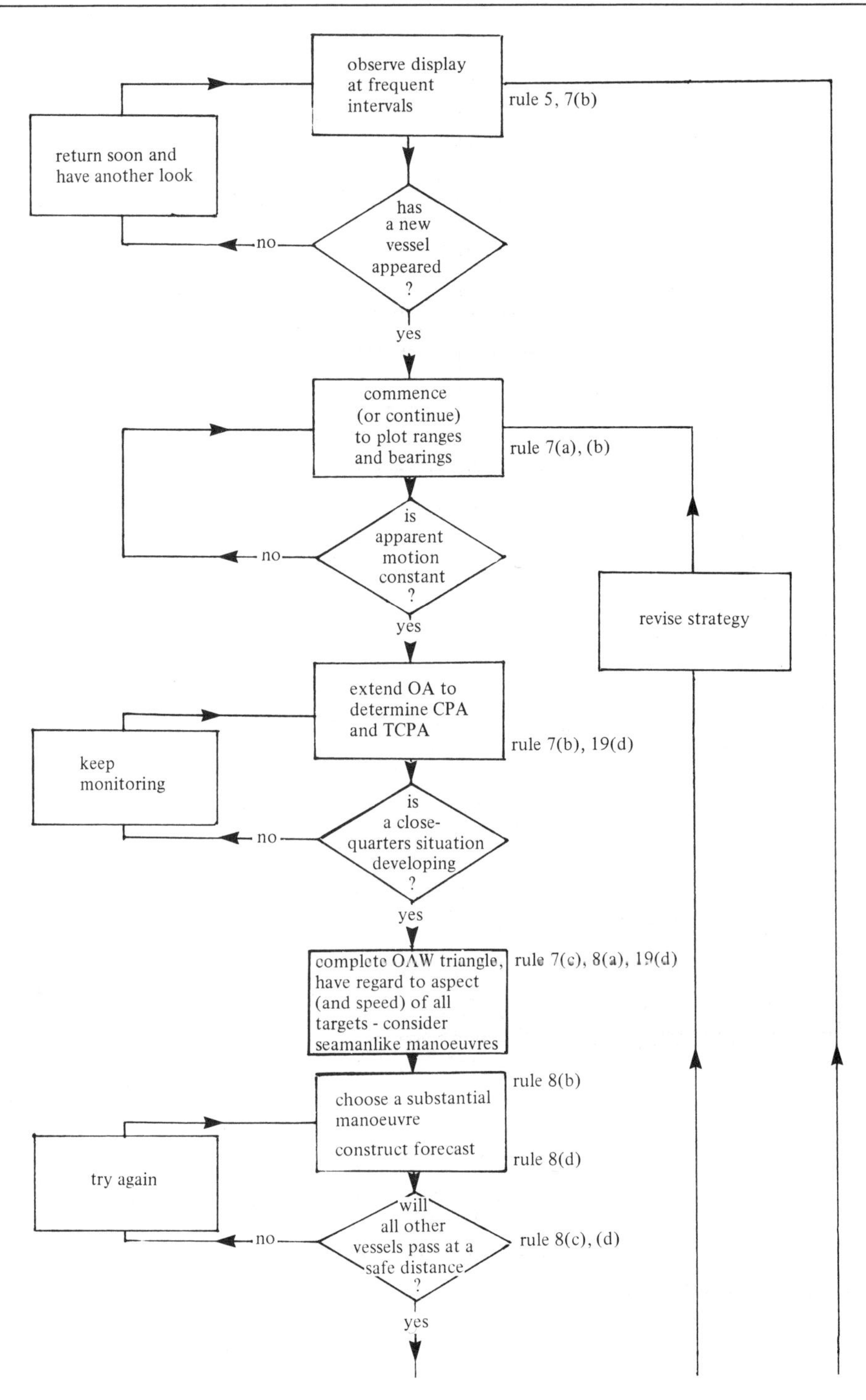
observe display at frequent intervals
rule 5, 7(b)
return soon and have another look
has a new vessel appeared ?
no
yes
commence (or continue) to plot ranges and bearings
rule 7(a), (b)
is apparent motion constant ?
no
yes
revise strategy
extend OA to determine CPA and TCPA
rule 7(b), 19(d)
keep monitoring
is a close-quarters situation developing ?
no
yes
complete OAW triangle, have regard to aspect (and speed) of all targets - consider seamanlike manoeuvres
rule 7(c), 8(a), 19(d)
choose a substantial manoeuvre
rule 8(b)
construct forecast
rule 8(d)
try again
will all other vessels pass at a safe distance ?
rule 8(c), (d)
no
yes

(continued)

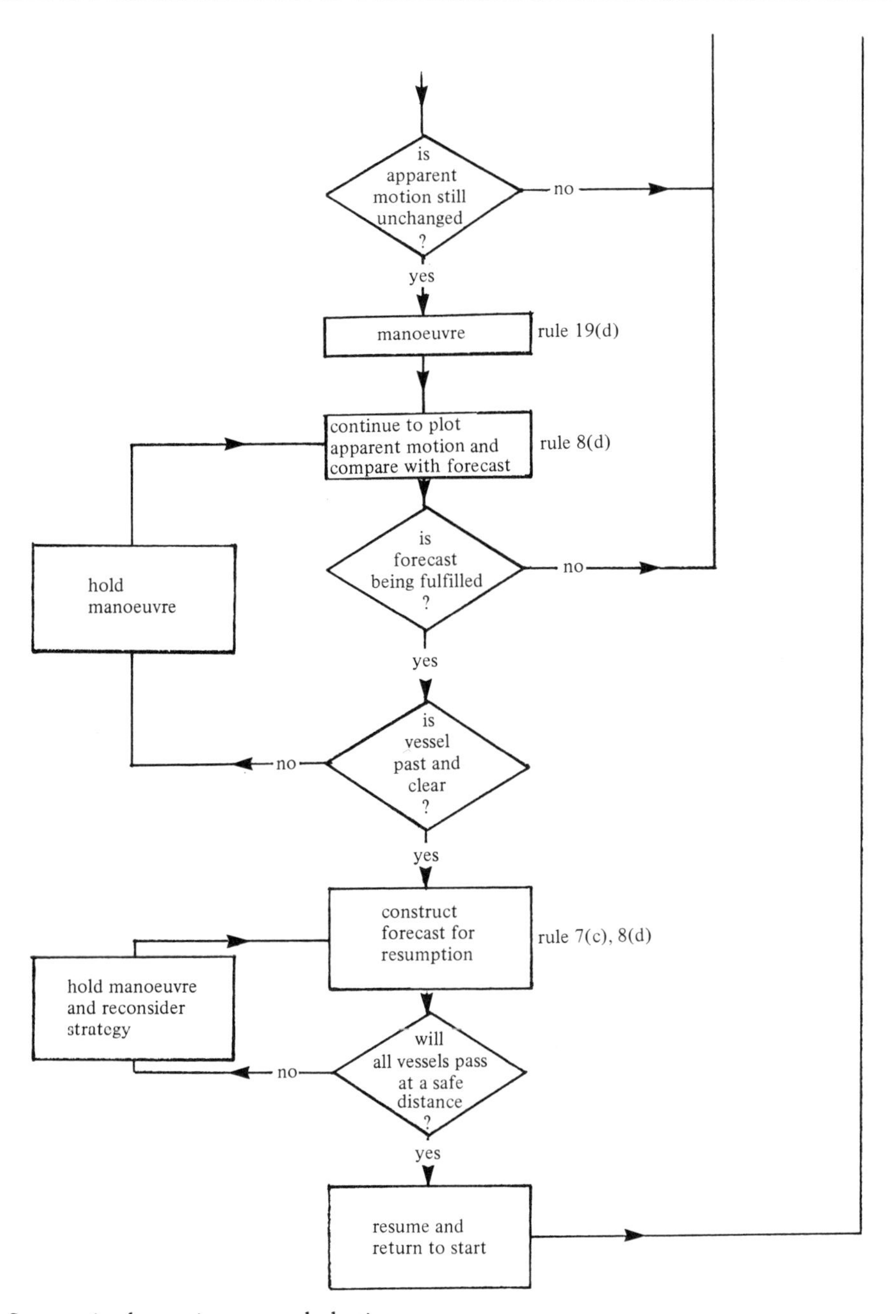

Figure 22.1 Systematic observation, manual plotting

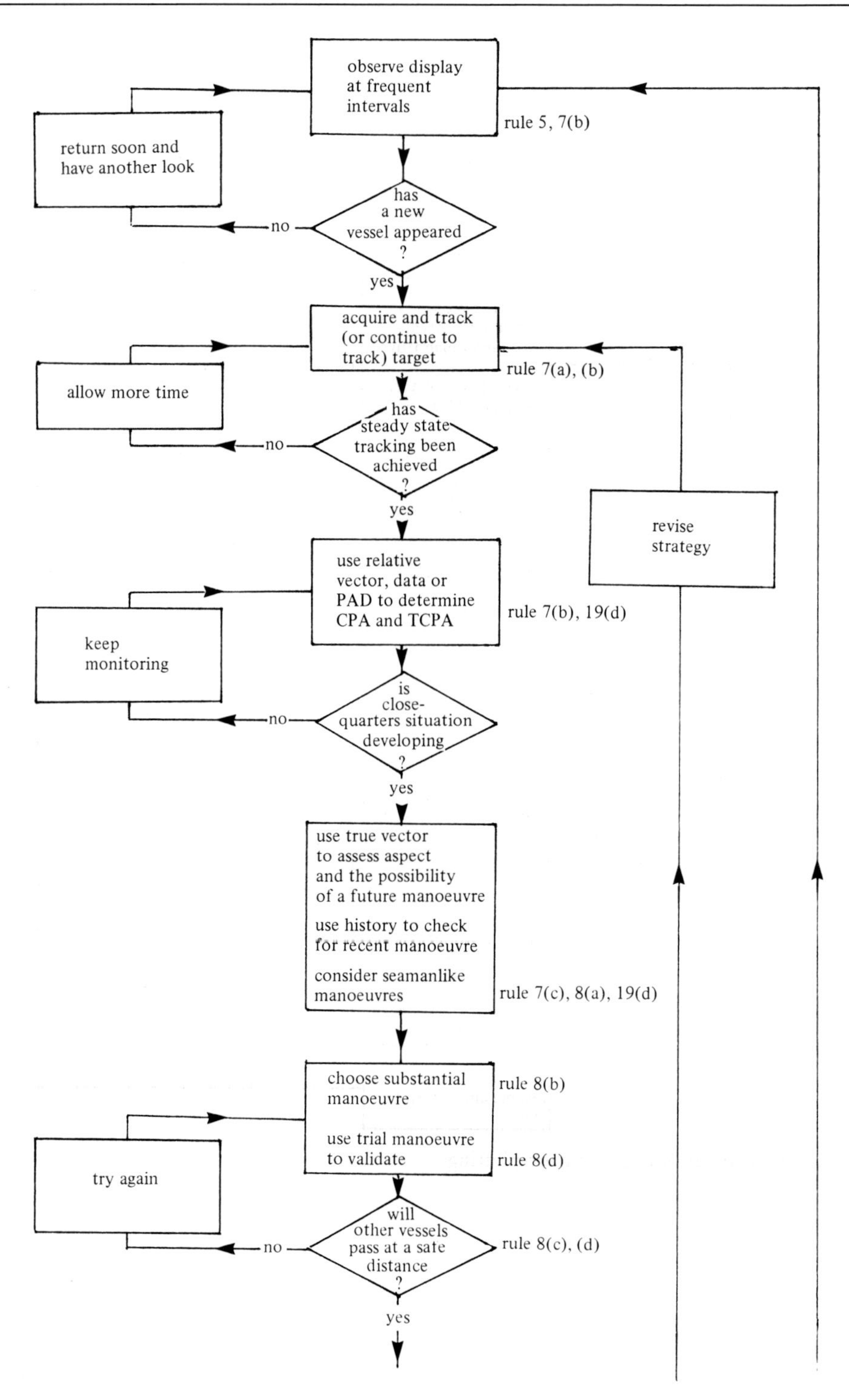
observe display
at frequent
intervals
rule 5, 7(b)
return soon and
have another look
has
a new
vessel appeared
?
no
yes
acquire and track
(or continue to
track) target
rule 7(a), (b)
allow more time
has
steady state
tracking been
achieved
?
no
yes
revise
strategy
use relative
vector, data or
PAD to determine
CPA and TCPA
rule 7(b), 19(d)
keep
monitoring
is
close-
quarters situation
developing
?
no
yes
use true vector
to assess aspect
and the possibility
of a future manoeuvre
use history to check
for recent manoeuvre
consider seamanlike
manoeuvres
rule 7(c), 8(a), 19(d)
choose substantial
manoeuvre
rule 8(b)
use trial manoeuvre
to validate
rule 8(d)
try again
will
other vessels
pass at a safe
distance
?
no
rule 8(c), (d)
yes

(continued)

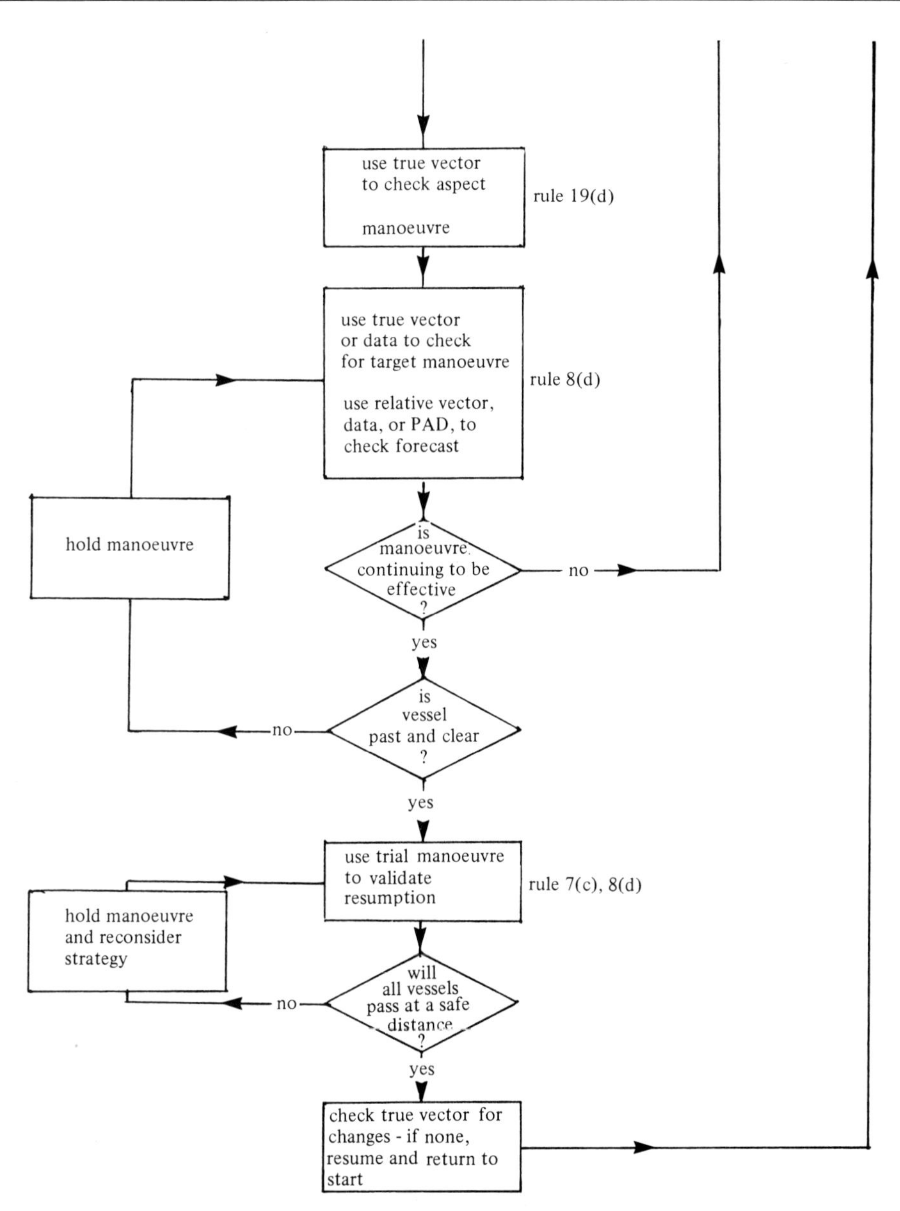

Figure 22.2 Systematic observation, automatic plotting

It should be remembered that the obligation to extract the data is universal and independent of the particular plotting technique employed (see Section 5.6). Clearly, the speed at which the data can be extracted is least where a paper plot is used and will increase progressively as more effective techniques are employed, culminating in the ability to quickly handle a large number of targets where computer assistance is available. In the context of rule 7 it is easy to defend the proposition that correct use of ARPA constitutes 'equivalent systematic observation of detected objects' (see Figure 22.2).

22.2.5 Development of a close-quarters situation

In cases where a target is detected by radar alone, rule 19(d) (see Section 22.3.1) places a specific obligation on the observer to determine if a close-quarters situation is developing. To comply with this requirement, the CPA and TCPA must be extracted. Having done this, it is the duty of the observer to decide if the predicted range at CPA constitutes a close-quarters situation.

The miss distance at which a close-quarters situation is considered to arise is a matter of individual judgement, and will depend on factors such as:

1 The geographical position
2 The handling capability of the ship
3 The density of traffic
4 The speeds of the vessels involved
5 The speed with which collision avoidance data can be extracted.

It must be appreciated that the other vessel(s) in any encounter may also be assessing the miss distance. Their perception of what constitutes a close-quarters situation may be quite different from that of the observing vessel. The unknown nature of the other vessel's handling characteristics is particularly relevant in this context.

Attempts have been made to define a close-quarters situation, but to date these have merely produced complex mathematical formulae which cannot compete, in the practical situation, with intelligent and seamanlike application of the above criteria.

If the decision is made that a close-quarters situation is not developing, no manoeuvre is necessary, but monitoring must continue to ensure early detection of any change in CPA.

22.2.6 Avoidance manoeuvres

If it is decided that a close-quarters situation is developing, the observer is required by rule 19(d) to take action to resolve the situation in ample time. Where the proposed action is an alteration of course, the rule requires that as far as possible, the following shall be avoided:

1 An alteration of course to port for a vessel forward of the beam, other than a vessel being overtaken
2 An alteration of course towards a vessel abeam or abaft the beam.

Clearly, implementation of 1 requires a knowledge of the heading of the other vessel. In general, any avoidance manoeuvre must be planned on a basis of a knowledge of the other vessel's true motion. This is necessary because it not only allows manoeuvring strategy to be based on the current conduct of other vessels, but also provides the observing vessel with the information needed to view the developing situation as seen by the other parties to the encounter. The latter allows evaluation of the possible manoeuvres open to the other vessels, and this must be taken into account when selecting an avoidance manoeuvre. In particular, consideration should be given to the significance of any recent manoeuvres which may, for example, indicate the possibility of resumption of a previous course.

Any avoiding action, not based on a knowledge of the true motion of all relevant targets, is unseamanlike and could certainly attract criticism as an assumption based on scanty radar information (rule 7(c)). Further, it would indicate failure to take account of paragraph 3.5 of Notice M.1158 (see Section 5.11.5), which states that choice of appropriate avoiding action is facilitated by knowledge of the other vessel's course and speed.

Rule 8(d) requires that action taken to avoid collision with another vessel shall be such as to result in passing at a safe distance. This rule applies to all states of visibility, but in restricted visibility the safe passing distance may well be greater than could be tolerated in good visibility. The plotting techniques which facilitate compliance with this rule are discussed fully in Sections 5.6, 5.8 and 5.9. Such techniques should also be used to ensure that the chosen manoeuvre does not result in a close-quarters situation with other vessels (rule 8(c)). It should be noted that good forecasting technique will frequently reveal the likelihood of targets manoeuvring to avoid each other. A further criterion for an acceptable manoeuvre is that it must be readily apparent to another vessel observing by radar (rule 8(b)). It is essential to remember that other vessels may be observing a single relative motion presentation and hence will be dependent on noting relative motion changes for warning of manoeuvres by other vessels. Thus, in judging whether or not an alteration is sufficiently large to be apparent, the relationship between the observing vessel's speed and that of the target must be considered. Further, it must be appreciated that even if the target vessel is equipped with radar, the plotting facilities may be very basic. Hence the rate at which he can extract data, and thus become aware of changes, may be slow and his ability to identify small changes may be particularly limited. (Thus rule 8(b) gives warning about a succession of small alterations.)

The possibility that a target vessel may not be plotting, or may not even have operational radar, must be continually kept in mind. The technique of calculating the course to steer for a specific passing distance must be used with extreme caution because it tends to promote small alterations, and this is contrary to the philosophy of the Collision Regulations. Further, there is the considerable probability that the CPA achieved will be less than that forecast.

22.2.7 *The effectiveness of the manoeuvre*

Rule 8(d) requires that the effectiveness of the action be checked until the other vessel is finally past and clear. This requirement is best satisfied by systematic monitoring of:

1 The true motion of all relevant targets to ensure early detection and identification of target manoeuvres;
2 The relative motion of all relevant targets to check for fulfilment of the forecast nearest approach.

This implies an initial plot and forecast, and not just a relative motion plot after completion of the manoeuvre by own vessel. The latter, while it might indicate that the target is passing clear, will not identify a target manoeuvre which in turn can affect the time of resumption.

22.2.8 *Resumptions*

When the target is finally past and clear, the decision may be made to resume course and/or speed.

As in the case of the avoidance manoeuvre, plotting techniques should be employed to verify that a safe passing distance will be achieved with respect to all relevant targets. It should be remembered that the resumption will be most obvious to other observing vessels if it is a single manoeuvre. The common practice of resuming course in steps by 'following a target round' will make it difficult for other observing ships to identify the manoeuvre positively and could be considered to be in contravention of the spirit, if not the letter, of rule 8(b). Before resumption, due consideration must be given to the possible reaction of other vessels. (When making alterations for navigational reasons, forecasting should also be used to assess the effect on the CPA of other vessels.)

22.2.9 *Conclusions*

In any potential collision situation in restricted visibility, the interpretation of radar information facilitates the determination and execution of action to avoid close-quarters situations. Traditionally the data was extracted by simple manual plotting but, with the passage of time, semi-automatic and automatic facilities have become progressively more available.

The navigator must be aware of the limitations placed on the procedure by the accuracy of the data (see Section 5.10). In all cases, the accuracy is such as to require large manoeuvres and large miss distances. Implicit reliance on the validity of small, non-zero passing distances should be avoided. The errors discussed in Section 5.10 clearly indicate that predicted passing distances of less than 1 mile should be treated with the utmost caution.

22.3 Extract from official publication

In Section 22.3.1, the paragraph numbering of the original publication has been retained.

22.3.1 International Regulations for Preventing Collisions at Sea (1972), as amended to the 1989 Amendment which came into force 19 April 1991

Part B: Steering and Sailing Rules

Contents of Part B

Section I: Conduct of vessels in any condition of visibility

Rule 5: Look-out

Every vessel shall at all times maintain a proper look-out by sight and hearing as well as by all available means appropriate in the prevailing circumstances and conditions so as to make a full appraisal of the situation and of the risk of collision.

Rule 6: Safe speed

Every vessel shall at all times proceed at a safe speed so that she can take proper and effective action to avoid collision and be stopped within a distance appropriate to the prevailing circumstances and conditions.

In determining a safe speed the following factors shall be among those taken into account:

(a) By all vessels:
  (i) The state of visibility
  (ii) The traffic density including concentrations of fishing vessels or any other vessels
  (iii) The manoeuvrability of the vessel, with special reference to stopping distance and turning ability in the prevailing conditions
  (iv) At night, the presence of background light such as from shore lights or from backscatter of her own lights
  (v) The state of wind, sea and current, and the proximity of navigational hazards
  (vi) The draught in relation to the available depth of water.

(b) Additionally, by vessels with operational radar:
  (i) The characteristics, efficiency and limitations of the radar equipment
  (ii) Any constraints imposed by the radar range scale in use
  (iii) The effect on radar detection of the sea state, weather and other sources of interference
  (iv) The possibility that small vessels, ice and other floating objects may not be detected by radar at an adequate range
  (v) The number, location and movement of vessels detected by radar
  (vi) The more exact assessment of the visibility that may be possible when radar is used to determine the range of vessels or other objects in the vicinity.

Rule 7: Risk of collision

(a) Every vessel shall use all available means appropriate to the prevailing circumstances and conditions to determine if risk of collision exists. If there is any doubt, such risk shall be deemed to exist.

(b) Proper use shall be made of radar equipment if fitted and operational, including long range scanning to obtain early warning of risk of collision and radar plotting or equivalent systematic observation of detected objects.
(c) Assumptions shall not be made on the basis of scanty information, especially scanty radar information.
(d) In determining if risk of collision exists, the following considerations shall be among those taken into account:
   (i) Such risk shall be deemed to exist if the compass bearing of an approaching vessel does not appreciably change
   (ii) Such risk may sometimes exist even when an appreciable bearing change is evident, particularly when approaching a very large vessel or a tow or when approaching a vessel at close range.

Rule 8: Action to avoid collision

(a) Any action taken to avoid collision shall, if the circumstances of the case admit, be positive, made in ample time and with due regard to the observance of good seamanship.
(b) Any alteration of course and/or speed to avoid collision shall, if the circumstances of the case admit, be large enough to be readily apparent to another vessel observing visually or by radar; a succession of small alterations of course and/or speed should be avoided.
(c) If there is sufficient searoom, alteration of course alone may be the most effective action to avoid a close-quarters situation provided that it is made in good time, is substantial and does not result in another close-quarters situation.
(d) Action taken to avoid collision with another vessel shall be such as to result in passing at a safe distance. The effectiveness of the action shall be carefully checked until the other vessel is finally past and clear.
(e) If necessary to avoid collision or allow more time to assess the situation, a vessel shall slacken her speed or take all way off by stopping or reversing her means of propulsion.
(f) (i) A vessel which, by any of these rules, is required not to impede the passage or safe passage of another vessel shall, when required by the circumstances of the case, take early action to allow sufficient sea room for the safe passage of the other vessel.
   (ii) A vessel required not to impede the passage or safe passage of another vessel is not relieved of this obligation if aproaching the other vessel so as to involve risk of collision and shall, when taking action, have full regard to the action which may be required by the rules of this part.
   (iii) A vessel the passage of which is not to be impeded remains fully obliged to comply with the rules of this part when the two vessels are approaching one another so as to involve risk of collision.

Rule 9: Narrow channels

(a) A vessel proceeding along the course of a narrow channel or fairway shall keep as near to the outer limit of the channel or fairway which lies on her starboard side as is safe and practicable.
(b) A vessel of less than 20 metres in length or a sailing vessel shall not impede the passage of a vessel which can safely navigate only within a narrow channel or fairway.
(c) A vessel engaged in fishing shall not impede the passage of any other vessel navigating within a narrow channel or fairway.
(d) A vessel shall not cross a narrow channel or fairway if such crossing impedes the passage of a vessel which can safely navigate only within such channel or fairway. The latter vessel may use the sound signal prescribed in rule 34(d) if in doubt as to the intention of the crossing vessel.
(e) (i) In a narrow channel or fairway when overtaking can take place only if the vessel to be overtaken has to take action to permit safe passing, the vessel intending to overtake shall indicate her intention by sounding the appropriate signal prescribed in rule 34(c)(i). The vessel to be overtaken shall, if in agreement, sound the appropriate signal prescribed in rule 34(c)(ii) and take steps to permit safe

passing. If in doubt she may sound the signals prescribed in rule 34(d).

(ii) This rule does not relieve the overtaking vessel of her obligation under rule 13.

(f) A vessel nearing a bend or an area of a narrow channel or fairway where other vessels may be obscured by an intervening obstruction shall navigate with particular alterness and caution and shall sound the appropriate signal prescribed in rule 34(e).

(g) Any vessel shall, if the circumstances of the case admit, avoid anchoring in a narrow channel.

Rule 10: Traffic separation schemes

(a) This rule applies to traffic separation schemes adopted by the organization and does not relieve any vessel of her obligation under any other rule.

(b) A vessel using a traffic separation scheme shall:

(i) proceed in the appropriate traffic lane in the general direction of traffic flow for that lane;

(ii) so far as practicable keep clear of a traffic separation line or separation zone;

(iii) normally join or leave a traffic lane at the termination of the lane, but when joining or leaving from either side shall do so at as small an angle to the general direction of traffic flow as practicable.

(c) A vessel shall, so far as practicable, avoid crossing traffic lanes, but if obliged to do so shall cross on a heading as nearly as practicable at right angles to the general direction of traffic flow.

(d) (i) A vessel shall not use an inshore traffic zone when she can safely use the appropriate traffic lane within the adjacent traffic separation scheme. However, vessels of less than 20 metres in length, sailing vessels and vessels engaged in fishing may use the inshore traffic zone.

(ii) Notwithstanding subparagraph (d) (i), a vessel may use an inshore traffic zone when en route to or from a port, offshore installation or structure, pilot station or any other place situated within the inshore traffic zone, or to avoid immediate danger.

(e) A vessel other than a crossing vessel or a vessel joining or leaving a lane shall not normally enter a separation zone or cross a separation line except:

(i) in cases of emergency to avoid immediate danger;

(ii) to engage in fishing within a separation zone.

(f) A vessel navigating in areas near the terminations of traffic separation schemes shall do so with particular caution.

(g) A vessel shall so far as practicable avoid anchoring in a traffic separation scheme or in areas near its terminations.

(h) A vessel not using a traffic separation scheme shall avoid it by as wide a margin as is practicable.

(i) A vessel engaged in fishing shall not impede the passage of any vessel following a traffic lane.

(j) A vessel of less than 20 metres in length or a sailing vessel shall not impede the safe passage of a power-driven vessel following a traffic lane.

(k) A vessel restricted in her ability to manoeuvre when engaged in an operation for the maintenance of safety of navigation in a traffic separation scheme is exempted from complying with this rule to the extent necessary to carry out the operation.

(l) A vessel restricted in her ability to manoeuvre when engaged in an operation for the laying, servicing or picking up of a submarine capable, within a traffic separation scheme, is exempted from complying with this rule to the extent necessary to carry out the operation.

Section III: Conduct of vessels in restricted visibility

Rule 19: Conduct of vessels in restricted visibility

(a) This rule applies to vessels not in sight of one another when navigating in or near an area of restricted visibility.

(b) Every vessel shall proceed at a safe speed adapted to the prevailing circumstances and conditions of restricted visibility. A power-driven vessel shall have her engines ready for immediate manoeuvre.

(c) Every vessel shall have due regard to the prevailing circumstances and conditions of restricted visibility when complying with the rules of section I of this part.

(d) A vessel which detects by radar alone the presence of another vessel shall determine if a close-quarters situation is developing and/or risk of collision exists. If so, she shall take avoiding action in ample time, provided that when such action consists of an alteration of course, so far as possible the following shall be avoided:
  (i) An alteration of course to port for a vessel forward of the beam, other than for a vessel being overtaken;
  (ii) An alteration of course towards a vessel abeam or abaft the beam.

(e) Except where it has been determined that a risk of collision does not exist, every vessel which hears apparently forward of her beam the fog signal of another vessel, or which cannot avoid a close-quarters situation with another vessel forward of her beam, shall reduce her speed to the minimum at which she can be kept on her course. She shall if necessary take all her way off and in any event navigate with extreme caution until danger of collision is over.

22.4 Bibliography

A. N. Cockcroft and J. N. F. Lameijer, *A Guide to the Collision Avoidance Rules* 4th edition, Butterworth-Heinemann, 1990.

Richard H. B. Sturt, M.A., *The New Collision Regulations*, Lloyds of London Press, 1977.

Richard A. Cahill, *Collisions and their Causes*, Fairplay, 1983.

23 Vessel traffic services

In 1948 the world's first harbour surveillance radar was installed, overlooking the approach to the Port of Liverpool. The facility was supplemented by the use of portable VHF radio equipment. From this beginning, a wide variety of vessel traffic services (VTS) has developed. Some facilities are to this day quite simple, being limited to the ability to broadcast routine general information. At the other extreme, highly complex traffic management centres exist. The latter may employ as many as thirteen VHF communication channels and thirty radars linked by a microwave network to one or more control centres. At such centres, computers process and analyse not only the radar signals but a host of other data relevant to the movement of traffic in the port.

The type of facility provided is to a great extent related to the role to which the VTS is dedicated. A VTS system might be required to perform functions ranging from the simple provision of routine information to the complex regulation of traffic. This will be discussed further in the remainder of this chapter. Vessel traffic services need not be limited to port and harbour areas but are also found associated with some (though by no means all) IMO traffic separation schemes (TSS). For example, in the case of the Dover Strait TSS, the Channel Navigation Information Service operates radar surveillance, broadcasts navigation information and co-ordinates a ship movement reporting scheme (MAREP) from the British side. A complementary service is operated by the French authorities from Cap Gris Nez (see Notice M.1449, in Section 23.9.3).

VTS systems have been developed for a variety of reasons. In some cases their establishment was promoted by heavy traffic conditions and/or regular adverse weather patterns. Alternatively, the catalyst may have been a serious casualty or the fear of such a casualty. The latter may be the case even where the level of traffic is relatively low if large vessels carrying noxious cargoes are involved. In general, the development will have been established by the relevant organization, on a perception of how best to meet its responsibility for marine safety in the particular waters over which it holds authority. It is thus not surprising that a variety of levels of VTS and a range of operational procedures are encountered worldwide, and even within the same country.

In order to promote a degree of international harmonization of operational procedures, a committee was set up to study the problem and make recommendations in the form of Guidelines for VTS. These were submitted to the International Maritime Organization. In some quarters the view was expressed that certain legal problems (such as liability and responsibility) must be resolved first. However, many others consider that the legal matters will have to be dealt with over a period of years and in the meantime advocated adoption of the Guidelines by IMO. The Guidelines were approved by the IMO Maritime Safety Committee and annexed to Resolution A578. The latter was adopted in November 1985 and recognizes that vessel traffic services would be further improved if operated in accordance with internationally-approved Guidelines. The full text of the Guidelines for VTS is too lengthy for inclusion here, but reference will be made in this chapter to the topics which are directly relevant to navigation control.

VTS has been, and still is, a subject which stimulates controversial debate. However, there is no doubt that it is here to stay. In Europe a concerted action project known as COST 301 was set up with the aim of improving safety of navigation in European waters by providing navigational assistance to ships through an integrated network of shore-based centres (VTS). Worldwide VTS is bound to be encouraged and stimulated by the IMO adoption of the Guidelines.

23.1 The functions of vessel traffic services

In suggesting the guidelines for VTS, no attempt was made to produce a rigid definition. It was felt that the following general description was more appropriate.

A vessel traffic service (VTS) is any service implemented by a competent authority primarily designed to improve the safety and efficiency of traffic and the protection of the environment. It may range from the provision of simple information messages to extensive management of traffic within a port or waterway. Given this broad description, a VTS might embrace some or all of the following functions:

1 Provision of routine information
2 Co-ordination of ship movement reports
3 Monitoring of compliance with established traffic rules
4 Provision of advice or guidance
5 Regulation of traffic

Fulfilment of all of the above functions requires a communications system, and surveillance is essential for all but the simplest information service. It would not be appropriate in this text to give a detailed treatment of the technical aspects of surveillance and communication systems, but some outline consideration is necessary before proceeding to detailed consideration of the listed functions of VTS.

23.2 Communications

In modern VTS systems a wide variety of communication systems may be found. These may include land line telephone, radiotelephony, telex, microwave data link and satellite communication (see Section 19.1.3). However, in almost every case it will be found that the principal method of communication between vessels within the VTS area and the vessel traffic centre (VTC) is speech, using radiotelephony in the marine VHF band. In some cases, additional use is made of the UHF band (see Section 10.1).

In some situations, visual signals may be used. While these may evoke memories of shapes swinging on the end of a harbour wall, it should be noted that modern lamp matrix systems exist. By choice of the pattern of lights illuminated, large letters or symbols can be produced. Such systems are, of course, dependent on clear visibility.

23.3 Surveillance

Radar is invariably the sensor used to carry out surveillance. In general, the systems are purpose built to achieve a very high degree of range and bearing discrimination. For long range surveillance, some discrimination may have to be sacrificed to obtain an adequate detection range. However, on short range scales, range discrimination of the order of 4 metres is often achieved. Horizontal beamwidth values of 0.25° are not uncommon, giving a bearing discrimination of about 8 metres at 1 mile range. In some cases the high resolution is achieved by using K band transmissions. (This corresponds with a wavelength of about 1.5 cm: see Section 10.1.) However, in most cases X band transmission (3 cm wavelength) is employed. In order to maximize bearing discrimination, the area may be divided into zones with a radar serving each, and the data relayed to the vessel traffic centre (VTC).

In simple systems, the surveillance may take the form of display observation and manual plotting, whereas in more complex systems, computers may be used to track targets, analyse the data and present a wide variety of traffic information in alphanumeric or graphical form. In complex systems, the computer may also manage other data such as general information relating to an identified vessel, berthing arrangements, cargo category, and agency details, to name but a few.

In any surveillance system, positive identification is extremely important and primary radar alone cannot do this. Normally, identification will be achieved by the vessel reporting its position, at some entry point to the system, by VHF radiotelephony to the VTC. This should, in most cases, allow the VTS operator to identify the radar echo of the reporting vessel. For this reason, it is very important that vessels navigating in the area of a VTS co-operate with the system, by reporting at the appropriate point(s) indicated on the chart and/or in the VTS handbook. Further, due care must be given to the precision with which the position is determined and reported (see also Section 23.5).

In some systems, VHF direction finding facilities are available and can be used to indicate a line of bearing on the VTS radar screen.

A possible future solution to the problem of identification is the use of a transponder on board the ship to be identified. Such a device, when interrogated by the shore-based radar, can transmit an identity signal which may be displayed on the VTS screen. Transponders are commonplace in air traffic control. In VTS applications, a few are in operation but they are by no means common.

23.4 Information services

In such a service, the VTC will disseminate general and particular information relevant to the area of its responsibility. The nature of such information will depend on the particular circumstance, but might include:

1 Weather, present and forecast
2 Tidal information
3 Scheduled movements of vessels
4 Actual movements of vessels
5 Navigation warnings
6 Information relating to pilotage, tugs and other port services
7 Specific traffic problems, e.g. large vessel swinging, vessel anchored in the channel, etc.

See also Section 23.9.1.1.

Many large vessel traffic centres broadcast this information on a dedicated VHF channel at scheduled times. In other cases it may be necessary for the vessel to call the VTC and request specific information.

The navigator should consult the Port Information Handbook, ALRS vol. 6, and the Sailing Directions for details of such service and schedules. There is a duty to listen to scheduled broadcasts and to keep a watch on the appropriate channel (see Section 19.3.1) in order to hear unscheduled broadcasts and messages from other vessels in the scheme which may assist in making a full appraisal of the traffic situation. It is essential, when listening in to messages transmitted by other vessels, that the vessel transmitting the message is identified beyond all doubt. Eavesdropping can provide valuable information but can also be potentially dangerous. In this context, it should be remembered that rule 7 of the Collision Regulations (see Sections 22.1.3 and 22.2.3) requires vessels to use *all* available means appropriate to the prevailing circumstances and conditions to determine if risk of collision exists (see also Section 19.1.8).

Where appropriate, the navigator should advise the VTC of any information concerning the position, movement or status of his vessel which might affect the overall safe movement of traffic. In all communications it must be remembered that, in the absence of confirmation, it cannot be assumed that a message has been received.

23.5 Ship reporting schemes

This general heading embraces the following three types of scheme which serve widely differing areas:

1 Oceanic schemes such as AMVER (see Notice M.1192, in Section 23.9.4), which were initiated to assist in the location of vessels in difficulty and to facilitate the organization of assistance from a knowledge of the location of other vessels in the same area

2 Schemes such as MAREP (see Notice M.1449, in Section 23.9.3), which provides for voluntary reporting by certain vessels in an area of international waters where IMO traffic routeing operates under rule 10 of the Collision Regulations
3 Schemes which operate in the area under the jurisdiction of a port authority.

In schemes of type 2 and 3, vessels are requested, or required, to report at specified positions. This is likely to include an initial position when entering the area and possibly subsequent positions as the vessel progresses through the area. The information requested (or required) will vary with authority, but might include:

1 Time, position, course, speed, draught, passage plan
2 ETA at the next reporting position
3 Information relating to cargo or defects (at the time of first contact) (see Notice M.988, in Section 23.9.5).

Whether required to or not, vessels should co-operate in such schemes in the general interest of safety, efficiency of traffic and the protection of the environment.

The Guidelines on VTS suggest that there should be a common language in every scheme. Where this is to be English, the communications should be in the format set out in the Standard Marine Navigational Vocabulary (see Notice M.1252, in Section 19.3.2). In particular, where positions are specified with respect to a charted object, the information should be given in the form of bearing and range *from* the object. It should, however, be noted that some VTS systems use more than one language and not all ports have adopted the IMO vocabulary.

23.6 Traffic separation schemes

As previously mentioned, in some traffic separation schemes in international waters, the countries which have coastlines adjacent to the scheme may operate a VTS. In addition to the information and reporting schemes already described, they may operate surveillance. A principal function of the surveillance is to monitor the traffic flow in the area of the TSS and hence detect vessels which are failing to comply with the traffic regulations. Such detection makes it possible to broadcast appropriate warnings to the whole area. Such broadcasts have the potential to warn the transgressor of the error of his ways and to alert the remainder of the traffic to the possible development of a dangerous situation.

Facilities may also exist, in the form of a patrol vessel or aircraft, visually to identify vessels which appear to be contravening the traffic regulations.

Irrespective of the presence or absence of VTS in any TSS adopted by IMO, the mariner has complete responsibility for compliance with all of the Collision Regulations but in particular with rule 10. (In schemes under other jurisdiction, local rules may apply. These may modify rule 10 and possibly other Steering and Sailing Rules. Such schemes may or may not be subject to surveillance.)

In general, the application of rule 10 is quite straightforward. However, two specific points are worthy of comment, namely the crossing of traffic lanes and the use of inshore traffic zones (see Notice M.1449, in Section 22.9.3; and Notice M.1448, in Section 23.9.6).

23.6.1 *Crossing traffic lanes*

When Rule 10 was first introduced it did not state specifically whether the crossing angle referred to heading or ground track and this gave rise to some controversy. Any ambiguity was removed by the 1987 amendment. As a result of this the rule now states that a vessel shall so far as practicable avoid crossing traffic lanes, but if obliged to do so shall cross on a heading as nearly as practicable at right angles to the general direction of traffic flow. This results in crossing the lane in minimum time irrespective of the tidal stream and presents clear encounters with vessels following the lanes.

23.6.2 *Use of inshore traffic zones*

When Rule 10 was first introduced it stated that inshore traffic zones should not normally be used by through traffic which could safely use the appropriate traffic lane within the adjacent traffic separation scheme. Some controversy arose over the meaning of the term 'through traffic'. The purpose of the 1989 amendment to Rule 10(d) (see Section 22.3.1) was to clarify the position. Notice M.1449 (see Section 23.9.3), which deals with the particular case of the Dover Strait, specifies the end-limits of the inshore zones and points out that they are charted. It is indicated that in the view of the Department of Transport, any vessel which commences its voyage from a location beyond one limit of either zone and proceeds to a location beyond the further limit of that zone, and is not calling at a port, pilot station, or destination or sheltered anchorage within that zone, should, if it can safely do so, use the appropriate traffic lane of the traffic separation scheme unless some abnormal circumstance exists in that lane. In this context, reduced visibility in the area is not considered by the Department as an abnormal circumstance warranting the use of the zone.

23.7 Provision of advice and guidance

In some VTS areas, advice and guidance, as opposed to information, is available. In some circumstances this will be volunteered by the VTC, and in others it may be solicited by vessels. The distinction between information and advice may be difficult to draw. In the Guidelines for VTS (see Section 23.9.1.1) a fairly wide list of items that might come under the heading of information are given, whereas the British Ports Association define 'information' quite specifically (see Section 23.9.2) and separately from 'advice'. The IMO Guidelines use the more general phrase 'navigational assistance service' and set out its meaning in broad general terms (see Section 23.9.1.2). Clearly there is some overlap, and it is by no means certain that all countries will interpret 'information' and 'advice' in the same way.

That which is information for one vessel may well constitute advice for another. For example, a VTC might call a vessel A and state that she appears to be to the south of the channel limit. This is clearly advice to vessel A but may well be considered to be information by other vessels. Should the VTC further inform vessel A that she would be better to steer more to port, this would then constitute guidance.

The ability of the VTC to offer advice or guidance will depend not only on the surveillance facilities but also on the qualifications and experience of the VTS operators.

23.7.1 *Influence of surveillance*

In the absence of surveillance no positional advice can be offered. Any directional guidance would be dependent on the vessel relaying its present position to the VTC and on the level of experience of the VTS operator.

Where surveillance is available, the precision with which positional advice and directional guidance can be given will be greatly influenced by the accuracy and discrimination of the shore-based radar system.

23.7.2 *Influence of the VTS operator*

The ability to give positional advice or directional guidance will depend on the background and experience of the VTS operator. A trained radar operator, using high resolution radar, can reasonably be expected to track the movement of a vessel and give positional advice within specified accuracy limits. This can be quite simply extended to the provision of the direction and distance to another position. However, a shore-based assessment of whether or not it would be possible or safe to navigate the particular vessel, in that direction for the specified distance, could only safely be made by an operator who had pilotage knowledge of the area, because a knowledge of the vessel's draught, handling characteristics, tidal streams and weather conditions would all be of significance.

23.7.3 *Use of advice and guidance*

The Guidelines for VTS encourage masters of vessels, navigating in an area for which VTS is provided, to make use of that service. Thus, if the master believes that advice or guidance from the shore will contribute to the safe passage of the vessel through the area, he should solicit this advice. However, it must be remembered that the function of the advice and guidance is to assist the navigation team in making the correct decision and *not* to make the decision for them. The advice from the shore must be seen as one element in the range of data available to the team, and the weighting attributed to this advice will be determined by the circumstances of the case.

The Guidelines stress that decisions concerning the navigation and manoeuvring of the vessel remain with the master (see Section 23.9.1, paragraph 3.3.2). If voluntary or compulsory pilotage exists in the VTS area it plays an important role in such VTS, and the Guidelines draw attention to the function of a pilot in such circumstances. (See 23.9.1 paragraph 3.3.3).

23.8 Regulation of traffic

In some ports a level of traffic regulation may be a normal feature of the VTS operation, while in others regulation may only be implemented in cases of emergency. The form which such regulation may take, the legislation whereby it is enforced and the division of responsibility will vary around the world. It is generally accepted that it may take some years for the legal aspects of the responsibility of masters, pilots and VTS operators to be defined internationally for the purposes of VTS.

The term 'regulation' is defined by the British Ports Association (BPA) as meaning 'giving an instruction, under the port authority's statutory powers, to a vessel which will necessitate an adjustment to, or confirmation of, its proposed movement within the VTS area' (see Section 23.9.2). The IMO Guidelines adopt a more descriptive approach. They explain, in general terms, that a traffic organization service is concerned with the forward planning of movements to prevent the development of dangerous situations and to provide for the safe and efficient movement of traffic within the VTS area (see Section 23.9.1, paragraph 4.6). Further, it indicates that this may be accomplished on the basis of sailing (passage) plans and may include:

1 Establishing and operating a system of traffic clearance and reports for specific manoeuvres and conditions, or establishing the order of movement;
2 Scheduling vessel movement through special areas such as where one-way traffic is established;
3 Establishing routes to be followed and speed limits to be observed;
4 Designating a place to anchor;
5 Organizing vessel movements by means of advice or instructions such as requiring a vessel to remain in, or proceed to, a safe position or other appropriate measure whenever the safety of life or protection of the environment or property warrants it.

Consideration of both the BPA strict definition and the IMO narrative approach indicates that it is envisaged that in certain circumstances decisions will be taken ashore, and that traffic will be required to comply with some overall tactical schedule.

In a simple case this might take the form of requiring a vessel to remain in her berth (or in a safe area outside the port) until another vessel has passed a critical point in the VTS area. In such circumstances, there should be no conflict between the responsibility of the master for the safety of his vessel and that of the VTC for effecting the safe regulation of the traffic within the area of the port.

Not all situations will be so simple. There may be some misgivings in cases where a VTC considers it necessary to give instructions to a vessel which is under way within the port area. The IMO Guidelines for VTS address this problem by counselling that care should be taken to ensure that VTS operations do not encroach upon the master's responsibility for the safe navigation of his vessel or disturb the traditional relationship between master and pilot (see Section 23.9.1, paragraph 2.1.5). The decisions concerning the actual navigation and manoeuvring of the vessel remain with the master. Neither the sailing plan, nor requested or instructed changes to the sailing plan, can supersede these decisions.

Where *instructions* are given to a vessel, they should not be dismissed or regarded lightly. In the event of a subsequent accident the ship's personnel will be called to account for their actions under the prevailing circumstances, and it would appear unlikely that an acceptable defence would be that the final authority rests aboard the ship.

It appears that it is the intention of the IMO Guidelines that the VTC *should not* give specific helm and engine manoeuvring instructions to any vessel. These are decisions for the master. If this is the case, instructions from the VTC must relate only to the general progress of the vessel and can thus only take the form of requirement for the vessel to either:

1 Remain within a safe area; or
2 Proceed to a safe area.

It must always be remembered that in any particular port the situation may well be governed by specific local legislation.

While the IMO Guidelines indicate that the proper authority should ensure that the VTS operations do not encroach on the master's responsibility for the safe navigation of the vessel, the master has a duty to proceed in such a way as to minimize the possibility of a conflict of responsibility arising within the VTS area. This is best achieved by careful planning and good communication (see also Section 19.1.8). In this respect the following factors are relevant:

1 Where traffic regulation is exercised, the published details should be carefully studied. Particular attention should be paid to critical areas in which a positive control or scheduling is known to, or is likely to, operate. In pilotage areas, this is a matter which should be discussed with the pilot at an early stage.
2 The VTC should be advised of the vessel's passage plan in good time, and ETAs at critical points should be clearly stated. This ensures that the VTC can implement any scheduling arrangements which may be necessary in the light of other plans already received.
3 Information should be sought from the VTC concerning the possibility of any scheduling arrangements which may be necessary to avoid the simultaneous arrival of vessels in critical areas, or because of weather or other conditions known to the centre.
4 Throughout the passage a proper listening watch should be kept on the appropriate VHF channel to ensure immediate reception of communications from the VTC, or other vessels within the area, which may require a reappraisal of the passage plan.
5 Where a report is addressed to the VTC, it must *not* be assumed that all vessels in the area are aware of the information contained therein. In some circumstances it may be best to communicate directly with other vessels, while in others it may be necessary to request the VTC to relay messages.

The question of traffic regulation is without doubt an issue which provokes lengthy and sometimes contentious discussion. However, there is little doubt that where it is seen as a co-operative venture, in which the shore organization and the vessels using the area liaise to produce a free but planned flow of traffic, the interests of safety at sea and the protection of the environment are well served.

23.9 Extracts from official publications

Within Sections 23.9.1 to 23.9.6, the paragraph numbering of the original publications has been retained.

23.9.1 IMO Resolution A578(14): Guidelines for VTS

23.9.1.1
4.4 Information service

An information service is a service provided by broadcasting information at fixed times, or at any other time if deemed necessary by the VTS centre, or at the request of a vessel, and may include:

Broadcasting information about the movement of traffic, visibility conditions or the intentions of other vessels, in order to assist all vessels, including small craft that are participating in the VTS only by keeping a listening watch;
Exchanging information with vessels on all relevant safety matters (notices to mariners, status of aids to navigation, meteorological and hydrological information, etc.);
Exchanging information with vessels on relevant traffic conditions and situations (movements and

intentions of approaching traffic or traffic being overtaken);

Warning vessels about hindrances to navigation such as hampered vessels, concentrations of fishing vessels, small craft, other vessels engaged in special operations, and giving information on alternative routeing.

23.9.1.2
4.5 Navigational assistance service

A navigational assistance service is a service given at the request of a vessel or, if deemed necessary, by the VTS centre, and may include assistance to vessels in difficult navigational or meteorological circumstances or in case of defects or deficiencies.

23.9.1.3

2.1.5 Care should be taken that VTS operations do not encroach upon the master's responsibility for the safe navigation of his vessel, or disturb the traditional relationship between master and pilot.

3.3.2 The decisions concerning the actual navigation and manoeuvring of the vessel remain with the master. Neither the sailing plan (see paragraph 5.3.1; not printed here), nor requested or instructed changes to the sailing plan, can supersede the decisions of the master concerning the actual navigation and manoeuvring of the vessel, if such decisions are required according to his judgement by the ordinary practice of seamen or by the special circumstances of the case.

3.3.3 If voluntary or compulsory pilotage exists in the VTS area, pilotage plays an important role in such VTS. The function of a pilot is to provide the master with assistance in manoeuvring his vessel; local knowledge both concerning navigation and national and local regulations; and assistance with ship/shore communications, particularly where there are language difficulties.

23.9.1.4
4.6 Traffic organization service

This is concerned with the forward planning of movements in order to prevent the development of dangerous situations and to provide for the safe and efficient movement of traffic within the VTS area, which may be accomplished on the basis of sailing plans. This service may include:

Establishing and operating a system of traffic clearance and reports for specific movements and conditions, or establishing the order of movements; scheduling vessel movements through special areas such as those in which one-way traffic is established.

establishing routes to be followed and speed limits to be observed;

designating a place to anchor;

organizing vessel movements by means of advice or instructions, such as requiring a vessel to remain in or proceed to a safe position or other appropriate measure, whenever the safety of life or protection of the environment or of property warrants it.

23.9.2 VTS in British Ports (British Ports Association)

23.9.2.1
Definitions

Advice means a message containing information or suggested future action.

Information meaning a message containing factual information relating to navigation.

Regulation means giving an instruction, under the port authority's statutory powers, to a vessel which will necessitate an adjustment to or confirmation of its proposed movement within the VTS area.

23.9.3 UK DTp Merchant Shipping Notice M.1449: Navigation in the Dover Strait

Introduction

1 The Dover Strait and its approaches are among the busiest shipping lanes in the world and pose serious problems for the safety of navigation. The traffic separation scheme, its associated inshore traffic zones, the Ship Movement Reporting (MAREP) scheme and the Channel Navigation Information

Service (CNIS) have been designed to assist seafarers to navigate these waters in safety. The Department wishes, therefore, to emphasize the need for careful navigation in the area in acordance with the International Regulations for Preventing Collisions at Sea 1972 (as amended) and for use to be made of the MAREP scheme and the CNIS. Merchant Shipping Notice M.1448 contains guidance on the observance of traffic separation schemes in general. Details of the MAREP scheme and CNIS are contained in the Admiralty List of Radio Signals Vol. 6 and the Mariners' Routeing Guide, English Channel and Southern North Sea (BA Chart No. 5500).

2 The number of collisions in the Dover Strait and its approaches has declined since the introduction of the traffic separation scheme and its application becoming mandatory for all ships in 1977. Nevertheless the risk of collision is ever present and heightened if vessels do not comply with the requirements of the scheme, and Rule 10. Non-compliance subsequently causes an increase in 'end on' ship/ship encounters and heightened collision risks.

Inshore traffic zones

3 The French inshore traffic zone extends from Cap Gris Nez in the north to a line drawn due west near Le Touquet in the south. The English inshore traffic zone extends from a line drawn from the western end of the scheme to include Shoreham to a line drawn due south from South Foreland. These end-limits are charted.

4 Surveys have shown that a large proportion of ships in the EITZ ought not to have been there under any interpretation of Rule 10(d) of the Regulations. A vessel of less than 20 metres in length, a sailing vessel and vessels engaged in fishing may, under all circumstances, use the English and the French inshore traffic zones. With respect to other vessels, it is the Department's view, in line with the amendment to Rule 10(d) contained in IMO Resolution A678(16) [see M.1448] that where such a vessel commences its voyage from a location beyond one limit of either zone and proceeds to a location beyond the further limit of that zone and is not calling at a port, pilot station or destination or sheltered anchorage within that zone, it should, if it can safely do so, use the appropriate traffic lane of the traffic separation scheme unless some abnormal circumstance exists in that lane. In this context reduced visibility in the area is not considered by the Department as an abnormal circumstance warranting the use of the zone.

5 After careful consideration of traffic surveys in the area the Department is of the view that, in general, the interests of safety are best served by excluding from the EITZ as many vessels, other than those with a clear need or right to use it, as possible, Accordingly, the Department will consider action against SW-bound vessels in the EITZ (other than those exempted by Rule 10(d)) and NE-bound vessels proceeding to Continental ports. NE-bound vessels voyaging to the Thames or East Coast ports are required to use the north-bound lane of the scheme where they can safely do so. A ruling on whether in any particular case a master of a NE-bound vessel was justified on safety grounds in choosing to use the EITZ rather than the north-bound lane is for the Courts to decide in the light of individual circumstances. It should be noted that neither CNIS nor HM Coastguard has authority to interpret the Collision Regulations or grant permission for vessels to use the EITZ in contravention of Rule 10(d).

Passage planning/Crossing traffic lanes

6 Radar surveillance surveys show that many vessels proceeding from the NE Lane towards the Thames and East Coast ports use the MPC buoy as a turning point irrespective of the traffic present in the SW lane. Masters are reminded that crossing the lane in compliance with Rule 10(c) can be made anywhere between the Ridge and Sandettie Bank. In selecting the crossing point regard should be given to traffic in the SW lane and the need to avoid the development of risk of collision situations with such traffic.

Surveillance surveys also indicate that risk of collision increases if cross channel traffic, leaving Dover or the Calais approach channel, assume courses without due regard to the traffic situation in the adjacent lane. Vessels proceeding along the traffic lanes in meeting their obligations under Rules 15 and 16 are often observed making substantial course alterations and their actions are frequently complicated when bunching of traffic exists in their lane. Attention is therefore drawn to the need for cross channel traffic to consider this possible situation arising when passage planning and ultimately selecting the point where a lane is to be crossed so that the collision risk situations can be anticipated and are not allowed to develop.

Regulations for prevention of collisions – general

7 Use of the scheme in accordance with Rule 10 does not in any way alter the overriding requirement for vessels to comply with the other Rules of the Regulations. *In particular, vessels, other than those referred to in rule 10(k) and (l), do not by virtue of using the traffic lanes in accordance with Rule 10 enjoy any privilege or right of way that they would not have elsewhere.* In addition, vessels using the traffic separation scheme are not relieved of the requirement to proceed at a safe speed, especially in conditions of restricted visibility, or to make course and/or speed alterations in accordance with Rule 8.

Crossing traffic

8 Mariners are reminded that there is a concentration of crossing ferry traffic in the Strait. These vessels may make course alterations outside the lanes in order to cross them at right angles.

Rules 10(b)(ii) and 10(b)(iii)

9 In conclusion, the Department wishes to draw attention to Rule 10(b)(iii) which requires vessels normally to join and leave a traffic lane at the termination of the lane. This rule does not preclude a vessel from joining a lane from the side at a small angle to the general direction of traffic flow. Consequently, vessels bound SW from locations in the EITZ are advised to join the SW lane as soon as it is safe and practicable to do so. All vessels are advised to keep clear of boundary separation lines or zones in accordance with Rule 10(b)(ii); failure to observe this rule has been one cause of repeated damage to the CS4 buoy. This buoy is protected by a charted 'area to be avoided' by all vessels.

23.9.4 UK DTp Merchant Shipping Notice M.1192: Merchant Ship Position Reporting

1 Notice No. M.832 drew attention to the facilities provided by AMVER (Automated Mutual Assistance Vessel Rescue System) and ships were encouraged to co-operate in both AMVER and other position reporting systems by sending periodic position reports to the authorities operating the systems.

2 The following United Kingdom coast radio stations will accept AMVER reports from ships for transmission to the AMVER centre in New York:

Portishead Radio	Niton Radio
Anglesey Radio	North Foreland Radio
Cullercoats Radio	Hebrides Radio
Humber Radio	Portpatrick Radio
Ilfracombe Radio	Stonehaven Radio
Lands End Radio	Wick Radio

3 Ships fitted with satellite terminals can send messages direct to the AMVER centre at Southbury (USA) or via a British Earth Station using the INMARSAT satellite system. Otherwise communication from ship to shore for passing AMVER traffic should be by wireless telegraphy or radioteleprinter. Exceptionally, ships not fitted with wireless telegraph may use radiotelephone.

4 A brief description of the AMVER system is contained in Annual Summary Admiralty Notice to Mariners 4B; also in the separate section on ship reporting systems in parts I and II of the current Admiralty List of Radio Signals.

5 Addressees of this notice are encouraged to participate in AMVER and other position reporting systems, details of which are given in the above mentioned publications. All position reporting messages sent through authorized coast radio stations or INMARSAT are accepted free of cost to the ship concerned.

23.9.5 UK DTp Merchant Shipping Notice M.988: Reports to Harbour Authorities by Tankers Entering or Leaving United Kingdom Ports

1 The Government are introducing a number of measures, to take effect from 15 October 1981, intended to improve tanker safety in United Kingdom ports and their approaches. In particular, masters of certain oil, gas and chemical tankers of 1600 gross register tons and over will be required, in advance of entering a port, to notify the harbour master of specified information about the tanker and to complete a checklist concerning the status of the ship, its equipment and personnel for the information of the pilot and, if he so requests, of the harbour master.

23.9.6 UK DTp Merchant Shipping Notice M.1448: Observation of Traffic Separation Schemes

4 Procedure within a traffic lane

Rule 10(b) and (c). All vessels using a traffic lane must conform to the essential principles of routeing. If they are following the lane they must proceed in the general direction of traffic flow and if they are crossing it they must do so on a heading as nearly as practicable at right angles to that direction. Vessels should normally join or leave a traffic lane at its termination, however they may join or leave from either side of a lane provided they do so at as small an angle as possible to the general direction of traffic flow. The same procedure with certain exemptions, as stated in Rule 10(k) and (l), applies to vessels which are within a lane for purposes other than for passage through or across it, such as vessels engaged in fishing, if they are making way; it is appreciated that such vessels cannot always maintain a steady course and speed but their general direction of movement must be in accordance with this principle. Any substantial departure from this direction by any vessel is only allowed if it is required by overriding circumstances, such as the need to comply with other steering and sailing rules or because of extreme weather conditions. Particular attention is drawn to the requirement that vessels which must cross a traffic lane shall do so on a heading as nearly as practical at right angles to the direction of traffic flow. Steering at right angles keeps the time a crossing vessel is in the lane to a minimum irrespective of the tidal stream, and leads to a clear encounter situation with through vessels.

5 Inshore zones

Rule 10(d). Vessels other than those of less than 20 metres in length, sailing vessels, vessels engaged in fishing, and vessels en route to or from a destination within an Inshore Traffic Zone, should if it is safe to do so use the appropriate adjacent traffic lane. It does not preclude traffic under stress of weather from seeking protection of a weather shore within such a zone nor does it impose any specific behaviour on vessels within an inshore zone and traffic heading in any direction may be encountered. Within the context of this rule it is the Department's view that the density of traffic in a lane is not sufficient reason by itself to justify the use of an inshore zone, nor will the apparent absence of traffic in the inshore zone qualify as a reason for not complying with this Rule.

Index